VISUALIZING
Human Biology Lab Manual

JENNIFER ELLIE

VISUALIZING
Human Biology
Lab Manual

JENNIFER ELLIE

WILEY   JOHN WILEY & SONS, INC.

CREDITS

EXECUTIVE VP AND PUBLISHER   Kaye Pace
SENIOR ACQUISITIONS EDITOR   Rachel Falk
MARKETING MANAGER   Kristine Ruff
PRODUCTION MANAGER   Micheline Frederick
ASSISTANT EDITOR   Jenna Paleski
PHOTO MANAGER   Hilary Newman
PHOTO RESEARCHER   Terri Stratford
SENIOR ILLUSTRATIONS EDITOR   Sandra Rigby
CREATIVE DIRECTOR   Harry Newman
DESIGNER   Wendy Lai
SENIOR MEDIA EDITOR   Linda Muriello
INTERACTIVE PROJECT MANAGER   Daniela DiMaggio
EDITORIAL ASSISTANT   Jennifer Dearden
ART DEVELOPER   Elizabeth Morales
PRODUCTION SERVICES   Furino Production
Cover PHOTOS (Center) © Stockbyte/GettyImages; (bottom) Fawn V. Beckman

This book was set in Adobe Caslon by Silver Editions and printed and bound by Quad/Graphics. The cover was printed by Quad/Graphics.

This book is printed on acid-free paper.

The case stories that appear in this text are fictitious and not based on any specific real people or situations. The photographs used to illustrate the case stories do not depict the individuals described.

ISBN   978-0470-59149-9

Printed in the United States of America

V10006080_112918

PREFACE

Visualizing Human Biology Lab Manual engages students by focusing on the structure and function of each person's own unique body. Step-by-step visual guides are provided so that students can successfully complete each experiment in a timely manner. Visuals are used to teach and explain, not just illustrate, and readers with varied learning styles will be engaged. The applications of common laboratory techniques in science, medicine, and everyday life are also explored in each lab topic.

Visualizing Human Biology Lab Manual:

- Uses the unique **Visualizing the Lab** feature to provide **Step by Step** instructions and photos to help students successfully complete each lab

- Includes a comprehensive Preparation Guide with teaching tips, supply lists, supplier listings, and instructor support on the Instructor Companion site (www.wiley.com/college/ellie)

- Encourages students to grasp the big picture by relating each lab activity to real life conditions and their own connections to Biology

- Provides increased opportunities for critical thinking through the Active Learning Questions, Introductions, Exercises and Review Questions included in each of the 18 labs

- Can be used with **any Human Biology text**, but mirrors the table of contents of *Visualizing Human Biology* and key visuals from the text are used where appropriate

- *Closer Look At* boxes delve into the practical applications of the experiments performed in lab

- Gross anatomy activities include color photographs of anatomical models, in addition to color photographs of human organs for the accompanying review questions

- Case studies on disorders are provided in each organ system lab

- Histology activities include color photomicrographs to help students identify key structures

- URLs are provided within each lab to reliable sources of health information on the Internet

INTRODUCTION

This lab manual is devoted to the diverse population of students who are taking introductory courses on human biology. For some students, this course will serve as their only exposure to the natural sciences at a higher learning institution. For others, this course will serve as a stepping stone into the health professions or the natural sciences. As educators, we strive to meet the needs of our diverse student populations, particularly when students with varying career goals enroll in the same course. To meet the needs of diverse student populations, core biological concepts are addressed from the following perspectives:

- The perspective of a well-rounded citizen who understands the practical applications of controlled research studies and common laboratory techniques in the biological sciences.

- The perspective of a student who is learning basic medical terminology, as well as the principles behind common diagnostic exams.

- The perspective of a burgeoning health profession major who is getting ready to embark on rigorous coursework in his or her area of specialty.

Throughout the manual, students are exposed to the practical applications of the activities they perform during lab. As a result, students can easily relate the importance of key topics to their own lives, regardless of differences in chosen majors and career paths.

For more information, visit: www.wiley.com/college/ellie

Visualizing Human Biology Lab Manual is available as a standalone or in a customizable package with *Visualizing Human Biology* and your own materials, through the **Wiley Custom Select Program** (www.customselect.wiley.com). Please contact your Wiley representative for more information.

Reviewers of Visualizing Human Biology

Shazia Ahmed, *Texas Woman's University*
Dorothy Anthony, *Keystone College*
Chantilly Apollon, *City College of San Francisco*
David Bailey, *St. Norbert College*
Tamatha Barbeau, *Francis Marion University*
Laurie J. Bonneau, *Trinity College*
Jennifer R. Chase, *Northwest Nazarene University*
Marie Cook, *Georgian Court University*
Tracy Deem, *Bridgewater College*
Deborah Dodson, *Vincennes University*
Kari-Ann Draker, *Sir Sanford Fleming College*
Steven D. Fenster, *Ashland University*
Donald Glassman, *Des Moines Area Community College*
Leif Hembre, *Hamline University*
Linda Jensen-Carey, *Southwestern Michigan College*
Brad Kennedy, *Iowa Western Community College*
Debra Levinthal, *Roosevelt University*
Amy Liptak, *NHTI – Concord's Community College*

Mary Katherine Lockwood, *University of New Hampshire*
Philip J. McCrea, *McHenry County College*
Susan Meacham, *University of Nevada, Las Vegas*
Rachel J. Meyer, *Midland Lutheran College*
Nibedita Mitra, *University of Hartford*
Peter Mullen, *Florida Community College*
Kelly Neary, *Mission College*
Linda Peters, *Holyoke Community College*
Jean Revie, *South Mountain Community College*
Steven Revie, *South Mountain Community College*
Lori Ann Rose, *Sam Houston State University*
Stacy Seeley, *Kettering University*
Joshua Smith, *Missouri State University*
Leah Stands Lanier, *Virginia Military Institute*
Rick Stewart, *Fresno City College*
Robert Turnbull, *University of Southern Mississippi*
Stacy Vaughn, *Des Moines Area Community College*

Acknowledgments

First and foremost, I extend my heartfelt gratitude to the editorial and production team at John Wiley and Sons. It is with their vision, expertise, and tireless efforts that this book reached its full potential on a tight schedule. In particular, I would like to thank Rachel Falk and Jenna Paleski for their passion and dedication to the book, their innovative ideas, and – most importantly - their patience in answering a myriad of questions from the novice author! I sincerely thank Hilary Newman, in addition to Teri Stratford, for researching and managing the staggering number of photos in this lab manual. Sandra Rigby provided invaluable expertise while overseeing the illustration program, and I am grateful for her efforts. I am also grateful to our production manager, Jeanine Furino, for her tireless efforts in putting all of the pieces of the puzzle together. Special thanks are also extended to the following individuals at John Wiley and Sons: Kaye Pace, Kristine Ruff, Micheline Frederick, Harry Nolan, Wendy Lai, Linda Muriello, Daniela DiMaggio, and Jennifer Dearden.

I am extremely grateful to my colleagues at Wichita State University for their support and guidance throughout this journey. First and foremost, my gratitude goes to Fawn Beckman for her passion and dedication to the lab manual. Although she worked behind the scenes – optimizing experiments, overseeing the step-by-step photos, and even compiling the preparation guide – I could not have written this book without her. I thank the following individuals for their willingness to model in the step-by-step vignettes: Carrie Chambers, Derek Norrick, Matt Moore, Megan Simpson, Brian Kilmer, Pravin Wagley, Katie Coykendall, Bryauna Carr, Barbara Fowler, and Lindsey Drees. I sincerely thank my supervisor and departmental chair, Dr. William H. Hendry III, for supporting my decision to embark on this exciting project. Mary Jane Keith lent her expertise in anatomy to me on countless occasions, and Maria Martino provided professional and personal support throughout this entire journey. I thank Dr. Kent R. Thomas for his assistance on multiple labs, and in particular, for suggesting the sperm motility experiment. Dr. Karen Brown-Sullivan provided guidance on the evolution lab, as did Dr. Jessica Bowser on the cancer lab and Isabel Hendry on the DNA fingerprinting lab.

I extend my heartfelt gratitude to Ellie Skokan, MA for piquing my interest in biology as an undergraduate. The same is true for my graduate mentor, Dr. J. David McDonald, who nurtured my interest in molecular biology. On a final note, I would like to thank my brother, Michael McCoy, for his love and support.

Dedication

This book is dedicated to my students – past, present, and future – who make the time we spend together in the classroom such a valuable, interactive experience. You remind me how challenging it can be to learn the biological concepts for the first time, and you allow me to grow as an instructor as I strive to help you grow academically.

About the Author

Jennifer Ellie was born and raised in Wichita, Kansas, where she happily lives and works to this day. She obtained her BS in Biological Sciences from Wichita State University with an emphasis in Biochemistry. Following several years of work as a laboratory technician, Jennifer decided to pursue an MS in Biological Sciences. As a graduate teaching assistant, she quickly realized that her passion lies in the area of science education. Currently, she coordinates laboratory courses for the biology department at Wichita State University, where she also instructs Human Biology courses.

BRIEF TABLE OF CONTENTS

TABLE OF CONTENTS

LAB 1:
Using the Scientific Method in Everyday Life

By the End of This Lab, You Should Be Able to Answer the Following Questions:

- How do biologists incorporate the scientific method into their research studies?

- How do people use the scientific method in everyday life?

- What is the difference between a hypothesis and a theory?

- Why are control groups included in research studies?

- During experiments, what measures can be taken to eliminate (or minimize) human bias?

- Who should address ethical questions that arise from new technology?

INTRODUCTION

At some point in life, we all seek answers to questions about our bodies and the physical world. Often times these questions arise when things don't work as expected. For instance, the onset of crippling shoulder pain may prompt you to identify the source of this pain, find effective treatment options, and learn techniques to prevent similar injuries in the future. The answers we obtain about our health and the physical world come from a variety of different sources: the Internet, textbooks, health professionals, friends and family members, magazines, and so on. How do you know whether the information you receive is reliable or not? Understanding how credible, scientific conclusions are drawn gives you power to be critical of information in everyday life.

The **scientific method** is a logical series of steps used by scientists to answer questions about the physical world. Figure 1.1 outlines the basic steps involved in the scientific method. Before using the scientific method, researchers must determine whether their questions can be answered through experimentation. Questions about morality, the supernatural, and ethics cannot be tested within the physical world; for this reason, they fall outside of the realm of science.

It may be surprising to learn that people also use the scientific method in everyday life. Let's say a student named Morgan needs to call her brother after school. Morgan discovers her cell phone won't turn on after class, despite the fact that it worked earlier in the day. What is causing this problem? Morgan suspects her cell phone doesn't work because she forgot to charge the battery last night. This statement is a hypothesis that can be tested through experimentation. A **hypothesis** is simply one possible answer to the question or problem of interest. When Morgan returns home, she performs a simple experiment to test this hypothesis: she charges her cell phone battery. Afterward, Morgan examines her cell phone to find out whether her hypothesis was supported or rejected. Since her cell phone does indeed work again, Morgan's initial hypothesis was supported.

Morgan's experiment tested the effects of one variable—battery charging—on cell phone performance. It is preferable to test one variable at a time, as Morgan did during her experiment. Imagine that Morgan charged her cell phone battery *and* tested the battery's performance in a different cell phone. It would be hard to tell if the results were due to battery charging, the new cell phone, or a combination of the two variables.

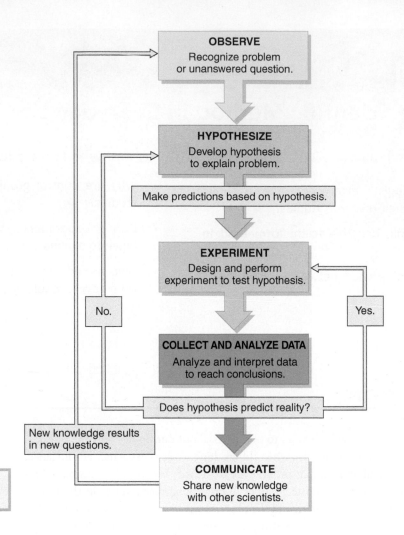

FIGURE 1.1
The Scientific Method

OBSERVE
Recognize problem
or unanswered question.

HYPOTHESIZE
Develop hypothesis
to explain problem.

Make predictions based on hypothesis.

EXPERIMENT
Design and perform
experiment to test hypothesis.

No.

Yes.

COLLECT AND ANALYZE DATA
Analyze and interpret data
to reach conclusions.

Does hypothesis predict reality?

New knowledge results
in new questions.

COMMUNICATE
Share new knowledge
with other scientists.

Write one question from everyday life that can be tested using the scientific method.

_____

_____

Write one question from everyday life that cannot be tested using the scientific method.

_____

_____

SCIENTIFIC THEORIES

The word *theory* has a much different meaning in science than it does in everyday life. Imagine a friend says to you, "My theory is Matt ignores Carrie because deep down, he still has feelings for her." In this situation, the word *theory* means your friend is speculating about a situation without having concrete evidence to support his belief. In science, the word *theory* is not thrown around lightly, since very few hypotheses stand the test of time and reach the level of a theory. A **scientific theory** is a unifying principle that has undergone extensive testing without being refuted. One example is the cell theory, which states that all living organisms are composed of one or more cells. Since the creation of the microscope in the late 1500s, almost every imaginable organism has been observed at high levels of magnification. Cells have been observed within every single creature, regardless of the microscope used or the organism tested.

On a final note, scientific theories are expressed as general statements, while hypotheses tend to be expressed as specific statements. The statement *mushrooms are composed of cells* is an example of a hypothesis; this statement is specific enough to be supported or rejected by the findings of one experiment. *All organisms are composed of cells*, on the other hand, is a general statement that encompasses the findings of thousands of experiments.

What is the difference between a hypothesis and a scientific theory?

_____

_____

_____

CONTROLLED SCIENTIFIC EXPERIMENTS

The scientific method is used by researchers who seek to understand human disease and advance medical treatments. To ensure that reliable results are collected, researchers examine two groups of patients simultaneously: the control group and the test group. Let's use a pharmaceutical study to illustrate the difference between these groups.

The effectiveness of a new migraine medication is being evaluated by a pharmaceutical company. During this study, patients in the **test group** are given the new migraine medication. Another group of patients—the **control group**—is not given any sort of migraine medication. Instead, the control group is given a substance with no medicinal value called a **placebo**. (Sugar pills are commonly used as placebos in pharmaceutical studies.) Patients in the control group and test group are treated in the same manner, with the exception of the variable being tested: migraine medication. The following **data**, or information, is collected from patients throughout the course of the study: migraine frequency, migraine severity, and adverse side effects. Researchers must compare data from the test group and control group to draw reliable conclusions from the experiment. Unless the frequency and/or intensity of migraines is significantly lower in the test group versus the control group, there is no concrete evidence to verify the effectiveness of this drug.

Sample size is another crucial consideration in research studies. **Sample size** refers to the number of data points—people, plants, bacterial cells, objects, etc.—that are evaluated during an experiment. In order to draw reliable conclusions from a study, the sample size must be large enough to accomplish the following tasks:

- Minimize the effects of natural variations among participants

- Minimize the effects of random experimental errors

- Detect any significant differences that exist among the groups tested

- Ensure the results of the study are replicable by those who wish to repeat the experiment

Throughout the course of the semester, we will routinely combine data to produce larger sample sizes for analysis. Each student must take precautions to minimize human bias, experimental errors, and deviations from the stated procedures, since these factors will ultimately affect the entire class.

During this lab period, you will use the scientific method to address questions about daily life. What makes bottled water different from tap water? Does hand washing really remove bacteria from the surface of your skin? As you analyze data and formulate conclusions, consider the sources of information people rely on in everyday life. How can people ensure that information from outside sources is reliable?

During this exercise, you will work in teams to assess the chemical content of bottled water and tap water. Distilled water will serve as the control, so it will undergo an identical testing procedure. Water quality test strips will measure the concentration of different chemicals in each sample; as a result, the data will be **quantitative** (numerical) in nature. Discuss the following questions as a class before beginning this experiment.

- How often do you drink bottled water?
- What are the advantages of drinking bottled water?
- What are the drawbacks to drinking bottled water?
- Do you think bottled water consumption is prevalent in the general public? How could you verify the answer to this question?

Create a hypothesis to test that compares the chemical content of tap water to bottled water. (Example: Tap water and bottled water have the same chemical content.)

_____

_____

Procedure

1. Collect three paper cups from the materials bench. Label one cup as *bottled water*, one cup as *tap water*, and one cup as *distilled water*.

2. Pour the appropriate water sample into each cup. **Note**: The water depth must match or exceed the length of the test strip.

3. Read the directions included with the water quality test strips prior to use.

4. Submerge one test strip into each sample as specified by the directions. Be sure to use a new test strip in each water sample.

5. Compare each test strip to the analysis chart on the test strip bottle. Record the results in Table 1.1.

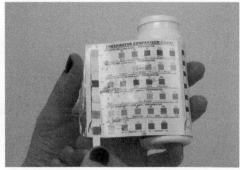

TABLE 1.1 Chemical Content of Bottled Water, Tap Water, and Distilled Water

Water Source	pH	Chlorine (ppm)	Nitrate (ppm)	Nitrite (ppm)	Hardness (ppm)
Tap Water Group Data:					
Class Average:					
Bottled Water Group Data:					
Class Average:					
Distilled Water Group Data:					
Class Average:					

* The abbreviation *ppm* stands for *parts per million*. One ppm is equivalent to one milligram (mg) per liter (L) of water.

Review Questions for Exercise 1.1

Was your hypothesis supported or rejected by the results of this experiment? If necessary, revise your hypothesis in the space provided below.

_____

_____

_____

How do your results compare with those obtained by other groups?

_____

_____

_____

What are potential sources of error in this experiment? What improvements (if any) could be made to this experiment?

_____

_____

_____

A CLOSER LOOK AT DRINKING WATER STANDARDS

The United States Environmental Protection Agency (EPA) works with partners at the state and local levels to enforce the Safe Drinking Water Act. To comply with EPA guidelines, which are in place to protect human health, public water systems rigorously monitor the levels of contaminants in tap water. To learn more about the contaminants that are regulated by EPA, you can access the National Primary Drinking Water Regulations at **www.epa.gov/safewater/consumer/pdf/mcl.pdf**.

The Food and Drug Administration (FDA) regulates the contents and safety of bottled water. Bottled water manufacturers are monitored by the FDA to ensure their products contain permissible levels of contaminants. To learn more about bottled water regulations, you can access **http://www.fda.gov/Food/FoodSafety** and click on the Product-Specific Information link.

EXERCISE 1.2 Comparing the Taste and Smell of Different Water Samples

During this exercise, you will work in pairs to determine whether tap water, bottled water, and distilled water differ in taste and smell. **Qualitative data**—descriptive, subjective data—will be collected during this experiment. Taste and smell will be ranked using the following descriptive terms: *pleasant*, *neutral*, or *foul*. Determine as a class what criteria will produce a *pleasant*, *neutral*, or *foul* ranking.

Pleasant Taste: _____

Neutral Taste: _____

Foul Taste: _____

Create a hypothesis to test that compares how tap water and bottled water taste.

_____

_____

Pleasant Smell: _____

Neutral Smell: _____

Foul Smell: _____

Create a hypothesis to test that compares how tap water and bottled water smell.

_____

_____

Procedure

1. Collect three paper cups from the materials bench. Label one cup as *bottled water*, one cup as *tap water*, and one cup as *distilled water*.

2. Pour the appropriate water sample into each cup.

3. Determine who will serve as the taster and the researcher. The **taster** places a blindfold over his or her eyes prior to testing. The **researcher** hands water samples to the taster and records data in Table 1.2.

4. **Researcher**: Randomly choose one water sample for testing. Hand this water sample to the taster, who will analyze its taste and smell. Record the results in Table 1.2.

5. Repeat step 4 with the remaining water samples.

6. Switch roles with your partner and repeat this experiment.

TABLE 1.2   Smell and Taste Data for Different Water Samples

Water Sample	Taste	Smell
Distilled Water (Control)		
Personal Data:		
Partner's Data:		
Mode for Entire Class:		
Bottled Water		
Personal Data:		
Partner's Data:		
Mode for Entire Class:		
Tap Water		
Personal Data:		
Partner's Data:		
Mode for Entire Class:		

* Mode = the response that occurred most frequently in the class

Was your hypothesis supported or rejected by the results of this experiment? If necessary, revise your hypothesis in the space provided below.

_____

_____

_____

How do your results compare with those obtained by other students?

_____

_____

_____

What steps were taken to minimize human bias during this experiment?

_____

_____

_____

What are potential sources of error in this experiment? What improvements (if any) could be made to this experiment?

_____

_____

_____

A CLOSER LOOK AT BIOETHICAL ISSUES

People purchase and consume bottled water for various reasons, including convenience, taste, and smell. One question that may arise during this experiment relates to the effects of bottled water consumption on the environment. Do the merits of bottled water outweigh the effects of pollution? While scientists can assess the impact of pollution on various environmental parameters, they cannot answer ethical questions for society. Society as a whole must determine whether the drawbacks of bottled water outweigh any perceived benefits.

The Centers for Disease Control and Prevention (CDC, www.cdc.gov) is a branch of the United States Department of Health and Human Services. In 1998, the CDC launched a campaign called *An Ounce of Prevention* that continues to this day. The goal of *An Ounce of Prevention* is to inform the general public and health educators about strategies for minimizing the spread of infectious disease. Frequent hand washing is stressed by this campaign as a simple, low-cost way to prevent illness and disease transmission. While not all bacteria are harmful to our health—in fact, certain strains of bacteria are quite beneficial to us—we can pick up **pathogens** (disease-causing microorganisms) on our skin as we touch objects throughout the day.

During this exercise, you will determine whether hand washing actually decreases the number of bacteria that are living on the surface of your skin. Since the results will be pooled for analysis, variations must be minimized in the hand washing procedure. Before you perform this experiment, brainstorm as a class to develop a uniform hand washing procedure. Consider the following questions while formulating a procedure:

- How long will hand washing be performed?
- What temperature of water will be used? (Are different temperatures of water available at the laboratory sink?)
- Will plain soap, antibacterial soap, or alcohol foam be used in addition to water?
- Will you dry your hands after they are washed? If so, how?

Hand Washing Procedure:

_____

_____

_____

_____

_____

Do you think hand washing removes a significant number of bacteria? Formulate a hypothesis to test during this experiment.

_____

_____

Procedure

1. Collect one Petri dish that contains sterile TSA agar. This medium contains the nutrients required for bacterial growth. **Note**: Do not remove the lid until instructed to do so!

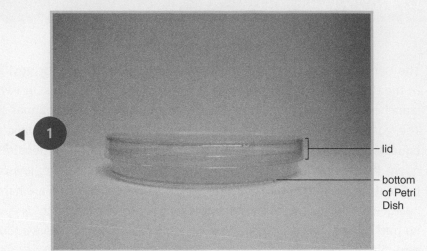

— lid

— bottom of Petri Dish

Petri Dish Filled with TSA Agar

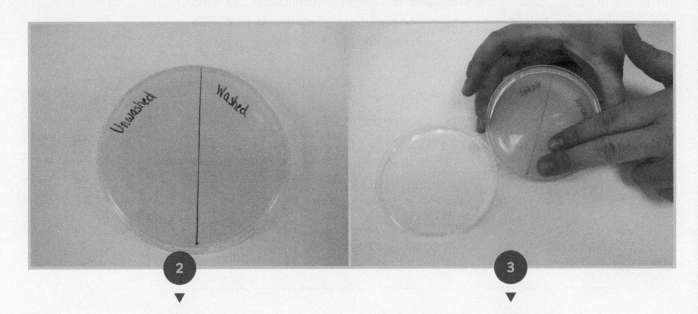

2. Using a wax pencil or a permanent marker, draw a line on the **bottom** of the Petri dish to divide it in half. Label one side as *unwashed* and the other side as *washed*. Be sure to write your name on the bottom of the Petri dish, as well.

3. Remove the lid from the agar plate, and smear two fingers across the *unwashed* side of the plate.

4. Place the lid back on the Petri dish, and wash your hands as specified by the class.

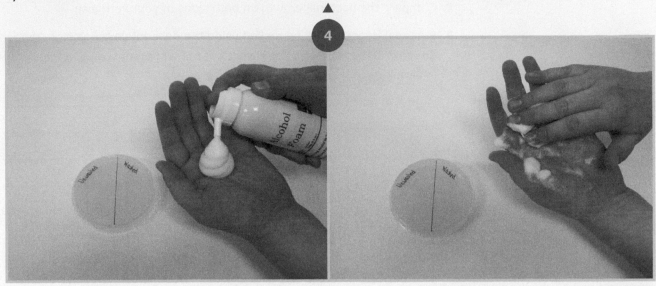

4

5. Remove the lid from the agar plate, and smear the same two fingers across the *washed* side of the plate.

◀ **5**

bacterial colony

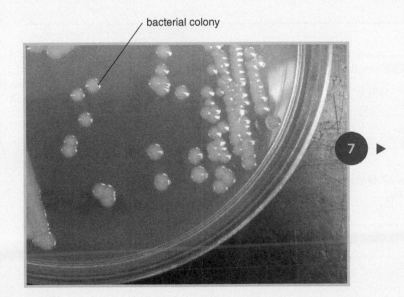

6. Place the lid back on the agar plate.

7. Place the Petri dish in the location specified by your instructor. Your plate must incubate for 24–48 hours so bacteria can divide and produce visible growths called **colonies**.

7 ▶

Next Lab Period

8. Compare the bacterial growth on both sides of your Petri dish.

9. Bacterial densities will be ranked using the descriptive terms *high*, *moderate*, and *low*. Determine as a class what criteria will produce a *high*, *moderate*, or *low* ranking.

   High Bacterial Density (+++): _____

   Moderate Bacterial Density (++): _____

   Low Bacterial Density (+): _____

   No Bacterial Growth (−): _____

10. Rank the bacterial densities observed on the *washed* and *unwashed* sides of your Petri dish. Record this data in Table 1.3.

11. Count the number of colonies growing on the *washed* and *unwashed* sides of your Petri dish. Record this data in Table 1.3.

TABLE 1.3 Bacterial Growth from Washed and Unwashed Hands

Side of Plate	Bacterial Density on Your Plate (+++, ++, +, or −)	Mode for Entire Class (+++, ++, +, or −)	Number of Colonies Growing on Your Plate	Average for Entire Class (units = colonies)
Unwashed				
Washed				

Review Questions for Step by Step 1.3

Was your hypothesis supported or rejected by the results of this experiment? If necessary, revise your hypothesis in the space provided below.

_____

_____

How do your results compare with those obtained by other students?

_____

_____

In regards to bacterial density, do you think qualitative or quantitative results render the highest accuracy? Why?

_____

_____

What are potential sources of error in this experiment? What improvements (if any) could be made to this experiment?

_____

_____

REVIEW QUESTIONS FOR LAB 1

1. Can the following statements be tested using the scientific method? Why or why not?

   It is unethical to drink bottled water since plastic bottles accumulate in land-fills. _____

   _____

   Bottled water contains fewer impurities than tap water.

   _____

   _____

2. How does a hypothesis differ from a scientific theory?

   _____

   _____

   _____

3. _____ data is numerical in nature, while _____ data is descriptive in nature.

4. What conclusions can be drawn about a hypothesis following a single experiment?

   _____

   _____

   _____

5. True or False? Scientists alone should address ethical issues that arise from new technology.

6. What is the difference between a test group and a control group?

   _____

   _____

   _____

7. A _____ is a substance with no medicinal value, such as a sugar pill.

8. Are the following statements considered hypotheses or theories? Why?

   All living organisms are composed of cells.

   _____

   _____

   Celery is composed of cells.

   _____

   _____

9. How does sample size affect the results of an experiment? Use examples from this lab period to support your answer.

   _____

   _____

   _____

   _____

10. How might human bias distort the results of a scientific experiment? What measures can be taken to eliminate (or minimize) human bias?

   _____

   _____

   _____

   _____

LAB 2:
Observing Cells with Light Microscopy

By the End of This Lab, You Should Be Able to Answer the Following Questions:

- Why are microscopes used to view tissues and other biological specimens?

- What are some practical applications of microscopy?

- What is the proper way to carry, focus, and store a microscope?

- What is a wet mount? How is a wet mount created?

- What type of microscope is used to view the three-dimensional surface of a specimen?

- How do cheek cells differ in appearance from skin cells? What similarities exist between them?

- When viewing a specimen through a microscope, how do you calculate the specimen's magnification level?

INTRODUCTION

Imagine you work in a clinical laboratory that specializes in diagnosing cancer. Every day, you examine **biopsies** (tissue specimens) from patients and look for any signs of disease. The biopsies shown in Figure 2.1 have just been shipped to your lab for analysis; how can you tell whether these specimens appear normal or abnormal? Since cells are too small to be seen with the naked eye, you rely on a microscope to magnify these specimens for analysis. (The cells in Figure 2.1 have already been magnified with a microscope.) Once the specimens are in focus, you draw on your knowledge of typical and atypical cell structure to detect any signs of disease. Within a matter of minutes, you conclude that the first patient's biopsy appears normal. However, the second patient's biopsy contains cells that appear to be cancerous.

Throughout the course of this semester, you will use microscopes to view cells, tissues, and organs in greater detail. In some cases, the characteristics of healthy tissue cells will be examined. Other times, microscopy will be used to examine the physical characteristics of diseased cells. During this lab, you will learn the proper way to handle and focus two types of microscopes: a dissecting microscope (also known as a

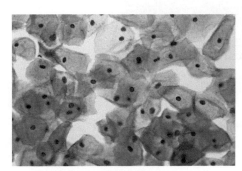

Patient 1

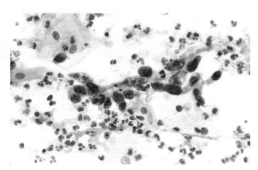

Patient 2

FIGURE 2.1. **Tissue Biopsies Observed through a Microscope**

stereomicroscope) and a compound light microscope. Since microscopy will be used repeatedly throughout the semester, make sure you are comfortable with these techniques by the end of the lab period.

Dissecting microscopes and compound microscopes differ from one another based on structure and function. Nevertheless, they do share several features in common:

- The specimen of interest (a microscope slide, an organism, etc.) is set on the **stage** of the microscope for viewing.

- The **light source** focuses visible light onto the specimen.

- Glass lenses magnify the specimen of interest. **Ocular lenses** are located in the eyepieces; the prefix *ocul-* actually means *eye* in Latin. **Objective lenses** point toward the object (specimen) of interest.

- **Adjustment knobs** bring the specimen into sharp focus.

A typical dissecting microscope is shown in Figure 2.2, and a typical compound microscope is shown in Figure 2.3. Microscope models differ from one another based on their parts and capabilities, so the microscopes owned by your university may differ slightly from these examples. Ask your instructor if you have questions regarding the unique features of your microscope model.

A CLOSER LOOK AT MICROSCOPY

If you would like to watch a demonstration video on microscopy prior to lab, you can access the *How to Use a Microscope* video available on the book companion site at www.wiley.com/college/ellie.

DISSECTING MICROSCOPES

A dissecting microscope shows the three-dimensional surface of a specimen in greater detail. The objects viewed through a dissecting microscope—organs, insects, leaves, etc.—are generally too large or too thick to be viewed through a compound microscope. As the name implies, dissections can also be performed under this microscope, when desired. The magnification range of a dissecting microscope (typically 5x-50x) is much lower than the magnification range of a compound microscope. For this reason, cells cannot be distinguished in most cases. The basic parts of a dissecting microscope are labeled on Figure 2.2, and the steps for proper usage are summarized next to the figure.

1. Using two hands, carry a dissecting microscope to your workstation. One hand grips the arm of the microscope, while the other hand holds the base of the microscope.

2. Use lens paper to clean the microscope lenses prior to use. **Note: Only use lens paper to clean the microscope lenses. Other materials (paper towels, Kleenex, shirt sleeves, etc.) can scratch the lenses.**

3. Place the specimen of interest (a microscope slide, a leaf, a bug, etc.) onto the stage.

4. Plug the microscope into the power outlet. Turn on the light source and point it toward the specimen.

5. Rotate the zoom control knob and bring the low-power objective lens (example: 0.5x) into the viewing field.

6. View the specimen by looking through the oculars.

7. Slowly turn the focus knob to bring the specimen into sharp focus.

8. To view the specimen at a higher level of magnification, rotate the zoom control knob and bring the high-power objective into the viewing field. Bring the specimen into sharp focus, as needed, using the focus knob.

9. When you are finished viewing the specimen, turn off the light source and clean off the stage. Use lens paper to clean off the microscope lenses.

10. Using two hands, carry the dissecting microscope back to its storage cabinet.

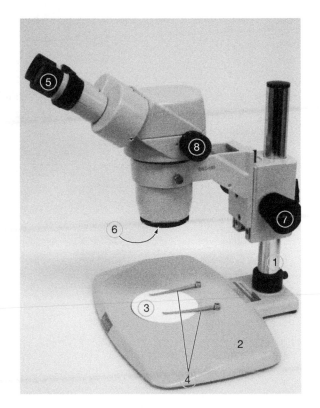

FIGURE 2.2 **Parts of a Typical Dissecting Microscope**
1. Arm
2. Base
3. Stage
4. Stage Clips
5. Oculars
6. Objective Lenses
7. Focus Knob
8. Zoom Control Knob

COMPOUND LIGHT MICROSCOPES

Biologists and medical professionals regularly use compound microscopes to observe thin, two-dimensional tissue sections. The basic parts of a compound microscope are labeled on Figure 2.3, and the steps for proper usage are summarized under the figure. The magnification power of a compound microscope (typically 40x-1000x) far exceeds the magnification power of a dissecting microscope. For this reason, individual cells can easily be distinguished.

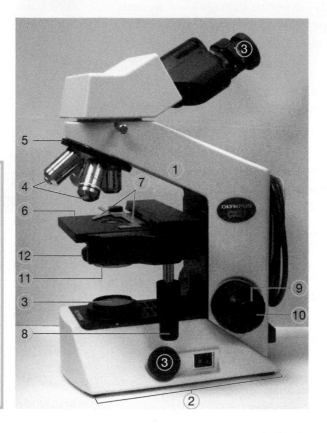

FIGURE 2.3   Parts of a Typical Compound Light Microscope
1. Arm
2. Base
3. Oculars
4. Objective Lenses
5. Revolving Nose Piece
6. Stage
7. Stage Clips
8. Stage Control Knobs
9. Course Focus Knob
10. Fine Focus Knob
11. Condenser
12. Iris Diaphragm

1. Using two hands, carry a compound microscope to your workstation. One hand grips the arm of the microscope, while the other hand holds the base of the microscope.

2. Use lens paper to clean the microscope lenses prior to use. **Note: Only use lens paper to clean the microscope lenses.** Other materials (paper towels, Kleenex, shirt sleeves, etc.) can scratch the lenses.

3. Plug the microscope cord into the power outlet, and turn on the light source. While viewing microscope slides, you can adjust light intensity (and therefore contrast) with the iris diaphragm.

4. Make sure the lowest power objective (example: 4x objective) is located directly above the hole in the stage. If needed, turn the revolving nosepiece to bring this objective into position.

5. Secure the microscope slide on the stage. If your microscope has a manual stage, then the stage clips are placed on *top* of the slide. If your microscope has a mechanical stage, then the stage clips grip the *sides* of the slide.

6. Center the microscope slide on the stage. If you are using a mechanical stage, use the stage control knobs to center the microscope slide.

7. *Carefully watch this step from the side of the microscope.* Turn the course focus knob to lift the stage toward the objective. Move the slide as close as possible to the objective, but *do not allow the slide to collide with the objective.*

8. *View your specimen by looking through the oculars from this point forward.* Using the course focus knob, move the slide *away* from the objective until the image comes into rough focus.

9. Slowly turn the fine focus knob and bring the image into sharp focus. **Only use the fine focus knob to sharpen images from this point forward.**

10. Scan the microscope slide in a pattern similar to the one shown below. If your microscope has a mechanical stage, use the stage control knobs to move the slide.

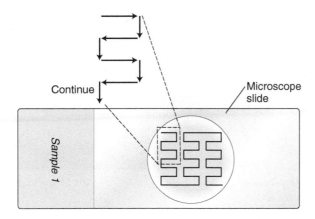

11. Center the object of interest before moving to a higher level of magnification.

12. Turn the nosepiece and position the next objective (example: 10x objective) directly above the light beam traveling through the slide. Since your microscope is **parfocal**, the image should still be in focus and require little (if any) adjustment. Bring the image into sharp focus using the fine focus knob, if needed.

13. Center the object of interest before moving to a higher level of magnification.

14. Turn the nosepiece and position the next objective (example: 40x objective) directly above the light beam traveling through the slide. Bring the image into sharp focus using the fine focus knob, if needed. **The 100x objective lens (if present) will not be used during Lab 2.**

15. When you are finished observing the microscope slide, turn the nosepiece and bring the lowest power objective (example: 4x objective) back into the viewing field.

16. Remove the microscope slide from the stage. If you are using a mechanical stage, move the slide away from the objectives prior to removal.

17. Turn off the light source, unplug the microscope, and wrap its cord around the cord holder (if present) or base. Use lens paper to clean the microscope lenses.

18. Using two hands, carry the microscope back to its storage cabinet.

FIELD OF VIEW AND TOTAL MAGNIFICATION POWER

You may wonder why a wide range of magnification levels are available on microscopes. Why isn't the highest level of magnification always desired? One answer to this question deals with the relationship between magnification power and the field of view. As the magnification of an object increases, the field of view inevitably decreases. In other words, less of an object can be viewed in one field as the magnification power gets higher. Figure 2.4 shows the relationship between magnification power and field of view.

FIGURE 2.4 **Relationship between Magnification Power and Field of View**

7.5x Magnification

20x Magnification

The magnification power of the ocular lens is written on the eyepiece, while the magnification power of each objective lens is written on the side of the objective. The total magnification of a specimen can be calculated using the following equation:

$$\text{Magnification of Ocular Lens} \times \text{Magnification of Objective Lens} = \text{Total Magnification Power}$$

For instance, a specimen is magnified one hundred times (100x) when the 10x ocular lens is used in conjunction with the 10x objective lens.

Review Questions

Imagine you are viewing cheek cells that are magnified by the 4x objective lens and the 10x ocular lenses. What is the total magnification of these cheek cells?
_____

To view bacteria clearly, they must be magnified one thousand times (1000x) larger than their original size. Which objective lens should be used in conjunction with the 10x ocular lenses? _____

DEPTH OF FIELD

Since compound microscopes create two-dimensional images, only one plane (or layer) of a specimen is in sharp focus at any given time. The *thickness* of this plane—the depth of field—is dependent on the magnification power being used. As shown in

Figure 2.5, an indirect relationship exists between depth of field and magnification power. In other words, as the magnification power increases, the depth of field decreases in size.

40x Magnification

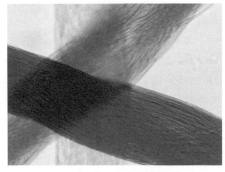

100x Magnification

FIGURE 2.5 Relationship between Depth of Field and Magnification Power

INVERSION OF MICROSCOPIC IMAGES

Finally, it is worth noting that images are inverted *and* flipped upside down as light travels through the microscope; Figure 2.6 shows an example of this phenomenon. To observe this phenomenon during lab, pay attention to the orientation of the specimen—and the orientation of the magnified image—as you view each microscope slide. When you move a microscope slide to the left, which way does the image appear to move? When you move a microscope slide up, which way does the image appear to move?

No Magnification

40x Magnification

FIGURE 2.6 Inversion of Microscopic Images

During this exercise, you will observe the outermost layer of your skin at different levels of magnification. Although your hand can fit on the stage of a dissecting microscope, the same is not true on a compound microscope! For this reason, a **wet mount** of your skin cells will be created on a glass microscope slide. To create a wet mount, the specimen of interest (skin cells, hair, pond water, etc.) is suspended in liquid between the microscope slide and a thin cover slip. Since the specimen is suspended in liquid, it will not dry out during observation. If the specimen contains translucent cells, then a liquid stain (example: methylene blue) may also be used to increase contrast.

Procedure

1. Choose a patch of skin on your hand to observe during this exercise. Draw a sketch of your skin in Table 2.1 as it appears to the naked eye.

2. Observe the same patch of skin using the low-power objective lens on a dissecting microscope. Draw a sketch of the skin detail observed in Table 2.1. Refer to Figure 2.2, as needed, to review the parts of the dissecting microscope and steps for proper usage.

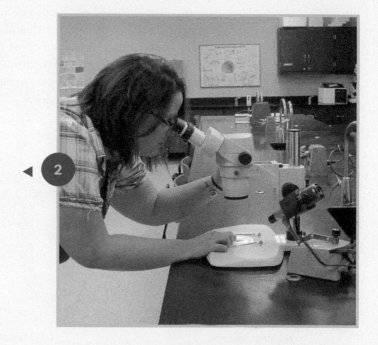

3. Observe the same patch of skin using the high-power objective lens on the dissecting microscope. Draw a sketch of the detail observed in Table 2.1.

4. Collect a clean microscope slide and a piece of double-stick tape that is approximately one inch long.

5. Attach the piece of double-stick tape onto the center of the microscope slide.

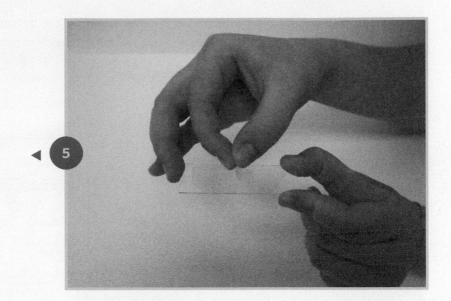

6. Wash your hands with soap to remove any oils or moisturizers.

7. Dry your hands.

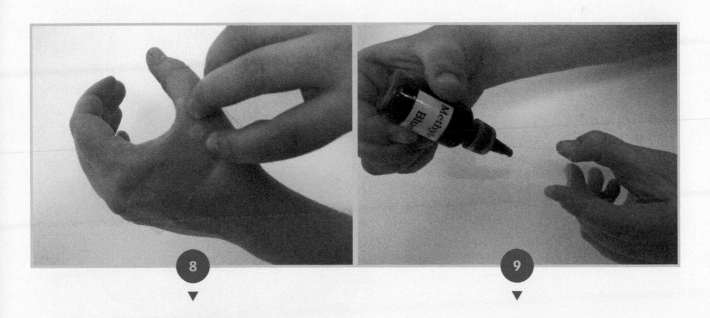

8. Gently press the microscope slide onto your hand. During this step, the double-stick tape should come in direct contact with your skin.

9. Place a small drop of methylene blue stain directly on top of the tape.

10. Starting with a 45° angle, gently lower a cover slip onto the microscope slide. *By the end of this process, the stain should be sandwiched between the cover slip and the slide.*

▲
10

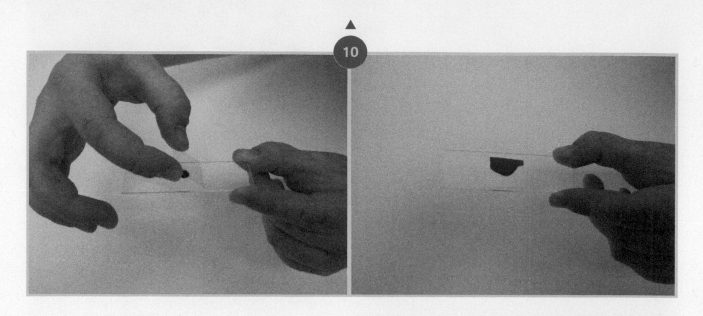

11. View your wet mount through a compound microscope. Refer to Figure 2.3, as needed, to review the parts of the compound microscope and steps for proper usage.

◄ 11

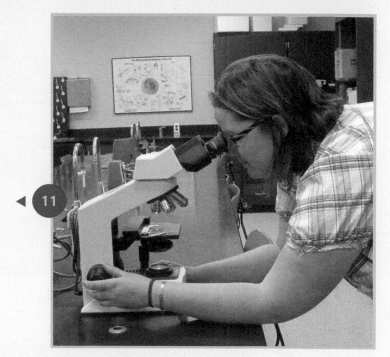

12. Draw a sketch of your skin cells in Table 2.1 at each level of magnification.

13. Place your microscope slide in a biohazard bag for disposal.

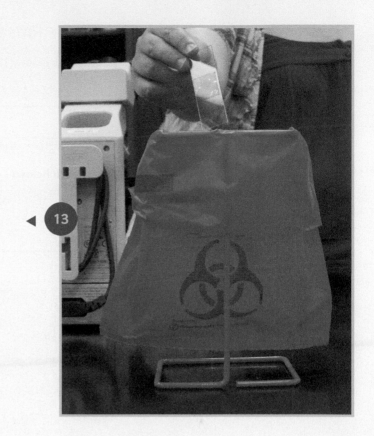

13

TABLE 2.1 Skin Observed at Different Levels of Magnification

Naked Eye	Dissecting Microscope Low-Power Objective Lens Total Magnification = ____x	Dissecting Microscope High-Power Objective Lens Total Magnification = ___x
Compound Microscope Low-Power Objective Lens Total Magnification = ___x	**Compound Microscope** Medium-Power Objective Lens Total Magnification = ___x	**Compound Microscope—** High-Power Objective Lens Total Magnification = ___x

Which photograph below (1 or 2) was taken through a dissecting microscope? How can you tell?

_____

_____

Which photograph below (1 or 2) was taken through a compound microscope? How can you tell?

_____

_____

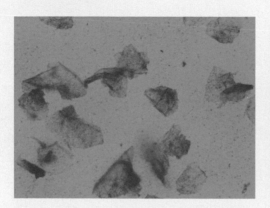

Skin Sample 1 (400x Magnification)

Skin Sample 2 (7.5x Magnification)

While observing your skin through a microscope, what happened to the field of view as magnification increased?

_____

_____

Visualizing THE LAB

STEP BY STEP 2.2 Observing Hair with Light Microscopy

1. Collect a hair strand from your head or arm. Draw a sketch of your hair strand in Table 2.2 as it appears to the naked eye.

2. Observe your hair strand using the low-power objective on a dissecting microscope. Draw a sketch of the detail observed in Table 2.2.

3. Observe your hair strand using the high-power objective on a dissecting microscope. Draw a sketch of the detail observed in Table 2.2.

4. Place a small drop of water on the center of a clean microscope slide.

5. Place your hair strand into the water drop.

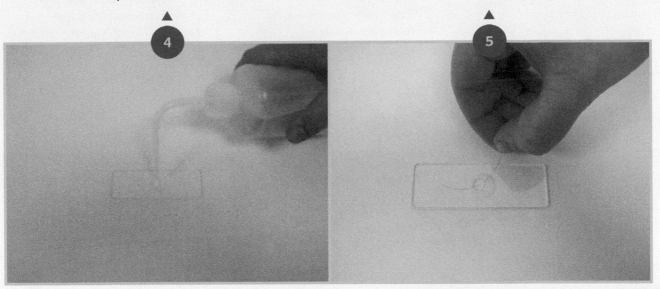

6. Starting with a 45° angle, gently lower a cover slip onto the microscope slide. *Your hair strand should be sand-wiched between the cover slip and slide by the end of this step.*

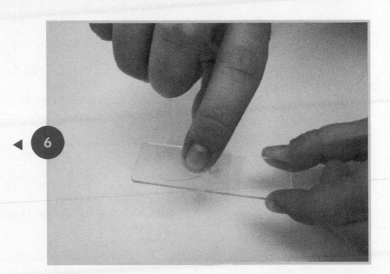

7. View your wet mount through a compound microscope. Refer to Figure 2.3, as needed, to review the parts of the compound microscope and steps for proper usage.

8. Draw a sketch of your hair strand in Table 2.2 at each level of magnification.

9. Compare your hair strand with that of students who have other colors or textures of hair.

TABLE 2.2   Hair Strand Observed at Different Levels of Magnification

Naked Eye	Dissecting Microscope Low-Power Objective Lens Total Magnification = _____x	Dissecting Microscope High-Power Objective Lens Total Magnification = _____x
Compound Microscope Low-Power Objective Lens Total Magnification = _____x	Compound Microscope Medium-Power Objective Lens Total Magnification = _____x	Compound Microscope High-Power Objective Lens Total Magnification = _____x

Review Questions for Step by Step 2.2

Which photograph below (1 or 2) was taken through a dissecting microscope? How can you tell?

_____

_____

Which photograph below (1 or 2) was taken through a compound microscope? How can you tell?

_____

_____

Hair Strand 1 (7.5x Magnification)

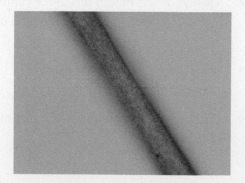

Hair Strand 2 (200x Magnification)

Why do you think methylene blue was added to your skin cells but *not* to your hair strand?

_____

_____

1. Collect a clean microscope slide from the materials bench.

2. Place a small drop of methylene blue stain onto the center of the slide.

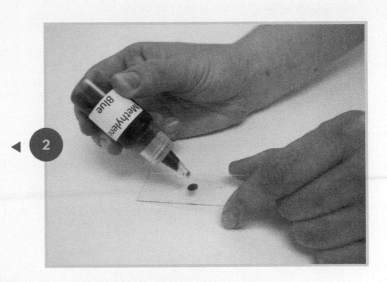

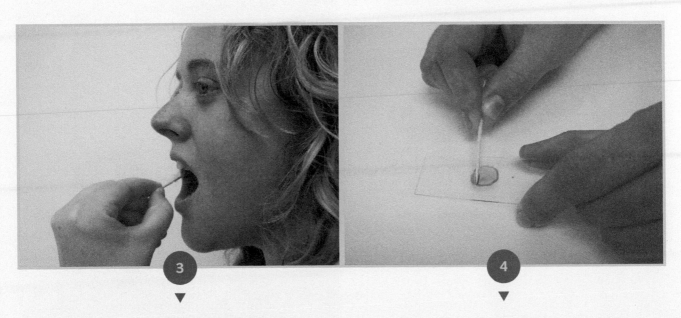

3. Use a toothpick to gently scrape the inside of your cheek.

4. Swirl the toothpick into the methylene blue stain on your microscope slide.

5. Place your used toothpick in a biohazard bag for disposal.

6. Starting with a 45° angle, gently lower a cover slip onto the microscope slide. *By the end of this process, the stain should be sandwiched between the cover slip and the slide.*

◀ 6

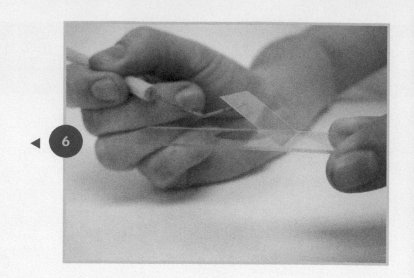

7. View your wet mount through a compound microscope. Refer to Figure 2.3, as needed, to review the parts of the compound microscope and steps for proper usage.

8. Draw a sketch of your cheek cells in Table 2.3 at each level of magnification.

9. Place your wet mount in a biohazard bag for disposal.

◀ 9

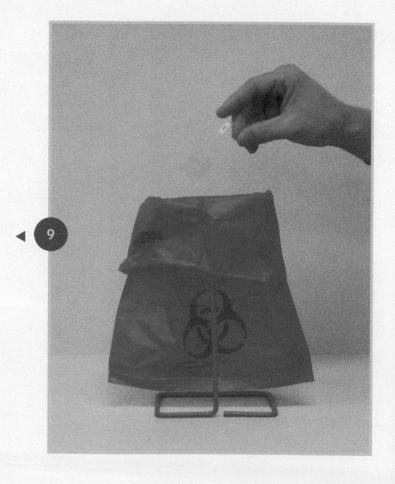

TABLE 2.3   Cheek Cells Observed at Different Levels of Magnification

Compound Microscope Low-Power Objective Lens Total Magnification = _____ x	Compound Microscope Medium-Power Objective Lens Total Magnification = _____ x	Compound Microscope High-Power Objective Lens Total Magnification = _____ x

Review Questions for Step by Step 2.3

Which image below (1 or 2) shows cheek cells? How can you tell?

_____

_____

Which image below (1 or 2) shows skin cells? How can you tell?

_____

_____

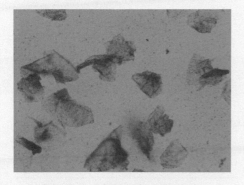
Specimen 1 (400x Magnification)

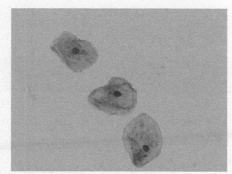
Specimen 2 (400x Magnification)

What similarities exist between your cheek cells and skin cells?

_____

_____

EXERCISE 2.4 Observing Prepared Tissue Slides

1. Choose one prepared microscope slide to observe from the collection provided by your instructor.
2. View this slide through a compound microscope. Refer to Figure 2.3, as needed, to review the parts of the compound microscope and steps for proper usage.
3. Draw a sketch of the detail observed at each magnification level in Table 2.4.

TABLE 2.4 Tissue Observed through a Compound Microscope

Low-Power Objective Lens Total Magnification = _____x	Medium-Power Objective Lens Total Magnification = _____x	High-Power Objective Lens Total Magnification = _____x

REVIEW QUESTIONS FOR LAB 2

1. What is the proper way to dispose of human specimens, such as wet mounts of cheek cells?

____

____

____

2. When you move a microscope slide to the left, which way does the image *appear* to move? ______________. When you move a microscope slide up, which way does the image *appear* to move? ______________

3. Label the ocular lens, objective lenses, and stage on the diagram below.

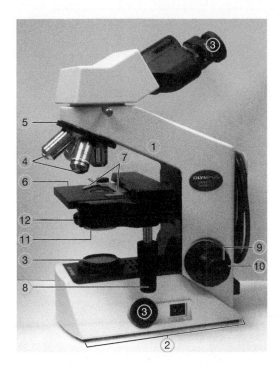

4. As the magnification power of a microscope increases, the field of view _____ in size.

5. Describe the proper way to carry a microscope.

   _____

   _____

   _____

6. Imagine you are viewing skin cells on a microscope with the 10x objective lens and 10x ocular lenses. What is the total magnification of these skin cells? _____

7. Describe the basic steps used to create a wet mount.

   _____

   _____

   _____

   _____

   _____

8. The specimen of interest (microscope slide, insect, etc.) is placed on the _____ of the microscope for viewing.

9. Which microscope (dissecting or compound) allows you to view individual cheek cells? _____

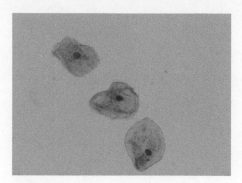

10. Which microscope (dissecting or compound) allows you to view the three-dimensional surface of your skin? _____

LAB 3:
Observing Body Tissues

By the End of This Lab, You Should Be Able to Answer the Following Questions:

- What levels of organization exist within the human body?

- How do the four major tissue classes (epithelial tissue, connective tissue, muscle tissue, and nervous tissue) differ in structure, function, and microscopic appearance?

- What physical characteristics are used to classify epithelial tissues?

- How can smooth, skeletal, and cardiac muscle be distinguished on microscope slides?

- How do neurons differ in appearance from muscle cells and epithelial cells? How do these structural differences relate to their differences in function?

- Why are tendons, blood, and compact bone all considered connective tissues?

INTRODUCTION

Would you believe there are more cells in your body than stars in the Milky Way galaxy? One thousand times more cells, to be exact! The Milky Way galaxy contains approximately 100 billion stars, while the average adult human body contains 100 trillion cells. What is the advantage of maintaining such a large number of cells? Several logical answers exist to this question. For instance, the death of one cell won't jeopardize your body if trillions of additional cells are present. Each cell can also focus on performing a specific task for the good of the entire body. Skin cells, for instance, can protect underlying cells from dehydration, infection, and injury. Muscle cells can contract to move the body itself and substances within the body. **Neurons** (nerve cells) can establish communication routes between distant organs. Bone cells can protect delicate internal organs and support the weight of the entire body.

Organization is essential inside of the body, just as organization is essential in other areas of life. Figure 3.1 illustrates the levels of organization that exist throughout the human body. **Cells** are the basic unit of life for all living organisms, including human beings. Each cell contributes to the well-being of the body by devoting itself to a specific task; the function performed by each cell depends on its structure and location within the body. The cells that make up your body are quite similar to employees who work for a large corporation. At work, employees use their specialized skills to help sustain the company's viability in the market. Regardless of each person's day-to-day responsibilities, the ultimate goal for each employee is to generate revenue and keep the company in business. Likewise, the ultimate goal for each cell is to help maintain **homeostasis** (stable conditions within the body, regardless of changes in the outside environment) and keep the body alive.

Employees with similar skill sets often work together as teams within a company. In the same way, tissue cells work together as teams to complete tasks in the body: the shortening of a muscle, the production of a hormone, the destruction of disease-causing microorganism, and so on. A **tissue** is simply a group of cells with uniform structure and function. Four major tissue classes construct the human body from head to toe: epithelial tissue, muscle tissue, nervous tissue, and connective tissue. **Organs** such as the heart, lungs, and kidneys are constructed out of many different tissue types. Ultimately, organs also

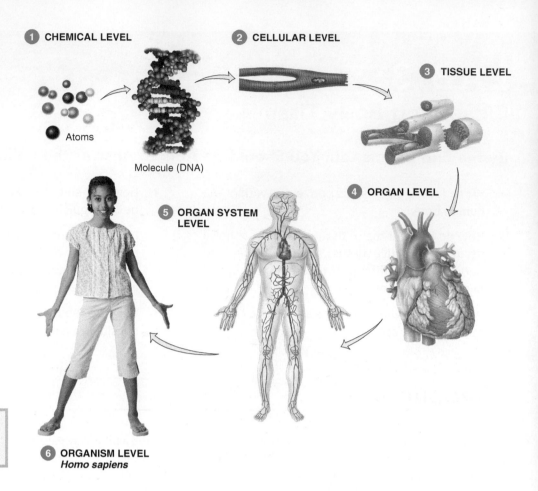

FIGURE 3.1 Levels of Organization in the Human Body

work together as teams (**organ systems**) to help maintain homeostasis. The respiratory system, for instance, exchanges gases with the environment. The lungs must work with other respiratory organs, such as the trachea and nose, to successfully complete this task.

During Lab 2, light microscopy was used to view skin cells, cheek cells, and hair strands in greater detail. During this lab, light microscopy will be used to examine the structure of various tissue types. As you examine each microscope slide, note the characteristics that make each tissue type unique. In addition, note the **histology** (tissue structure) of normal tissue samples: cell shape, cell arrangement, appearance of the nucleus, etc. To identify **pathology** (the disease state) in tissue biopsies, health professionals must first understand the characteristics of healthy tissue cells.

EPITHELIAL TISSUE

The term *epithelium* literally means a tissue layer found on top of (*epi-*) other cell layers (*-thelium*). The cheek cells and skin cells you observed during Lab 2 belong to this tissue category. Given their location, many epithelial cells are exposed to factors from the outside environment: ultraviolet light, ingested foods, inhaled air, etc. Epithelial tissue also lines **cavities** (hollow spaces) within organs such as the heart and the uterus. The epithelial tissue in your body has three primary functions: protection, absorption of substances, and **secretion** (release) of substances.

Epithelial tissue is classified based on cell shape and cell arrangement. Figure 3.2 illustrates the different shapes and arrangements of epithelial cells within the body.

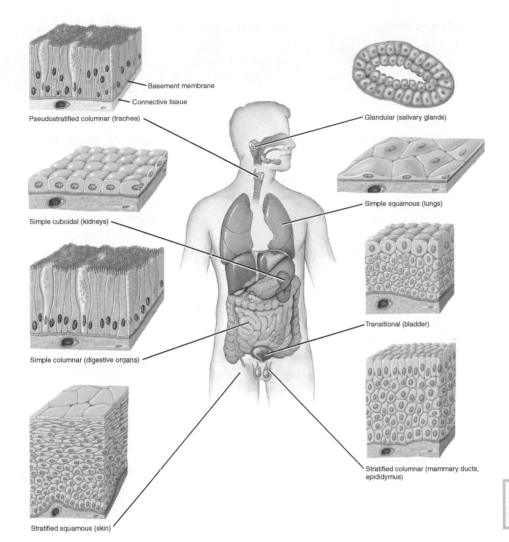

Basement membrane
Connective tissue
Pseudostratified columnar (trachea)

Glandular (salivary glands)

Simple squamous (lungs)

Simple cuboidal (kidneys)

Transitional (bladder)

Simple columnar (digestive organs)

Stratified columnar (mammary ducts, epididymus)

Stratified squamous (skin)

FIGURE 3.2 **Classification of Epithelial Tissue**

- **Squamous** cells are flat and thin like scales

- **Cuboidal** cells are shaped like cubes

- **Columnar** cells are shaped like columns

- **Transitional** cells can change shape to accommodate expansion or contraction

- **Simple** epithelium consists of a single cell layer

- **Stratified** epithelium consists of multiple cell layers (**strata** means layers)

- **Pseudostratified** epithelium may appear stratified at first glance, but closer observation reveals it is only composed of one cell layer

As with all tissues, the structure of epithelial tissue is directly related to its function. Simple squamous epithelium, for instance, is thin enough to allow the exchange of gases, nutrients, and wastes. For this reason, air sacs in the lungs *and* capillary beds in the cardiovascular system are composed of this type of tissue. Since the cells in transitional epithelium change shape, they accommodate expansion or contraction of the urinary bladder. Columnar and cuboidal epithelium are well-suited for producing, secreting, and absorbing substances. (A **gland** is a group of epithelial cells that produces a substance—such as tears, hormones, sweat, or mucus—and releases this substance into the surrounding environment.) As you observe different types of epithelium in the following exercise, think about how the structure of each tissue allows it to perform a specific task in the body.

To understand how a tissue functions, it is helpful to view its structure at the microscopic level. During this exercise, you will examine prepared microscope slides of epithelial tissue. Refer to Lab 2, as needed, to review the steps for proper usage of a compound microscope. Compare your focused microscope slides with the images provided below, and answer the review questions that accompany each slide.

Are the epithelial cells in Figure 3.3 squamous, cuboidal, or columnar in shape?

_____

Does Figure 3.3 show simple, stratified, or pseudostratified epithelium? How can you tell?

_____

_____

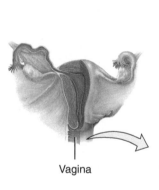

Vagina

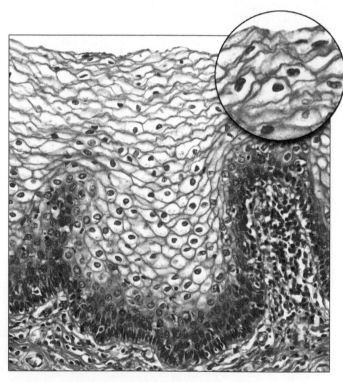

FIGURE 3.3   Epithelium Lining the Vagina

The outermost layer of your skin (the epidermis) is quite similar in structure to the epithelium that lines the vagina. Why is this type of epithelium a good barrier for protection?

_____

_____

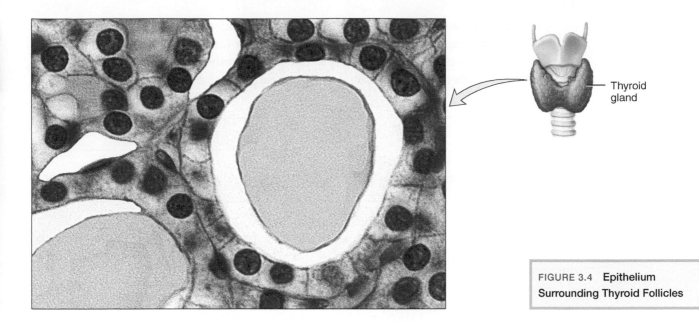

FIGURE 3.4 **Epithelium Surrounding Thyroid Follicles**

Are the epithelial cells in Figure 3.4 squamous, cuboidal, or columnar in shape?

_____

Does Figure 3.4 show simple, stratified, or pseudostratified epithelium? How can you tell?

_____

_____

Metabolic hormones are produced, stored, and secreted by thyroid epithelial cells. Why is this cell shape well suited for production, storage, and secretion of substances?

_____

_____

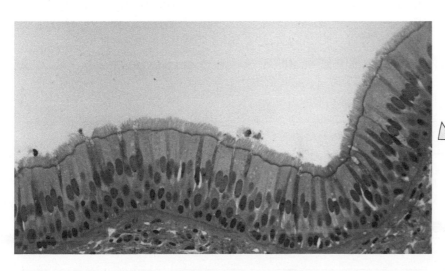

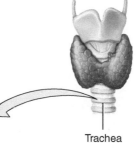

Trachea

FIGURE 3.5 **Epithelial Tissue Lining the Trachea**

The surface of a **ciliated** epithelial cell is covered with small, hair-like projections. When cilia beat together in unison, objects (or fluids) in the outside environment are moved in specific directions. Label cilia on the epithelial cells in Figure 3.5. What role do you think cilia play in the respiratory tract?

_____

_____

Are the epithelial cells that line the trachea squamous, cuboidal, or columnar in shape? _____

Does Figure 3.5 show simple, stratified, or pseudostratified epithelium? How can you tell?

_____

_____

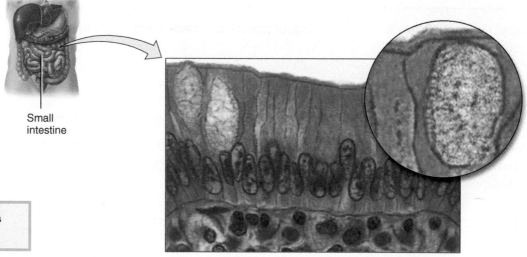

Small intestine

FIGURE 3.6 **Epithelial Cells Lining the Small Intestine**

Are the epithelial cells in Figure 3.6 squamous, cuboidal, or columnar in shape?

_____

Does Figure 3.6 show simple, stratified, or pseudostratified epithelium? How can you tell?

_____

_____

Digested nutrient molecules are absorbed by the epithelial cells lining the small intestine. Why is this type of epithelium well suited for absorption?

_____

_____

MUSCLE TISSUE

Muscle tissue is involved in a diverse array of bodily activities: breathing, regulation of blood pressure, facial expressions, heartbeats, heat production, movement, and even pushing substances through the body. All of these actions rely on the **contraction** (shortening) and **relaxation** (lengthening) of muscle cells in different parts of the body. (The terms *muscle cell*, *muscle fiber*, and *myocyte** can be used interchangeably.) Three distinct types of muscle tissue are found throughout the human body: skeletal muscle, cardiac muscle, and smooth muscle. Skeletal muscle and cardiac muscle are named based on their locations in the body, while smooth muscle is named for its appearance under the microscope.

Skeletal muscle is primarily involved in breathing, facial expressions, and body movements. As the name implies, most skeletal muscles are attached directly to bones. Skeletal muscle is under voluntary control, and **striations** (alternating light and dark bands) are seen in skeletal muscle fibers under the microscope. As illustrated in Figure 3.7, multiple nuclei are also found at the periphery of each muscle fiber. For this reason, skeletal muscle fibers are **multinucleated**.

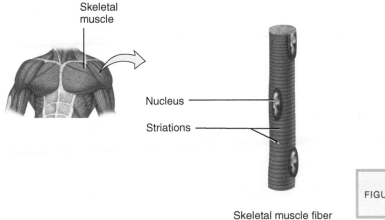

Skeletal muscle

Nucleus

Striations

Skeletal muscle fiber

FIGURE 3.7   **Skeletal Muscle**

Cardiac muscle is located exclusively within the heart, and its contraction propels blood through adjoining arteries. Unlike skeletal muscle, cardiac muscle is under involuntary control; this means the heart will beat whether or not we consciously think about it. Figure 3.8 illustrates the general structure of cardiac muscle fibers. Striations are present in cardiac muscle, just as in skeletal muscle. However, each cardiac muscle fiber only contains one nucleus; for this reason, cardiac muscle fibers are **uninucleated**. One unique feature of cardiac muscle is the presence of **intercalated discs**, which are found between (*inter-*) adjacent muscle fibers. Intercalated discs help synchronize the contraction of the heart.

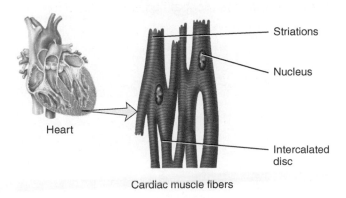

Heart

Striations

Nucleus

Intercalated disc

Cardiac muscle fibers

FIGURE 3.8   **Cardiac Muscle**

---

* The prefix *myo-* is commonly used in medical terminology when referring to muscle, and the suffix *–cyte* is commonly used when referring to a cell.

Smooth muscle (Figure 3.9) is located within tube-like structures such as the digestive tract, arteries, and veins. Wave-like contractions of smooth muscle push substances through these tubes, such as food and blood, respectively. Smooth muscle also regulates blood pressure by narrowing or dilating blood vessels. Striations are not seen in smooth muscle fibers, as implied by their name. However, smooth muscle is uninucleated and under involuntary control, just like cardiac muscle.

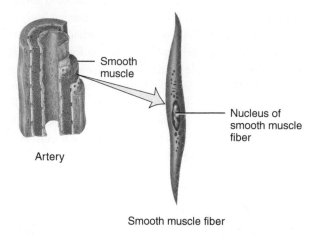

Smooth muscle

Nucleus of smooth muscle fiber

Artery

Smooth muscle fiber

FIGURE 3.9 **Smooth Muscle**

EXERCISE 3.2 Identifying Muscle Tissue on Microscope Slides

During this exercise, you will examine prepared microscope slides of smooth muscle, cardiac muscle, and skeletal muscle. Compare your focused microscope slides with the images provided below, and answer the review questions that accompany each slide.

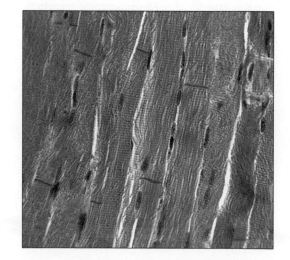

FIGURE 3.10 **Identification of Muscle Tissue**

What type of muscle (skeletal, cardiac, or smooth) is shown in Figure 3.10?

_____

Are striations present? _____

Are intercalated discs present? _____ If so, label an intercalated disc.

Label a nucleus on Figure 3.10. Are these muscle fibers uninucleated or multinucleated? _____

Where is this type of muscle found within the body? What happens when it contracts?

_____

_____

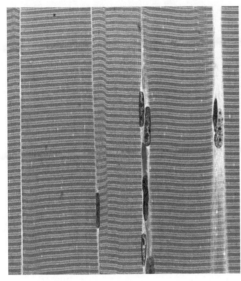

FIGURE 3.11 **Identification of Muscle Tissue**

What type of muscle (skeletal, cardiac, or smooth) is shown in Figure 3.11?

_____

Are striations present? _____

Are intercalated discs present? _____ If so, label an intercalated disc.

Label a nucleus on the image above. Are these muscle fibers uninucleated or multinucleated? _____

Where is this type of muscle found within the body? What happens when it contracts?

_____

_____

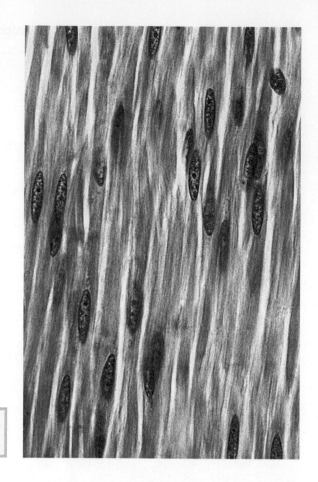

FIGURE 3.12   Identification of Muscle Tissue

What type of muscle (skeletal, cardiac, or smooth) is shown in Figure 3.12?

_____

Are striations present? _____

Are intercalated discs present? _____ If so, label an intercalated disc.

Label a nucleus on the image above. Are these muscle fibers uninucleated or multinucleated? _____

Where is this type of muscle found within the body? What happens when it contracts?

_____

_____

NERVOUS TISSUE

Your brain, spinal cord, and nerves are constantly monitoring conditions inside and outside of the body. If a problem arises that may jeopardize homeostasis, these organs work together to identify the problem, figure out a solution, and implement this plan in a timely manner. Since these organs all belong to the nervous system, it makes sense that they all contain nervous tissue. Some cells in nervous tissue are involved in communication, while others play a supporting role in the process. **Neurons** use electricity and chemicals to send messages throughout the body. Figure 3.13 shows the unique structure of a typical neuron.

Neuroglial cells support neurons in several ways; they provide nourishment to neurons, protect neurons from infection, and electrically insulate neurons. The term *neuroglia* literally means the glue (*-glia*) that holds the nervous system (*neuro-*) together. Neurons are much larger than neuroglial cells, but typically, neuroglial cells greatly outnumber neurons.

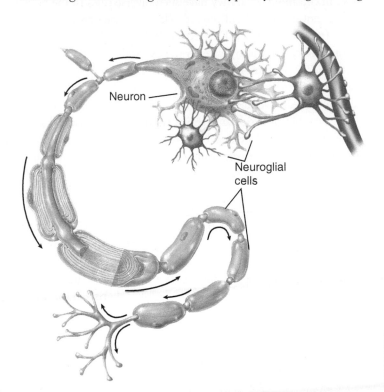

Neuron

Neuroglial cells

FIGURE 3.13   Nervous Tissue

EXERCISE 3.3   Identifying Nervous Tissue on Microscope Slides

During this exercise, you will examine a prepared microscope slide of nervous tissue. Compare your focused microscope slide to Figure 3.14, and label a neuron and a neuroglial cell on this figure.

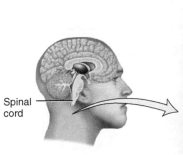

Spinal cord

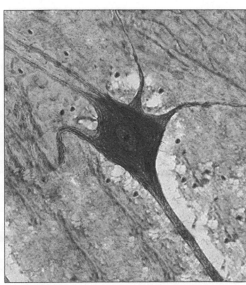

FIGURE 3.14   Nervous Tissue

CONNECTIVE TISSUE

Why are tendons, blood, and compact bone all considered connective tissues? You may wonder what these tissues have in common, since they have such diverse roles in the body. By definition, connective tissue is not composed entirely of cells, since a matrix of noncellular material separates the cells from one another. This **extracellular matrix** may be fluid, gel-like, or solid in nature. Protein fibers may also reside in the extracellular matrix, where they provide support, flexibility, and/or structural integrity.

Connective tissues are further classified into the following categories: fluid connective tissue, supportive connective tissue, loose fibrous connective tissue, and dense fibrous connective tissue. Blood is one example of **fluid connective tissue**, since blood cells are separated from one another by plasma. Cartilage and bone tissue, on the other hand, are classified as **supportive connective tissues**; cartilage contains a gel-like matrix, while bone tissue contains a solid matrix. Cartilage cells and bone cells are sequestered inside of small cavities called **lacuna** because the extracellular matrix is too harsh to sustain life. (Bone and cartilage will be discussed in greater detail during the skeletal system lab.)

Typically, tendons connect muscles to bones; this task requires a dense matrix of protein fibers for resilience and elasticity. Tendons are classified as **dense fibrous connective tissue** since they contain a high density of protein fibers. Fewer protein fibers are found in **loose fibrous connective tissues**, such as adipose (fat) tissue and areolar tissue. Areolar tissue is found throughout the body between epithelial tissue and muscle tissue; the word *areolar* tells us that tiny spaces are present in this tissue.

EXERCISE 3.4   Identifying Connective Tissue on Microscope Slides

During this exercise, you will examine prepared microscope slides of blood, hyaline cartilage, tendons, adipose tissue, compact bone, and areolar tissue. Compare your focused microscope slides with the images provided below, and answer the review questions that accompany each slide.

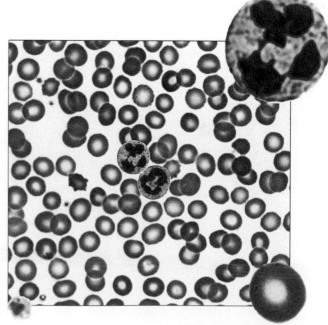

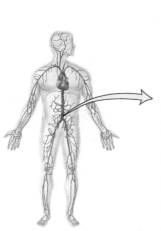

Blood in blood vessels

FIGURE 3.15   **Blood Smear**

Label a blood cell on Figure 3.15.

Label the extracellular matrix of blood on Figure 3.15.

Is blood classified as fluid, supportive, loose fibrous, or dense fibrous connective tissue? Why?

_____

_____

Based on the tasks blood performs in the body, do you think it is vital for blood to be fluid in nature? Why or why not?

_____

_____

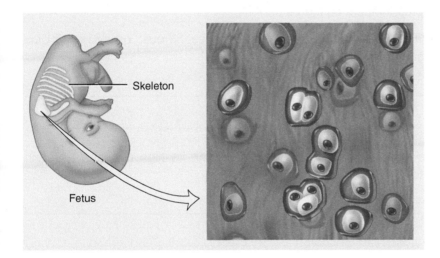

Skeleton

Fetus

FIGURE 3.16   Hyaline Cartilage

Label a chondrocyte (cartilage cell) and a lacuna on Figure 3.16

Label the extracellular matrix of hyaline cartilage on Figure 3.16.

Is hyaline cartilage classified as fluid, supportive, loose fibrous, or dense fibrous connective tissue? Why?

_____

_____

Hyaline cartilage is found in the ribcage, the knee, and the tip of the nose. How do you think its structure relates to its function in the body?

_____

_____

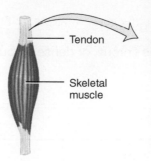

Tendon

Skeletal muscle

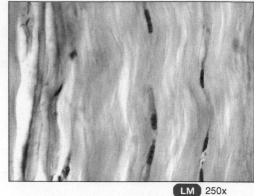

LM 250x

Sectional view of dense regular connective tissue of a tendon

FIGURE 3.17 **Tendon**

Label a fibroblast on Figure 3.17. A fibroblast is a cell that produces (-*blast*) protein fibers (*fibro-*).

Label several protein fibers on Figure 3.17.

Are tendons classified as fluid, supportive, loose fibrous, or dense fibrous connective tissue? Why?

_____

_____

How does the structure of a tendon—in particular, its high density of protein fibers—relate to its function in the body?

_____

_____

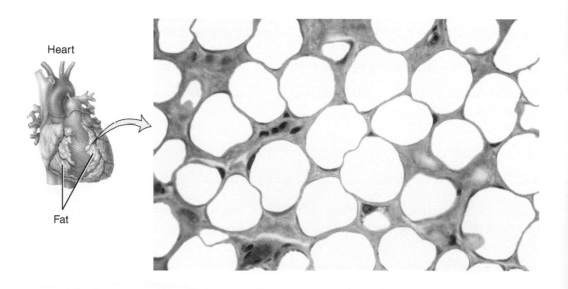

Heart

Fat

FIGURE 3.18 **Adipose Tissue**

Label an adipocyte (fat cell) on Figure 3.18.

Label a vacuole (a cavity in a cell) that is filled with fat droplets on Figure 3.18.

Is adipose tissue classified as fluid, supportive, loose fibrous, or dense fibrous connective tissue? Why?

_____

_____

Where is adipose tissue found in the body? How does its structure relate to its function?

_____

_____

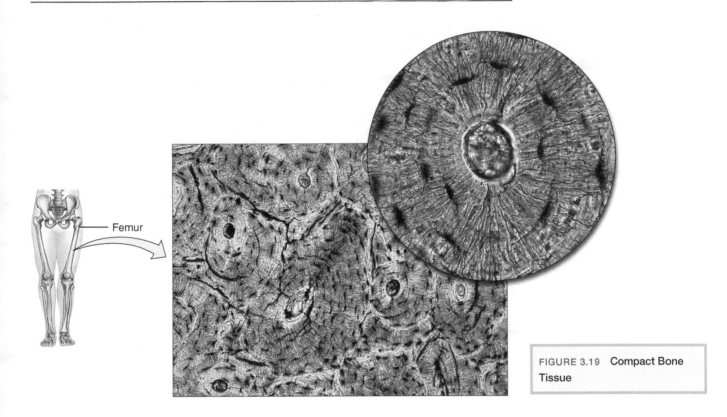

Femur

FIGURE 3.19   Compact Bone Tissue

Label a lacuna on Figure 3.19. (The word *lacuna* means *pit* in Latin.) Osteocytes (bone cells) live inside of lacunae, which look like dark pits embedded in the matrix.

Label the extracellular matrix of compact bone on Figure 3.19.

Is compact bone classified as fluid, supportive, loose fibrous, or dense fibrous connective tissue? Why?

_____

_____

How does the structure of compact bone relate to its function in the body?

_____

_____

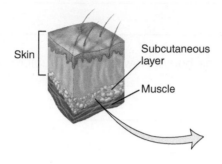

Skin

Subcutaneous layer

Muscle

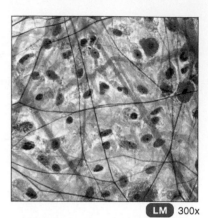

LM 300x

Sectional view of subcutaneous loose connective tissue

FIGURE 3.20 **Areolar Connective Tissue**

Label a cell and a protein fiber on Figure 3.20.

Is areolar tissue classified as fluid, supportive, loose fibrous, or dense fibrous connective tissue? Why?

_____

_____

Does areolar tissue contain a higher or lower density of protein fibers than tendons? How does this characteristic relate to its function in the body?

_____

_____

1. Fill in the missing levels of organization on the diagram below.

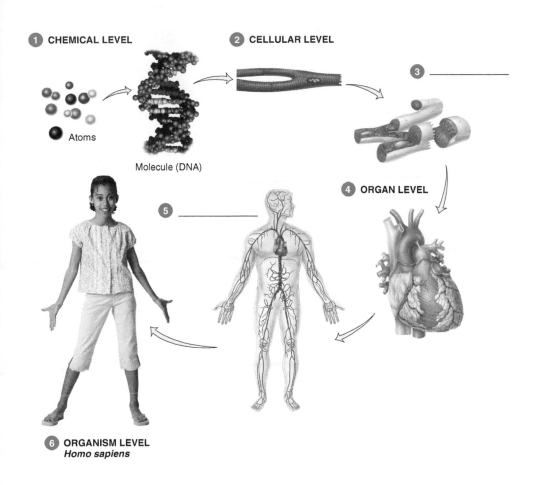

1 CHEMICAL LEVEL

Atoms

Molecule (DNA)

2 CELLULAR LEVEL

3 _____

4 ORGAN LEVEL

5 _____

6 ORGANISM LEVEL
Homo sapiens

2. What type of muscle tissue is found within the digestive tract? Is this type of muscle under voluntary or involuntary control?

_____

_____

3. _____ epithelium is depicted below.
   a. Simple squamous   c. Simple cuboidal
   b. Simple columnar   d. Stratified cuboidal

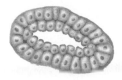

4. Why are tendons, cartilage, and adipose tissue all considered connective tissues?

    _____

    _____

5. Is cardiac, skeletal, or smooth muscle depicted below? How can you tell?

    _____

    _____

6. Name one gland in the human body. Are glands composed of muscle, nervous, connective, or epithelial tissue?

    _____

    _____

7-10. Label the pictures below as connective tissue, muscle tissue, epithelial tissue, or nervous tissue.

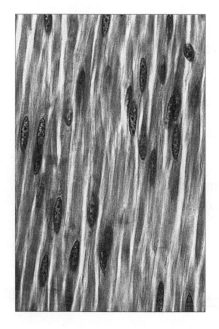

7. _____

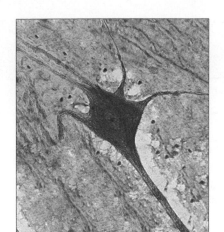

8. _____

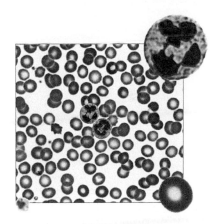

9. _____

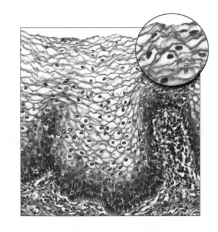

10. _____

LAB 4:
The Chemistry of Life

By the End of This Lab, You Should Be Able to Answer the Following Questions:

- What is an organic molecule?

- What roles do carbohydrates, proteins, lipids, and nucleic acids play in the human body?

- What are the basic subunits of carbohydrates, proteins, lipids, and nucleic acids?

- How do substances we use in everyday life, such as lotion, milk, and fabric softener, differ in composition from one another?

- What is the purpose of adding a chemical indicator to a test substance?

- Why are negative and positive controls included in experiments?

INTRODUCTION

It practically goes without saying that water is crucial for survival. Although humans can survive for several weeks without food, going two or three days without water can pose a serious threat to health. Water is by far the most abundant molecule in the body; this makes sense, given its involvement in so many bodily processes. The unique chemical properties of water help minimize rapid temperature changes in the body. A variety of other substances—such as sugars, salts, and proteins—can also dissolve in water and travel through the bloodstream to reach their target destinations. Although water is essential for life, it is *not* produced exclusively by living organisms. (Water is considered an **inorganic** molecule because it is composed solely of hydrogen and oxygen.) In contrast, most organic molecules would not exist on earth *unless* they were produced by living organisms. Starches, sugars, proteins, fats, oils, and DNA all fall into this category. (By definition, an **organic** molecule contains both carbon and hydrogen atoms.) Can you think of a source for oils, proteins, or carbohydrates that does not depend on organisms? Even fossil fuels are derived from the remains of perished organisms. Table 4.1 summarizes the properties associated with each class of organic molecules: carbohydrates, lipids, proteins, and nucleic acids.

During this lab period, your goal is to detect organic molecules in various household items such as foods, beverages, cosmetics, and cleaning products. **Chemical indicators** will be used to determine whether starches, sugars, proteins, lipids, and DNA are present or absent. If the chemical indicator does *not* change color by the end of the experiment, then the test results are negative. In other words, the molecule of interest (example: starch) was *not* detected in the substance you tested. If the indicator does change color in an expected manner, then the test results are positive. In other words, the molecule of interest (example: starch) was detected in the substance you tested.

TABLE 4.1   Classes of Organic Molecules

Class	Examples	Basic Subunits	Functions
Carbohydrates All atoms written out / Glucose monomer / Glycogen / Cellulose	• Sugars (monosaccharides and disaccharides) • Starches (polysaccharides) • Glycogen (polysaccharide)	Monosaccharides	• Immediate energy source for the body • Short-term energy storage (glycogen) in the liver
Lipids Head / Phosphate group / Glycerol / Tails / Polar heads / Nonpolar tails / Polar heads / Arrangement of phospholipids in a portion of a cell membrane / Chemical structure of a phospholipid	• Fats • Oils • Steroids (example: cholesterol)	Fatty Acids and Glycerol	• Energy source for the body • Long-term energy storage within adipose (fat) tissue • Insulation • Major component of cell membranes
Proteins ❶ Primary structure Polypeptide chain / ❷ Secondary structure / Alpha helix / Beta pleated sheet / Amino acids / Peptide bond / ❸ Tertiary structure / ❹ Quaternary structure / Polypeptide chain	• Enzymes • Transport proteins • Peptide hormones • Contractile proteins • Structural proteins • Defense proteins	Amino acids	• Enzymes speed up chemical reactions in the body • Transport proteins bind to specific substances and transport them throughout the body • Peptide hormones are chemical messages • Contractile proteins move the body itself or substances within the body • Structural proteins give strength and/or rigidity to certain tissues • Defense proteins protect the body from infectious disease

TABLE 4.1   Classes of Organic Molecules (*Continued*)

Class	Examples	Basic Subunits	Functions
Nucleic Acids	• DNA • RNA	Nucleotides	• DNA carries the instructions for building proteins • RNA is involved in converting the DNA code into proteins

To ensure reliable test results are collected, control tubes will be included in every assay. The **negative control** will only contain distilled water and the chemical indicator. Since pure water does not contain organic molecules, no color change is expected in the negative control tube.

If organic molecules are detected in the negative control tube, are your test results reliable? Why or why not?

_____

_____

What are some possible reasons why a positive result (a color change) might develop in a negative control tube?

_____

_____

A positive control will also be included in every assay. The **positive control** will contain a chemical indicator *and* the substance detected by this indicator. A color change is expected in every positive control tube.

If organic molecules are not detected in the positive control tube, are your test results reliable? Why or why not?

_____

_____

What are some possible reasons why a negative result (no color change) might develop in a positive control tube?

_____

_____

TESTING FOR ORGANIC MOLECULES IN HOUSEHOLD ITEMS

In the following exercises, you will test for the presence of sugars, starches, proteins, lipids, and DNA in two household items. Pick two substances to test from the choices provided by your lab instructor. Before you perform this experiment, formulate a hypothesis regarding the presence (or absence) of carbohydrates, lipids, proteins, and DNA in each substance. At the end of this experiment, the results will be analyzed to determine whether your hypotheses were supported or rejected.

Substance 1 _____

Hypothesis for Substance 1

_____

_____

Substance 2 _____

Hypothesis for Substance 2

_____

_____

If you choose a test substance that is not already in liquid form, then this substance must be homogenized with distilled water to produce a liquid. Twenty milliliters (20 mL) of each test substance is sufficient for completing the entire lab. Since the test results rely on your ability to detect color changes, dark substances may be difficult to analyze accurately during this experiment.

Benedict's reagent will be used to detect reducing sugars in your test substances. Glucose, fructose, lactose, and maltose are all examples of reducing sugars. (Sucrose is not a reducing sugar, so it will not be detected by this indicator.) Originally, Benedict's reagent is blue in color. When Benedict's reagent is boiled with reducing sugars, it changes color to green, yellow, orange, or red, depending on the amount of sugar present. Figure 4.1 shows positive and negative results for the Benedict's test.

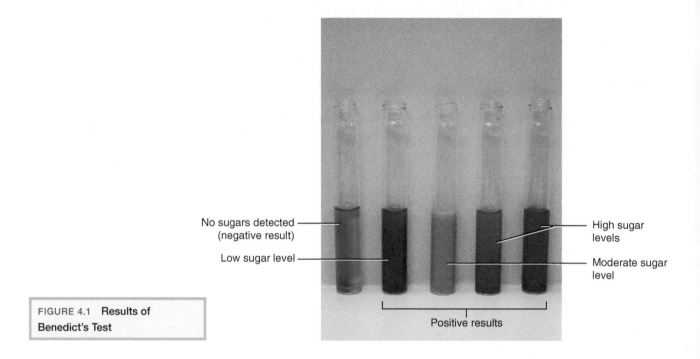

No sugars detected (negative result)

Low sugar level

High sugar levels

Moderate sugar level

Positive results

FIGURE 4.1   **Results of Benedict's Test**

CAUTION   Use care when handling Benedict's reagent to avoid potential irritation of the skin and eyes. In addition to wearing the proper safety gear—gloves and safety goggles—use this reagent in a well-ventilated area. If Benedict's reagent comes in direct contact with your skin or eyes, flush the area with water for at least 15 minutes. Notify your lab instructor of the situation to determine whether medical attention is necessary.

---

1. Collect four test tubes and rinse them thoroughly with distilled water prior to use.

2. Label the test tubes 1–4 and place them in a test tube rack. *Be sure to place labels near the top of each tube*.

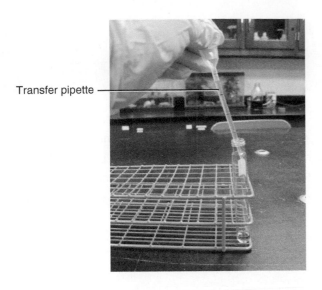

Transfer pipette

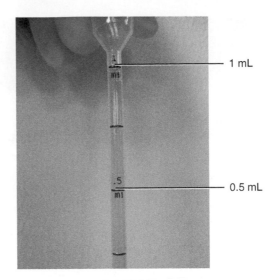

1 mL

0.5 mL

3. Use plastic transfer pipettes to fill each test tube with the contents specified by Table 4.2. **Use a separate transfer pipette with each solution to prevent cross-contamination.**

4. Thoroughly mix the contents of each test tube.

5. Carefully transfer each test tube to a boiling water bath, using a metal test tube holder or heat-resistant gloves.

6. Boil the test tubes for 3–5 minutes in the water bath. Observe any color changes that occur as these tubes boil.

7. Carefully remove each test tube from the boiling water bath, using a metal test tube holder or heat-resistant gloves.

8. Let the tubes cool to room temperature.

9. Examine the final color of each test tube, and record the results in Table 4.2.

10. Dispose of your solutions in the hazardous waste container provided by your instructor. **Do not pour any solutions down the sink without prior approval from your instructor.**

11. Thoroughly rinse the test tubes after use.

TABLE 4.2 Tube Contents and Results for Benedict's Test

Tube Number	Contents	Initial Color	Final Color (after boiling)	Results (+ or – for reducing sugars)
1 - Substance 1	1 mL of substance 1 2 mL Benedict's reagent			
2 - Substance 2	1 mL of substance 2 2 mL Benedict's reagent			
3 - Negative control	1 mL distilled water 2 mL Benedict's reagent			
4 - Positive control	1 mL of 3% glucose solution 2 mL Benedict's reagent			

The iodine test will be used to detect starch in your test substances. Originally, iodine reagent is amber in color. If starch is present, a purple or black color will develop inside of the test tube. Figure 4.2 shows positive and negative results for the iodine test.

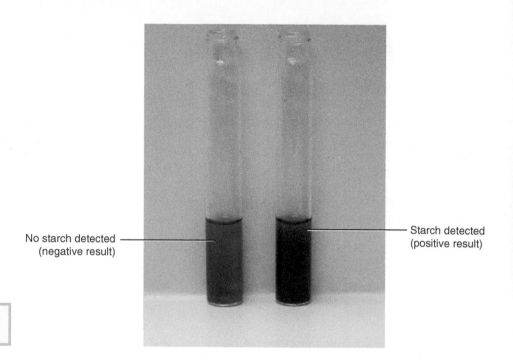

No starch detected (negative result)

Starch detected (positive result)

FIGURE 4.2 **Results of Iodine Test**

CAUTION Use care when handling iodine reagent to avoid clothing stains and potential irritation of the skin and eyes. In addition to wearing the proper safety gear—gloves and safety goggles—use this reagent in a well-ventilated area. If iodine reagent comes in direct contact with your skin or eyes, flush the area with water for at least 15 minutes. Notify your lab instructor of the situation to determine whether medical attention is necessary.

1. Collect four test tubes and rinse them thoroughly with distilled water prior to use.

2. Label the test tubes 1–4 and place them in a test tube rack. *Be sure to place labels near the top of each tube.*

3. Use plastic transfer pipettes to fill each test tube with the contents specified by Table 4.3. **Use a separate transfer pipette with each solution to prevent cross-contamination.**

4. Thoroughly mix the contents of each test tube.

5. Examine the final color of each test tube, and record the results in Table 4.3.

TABLE 4.3   Tube Contents and Results for Iodine Test

Tube Number	Contents	Initial Color	Final Color	Results (+ or − for starch)
1 - Substance 1	1 mL of substance 1 5 drops of Iodine reagent			
2 - Substance 2	1 mL of substance 2 5 drops of Iodine reagent			
3 - Negative control	1 mL distilled water 5 drops of Iodine reagent			
4 - Positive control	1 mL of 1% starch solution 5 drops of Iodine reagent			

6. Dispose of your solutions in the hazardous waste container provided by your instructor. **Do not pour any solutions down the sink without prior approval from your instructor.**

7. Thoroughly rinse the test tubes after use.

EXERCISE 4.3   Testing for Proteins with Biuret Reagent

Biuret reagent will be used to detect proteins in your test substances. Originally, Biuret reagent is blue in color. In the presence of **peptides** (short chains of amino acids) or proteins, Biuret reagent changes color to pink or purple, respectively. Figure 4.3 shows positive and negative results for the Biuret test.

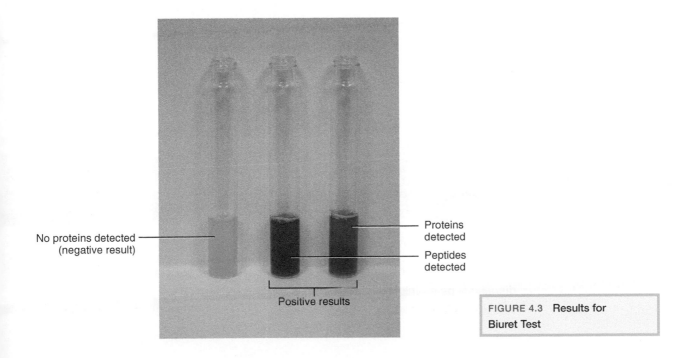

No proteins detected
(negative result)

Proteins detected

Peptides detected

Positive results

FIGURE 4.3   Results for Biuret Test

> **CAUTION** Biuret reagent contains sodium hydroxide, which is a caustic base. Use care when handling Biuret reagent to avoid irritation of the skin and eyes. In addition to wearing the proper safety gear—gloves and safety goggles—use this reagent in a well-ventilated area. **If Biuret reagent comes in direct contact with your skin or eyes, flush the area with water for at least 15 minutes. Notify your lab instructor of the situation to determine whether medical attention is necessary.**

1. Collect four test tubes and rinse them thoroughly with distilled water prior to use.

2. Label the test tubes 1–4 and place them in a test tube rack. *Be sure to place labels near the top of each tube.*

3. Use plastic transfer pipettes to fill each test tube with the contents specified by Table 4.4. **Use a separate transfer pipette with each solution to prevent cross-contamination.**

4. Thoroughly mix the contents of each test tube.

5. Examine the final color of each test tube, and record the results in Table 4.4.

6. Dispose of your solutions in the hazardous waste container provided by your instructor. **Do not pour any solutions down the sink without prior approval from your instructor.**

7. Thoroughly rinse the test tubes after use.

TABLE 4.4 Tube Contents and Results for Biuret Test

Tube Number	Contents	Initial Color	Final Color	Results (+ or – for proteins)
1 - Substance 1	1 mL of substance 1 2 mL Biuret reagent			
2 - Substance 2	1 mL of substance 2 2 mL Biuret reagent			
3 - Negative control	1 mL distilled water 2 mL Biuret reagent			
4 - Positive control	1 mL of 1% albumin solution 2 mL Biuret reagent			

EXERCISE 4.4   Testing for Lipids with the Sudan IV Test

Sudan IV stain will be used to detect lipids in your test substances. Sudan IV stain displays a pale pink color in aqueous solutions, but it stains lipid globules a bright red color. Figure 4.4 shows positive and negative results for the Sudan IV test.

No lipids detected — (negative result)

Lipids detected (positive result)

FIGURE 4.4 **Results of Sudan IV Test**

CAUTION Use care when handling Sudan IV stain to avoid irritation of the skin, throat, and eyes. In addition to wearing the proper safety gear—gloves and safety goggles—use this reagent in a well-ventilated area. Sudan IV stain contains acetone, so excessive inhalation should be avoided. **If Sudan IV stain does come in direct contact with the skin or eyes, flush the area with water for at least 15 minutes. Notify your lab instructor of the situation to determine whether medical attention is necessary.**

---

1. Collect four test tubes and rinse them thoroughly with distilled water prior to use.

2. Label the test tubes 1–4 and place them in a test tube rack. *Be sure to place labels near the top of each tube.*

3. Use plastic transfer pipettes to fill each test tube with the contents specified by Table 4.5. **Use a separate transfer pipette with each solution to prevent cross-contamination.**

4. Thoroughly mix the contents of each test tube.

5. Allow the tubes to sit undisturbed for five minutes. If lipids are detected by Sudan IV reagent, then red fat globules will rise to the top of the tube.

6. Examine the final color of each test tube, and record the results in Table 4.5.

7. Dispose of your solutions In the hazardous waste container provided by your instructor. **Do not pour any solutions down the sink without prior approval from your instructor.**

8. Thoroughly rinse the test tubes after use.

TABLE 4.5 Tube Contents and Results for Sudan IV Test

Tube Number	Contents	Initial Color	Final Color	Results (+ or − for lipids)
1 - Substance 1	1 mL of substance 1 3 mL of distilled water 10 drops of Sudan IV reagent			
2 - Substance 2	1 mL of substance 2 3 mL of distilled water 10 drops of Sudan IV reagent			
3 - Negative control	4 mL of distilled water 10 drops of Sudan IV reagent			
4 - Positive control	1 mL vegetable oil 3 mL of distilled water 10 drops of Sudan IV reagent			

Dische diphenylamine reagent will be used to detect DNA in your test substances. This reagent is originally clear, but it turns blue in the presence of DNA. Figure 4.5 shows positive and negative results for the Dische diphenylamine test; keep in mind that the color of your test substance may alter the color of a positive result.

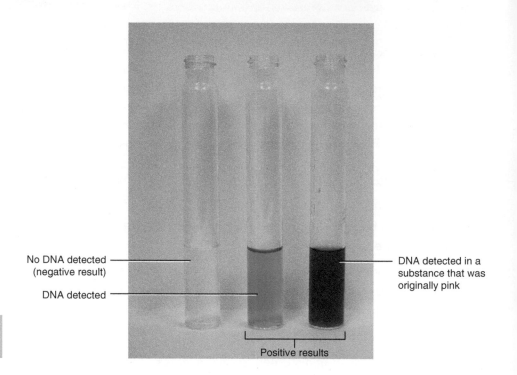

No DNA detected (negative result)

DNA detected

DNA detected in a substance that was originally pink

Positive results

FIGURE 4.5 **Results of Diphenylamine Test**

CAUTION Special care must be taken to avoid skin contact with diphenylamine reagent, since diphenylamine can be absorbed directly through the skin. In addition to wearing the proper safety gear—gloves and safety goggles—*use this reagent in a fume hood.* **If diphenylamine reagent comes in direct contact with the skin or eyes, flush with water for 15 minutes and seek medical attention.**

1. Collect four test tubes and rinse them thoroughly with distilled water prior to use.

2. Label the test tubes 1–4 and place them in a test tube rack. *Be sure to place labels near the top of each tube.*

3. Use plastic transfer pipettes to fill each test tube with the contents specified by Table 4.6. **Use a separate transfer pipette with each solution to prevent cross-contamination.**

4. Thoroughly mix the contents of each test tube.

5. Carefully transfer each test tube to a boiling water bath, using a metal test tube holder or heat-resistant gloves.

6. Boil the test tubes for 10 minutes in the water bath. Observe any color changes that occur as the tubes boil.

7. Carefully remove each test tube from the boiling water bath, using a metal test tube holder or heat-resistant gloves.

8. Let the tubes cool to room temperature.

9. Examine the final color of each test tube, and record the results in Table 4.6.

10. Dispose of your solutions in the hazardous waste container provided by your instructor. **Do not pour any solutions down the sink without prior approval from your instructor.**

11. Thoroughly rinse the test tubes after use.

TABLE 4.6   Tube Contents and Results for Dische Diphenylamine Test

Tube Number	Contents	Initial Color	Final Color	Results (+ or − for DNA)
1 - Substance 1	1 mL of substance 1 1 mL diphenylamine reagent			
2 - Substance 2	1 mL of substance 2 1 mL diphenylamine reagent			
3 - Negative control	1 mL distilled water 1 mL diphenylamine reagent			
4 - Positive control	1 mL strawberry solution 1 mL diphenylamine reagent			

CONCLUSIONS

Was your hypothesis regarding the composition of _____ (substance one) supported or rejected? Comment on the overall nature of this substance based on your results. If necessary, revise your initial hypothesis based on the results of this experiment.

_____

_____

_____

_____

Was your hypothesis regarding the composition of _____ (substance two) supported or rejected? Comment on the overall nature of this substance based on your results. If necessary, revise your initial hypothesis based on the results of this experiment.

_____

_____

_____

_____

Did any other students test the items you selected? If so, how do your results compare with those obtained by other students?

_____

_____

_____

_____

What are potential sources of error in this experiment? What improvements (if any) could be made to this experiment?

_____

_____

_____

_____

Did your control tubes produce the expected results? If not, what does this suggest about the reliability of your results?

_____

_____

_____

_____

A CLOSER LOOK AT FOOD COMPOSITION

One logical question that stems from this lab is related to nutrition. If you detected proteins in a weight-loss shake, for instance, *how much* protein is present per serving? How does this quantity relate to the amount of lipids and carbohydrates present? Nutritional labels are found on most food items, but many of us find it difficult to compare labels in the middle of a grocery store. If you would like to compare the composition of different food choices, you can access the National Nutrient Database online at http://www.nal.usda.gov. This database is provided by a scientific research agency called the Agricultural Research Service (ARS), which belongs to the United States Department of Agriculture (USDA). The goal of ARS is "to ensure that Americans have reliable, adequate supplies of high-quality food and other agricultural products."

If you would like to learn more about nutritional labels or the recommended dietary guidelines for children and adults, you can also access the USDA's Food and Nutrition Information Center at http://fnic.nal.usda.gov.

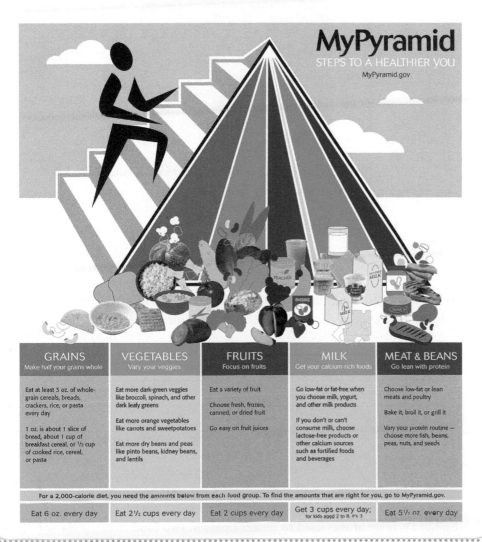

| GRAINS | VEGETABLES | FRUITS | MILK | MEAT & BEANS |
Make half your grains whole	Vary your veggies	Focus on fruits	Get your calcium-rich foods	Go lean with protein
Eat at least 3 oz. of whole-grain cereals, breads, crackers, rice, or pasta every day 1 oz. is about 1 slice of bread, about 1 cup of breakfast cereal, or ½ cup of cooked rice, cereal, or pasta	Eat more dark-green veggies like broccoli, spinach, and other dark leafy greens Eat more orange vegetables like carrots and sweetpotatoes Eat more dry beans and peas like pinto beans, kidney beans, and lentils	Eat a variety of fruit Choose fresh, frozen, canned, or dried fruit Go easy on fruit juices	Go low-fat or fat-free when you choose milk, yogurt, and other milk products If you don't or can't consume milk, choose lactose-free products or other calcium sources such as fortified foods and beverages	Choose low-fat or lean meats and poultry Bake it, broil it, or grill it Vary your protein routine — choose more fish, beans, peas, nuts, and seeds

For a 2,000-calorie diet, you need the amounts below from each food group. To find the amounts that are right for you, go to MyPyramid.gov.

Eat 6 oz. every day	Eat 2½ cups every day	Eat 2 cups every day	Get 3 cups every day; for kids aged 2 to 8, it's 2	Eat 5½ oz. every day

1. Match each subunit listed below with the proper class of organic molecules.

   a. Amino Acids   __ Nucleic Acids

   b. Glycerol and Fatty Acids   __ Proteins

   c. Nucleotides   __ Carbohydrates

   d. Monosaccharides   __ Lipids

2. Iodine reagent was used to detect starch in two food items: soda and pudding. The results from this experiment are shown below. Which tube (1 or 2) shows a negative result for the presence of starch? _____

Tube 1 ———   ——— Tube 2

3. _____ serve as long-term energy storage molecules in the body. Fats and oils belong to this class of organic molecules.

4. Which of the following is a nucleic acid?
   a. Starch   c. DNA
   b. Oil   d. Enzyme

5. What is the difference between a negative control and a positive control? Is a color change expected in each tube? Explain your answer.

   _____

   _____

   _____

   _____

6. _____ reagent is used to detect reducing sugars.
   a. Benedict's   c. Iodine
   b. Dische diphenylamine   d. Sudan IV

7. Sudan IV reagent was used to detect lipids in two food items: soda and salad dressing. The results from this experiment are shown on top of next page. Which tube (1 or 2) shows a positive result for the presence of lipids? _____

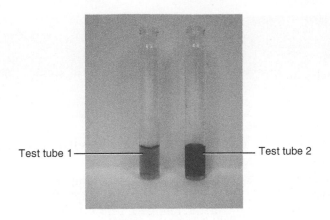

Test tube 1 — Test tube 2

8. _____ reagent is used to detect DNA in test substances.
   a. Benedict's reagent   c. Iodine reagent
   b. Dische diphenylamine reagent   d. Sudan IV stain

9. The Benedict's test was performed to detect reducing sugars in apple juice. Based on the results shown below, are reducing sugars present in apple juice? How can you tell?

   _____

   _____

10. Imagine you tested vanilla pudding for the presence of lipids, starch, sugars, proteins, and DNA. The following results were collected:

   Benedict's test → orange color after heating
   Iodine test → dark purple color
   Biuret test → purple color
   Sudan IV test → bright red color
   Dische diphenylamine test → blue color

   Based on your results, which organic molecules are present in this brand of pudding?

LAB 5:

The Digestive System

By the End of This Lab, You Should Be Able to Answer the Following Questions:

- Which organs come in contact with food as it travels through the gastrointestinal tract? What is the function of each of these organs?

- What accessory organs belong to the digestive system?

- What is the difference between mechanical digestion and chemical digestion?

- What does the suffix *–ase* denote in biology?

- What is the optimum temperature for enzyme activity in the human body?

- What is optimum pH for enzyme activity in the mouth? Stomach? Small intestine?

- Why are chemical indicators added to test tubes during enzymatic assays?

- What role does bile play in lipid digestion?

INTRODUCTION

Imagine you are rushing to the airport for a weeklong vacation with a friend. You prepared a sandwich to eat during the flight, but in a frenzy, you leave the sandwich on top of the kitchen counter. How do you think this sandwich will look by the time you return home? You'll definitely think twice about eating it, since it smells funny and has some mold growing on it! However, the sandwich still *looks* like a sandwich because of its shape, color, and overall consistency. Now, imagine your dog Fizzy devours the sandwich as soon as you leave the house. Within a matter of hours, this sandwich is digested as it travels through the mouth, pharynx (throat), esophagus, stomach, and small intestine. During this process, the sandwich changes in texture from a solid to a paste and then into a soupy mixture. The starch and meat are broken down along the way into monosaccharides and amino acids, respectively. Once these nutrient molecules hit the bloodstream, they can be delivered to Fizzy's cells for energy, growth, and repair.

A well-rounded diet provides the **micronutrients** (vitamins and minerals) and **macronutrients** (carbohydrates, proteins, and lipids) essential for survival. It is the job of the digestive system to **ingest** (take in) food, digest food into basic subunits, absorb nutrient molecules, and eliminate any nondigested remains. The digestive system is composed of the gastrointestinal (GI) tract and accessory organs that aid in the process of digestion. (The prefix *gastro-* is commonly used in medical terminology when referring to the stomach.) The GI tract is a muscular tube that extends from the mouth to the anus. Amazingly, this tube is approximately 30 feet (9 meters) long in adults, and the majority of this tube is coiled within the abdominal cavity.* Accessory organs such as the gallbladder, liver, and pancreas lie outside of the GI tract; nevertheless, they are connected to the GI tract by ducts. A **duct** is a tube that transfers fluid from its site of production and/or storage to its final destination. Figure 5.1 illustrates the organs in the GI tract, as well as the major accessory organs involved in digestion.

---

*Source for length of alimentary canal: Gray, Henry F.R.S. *Gray's Anatomy*. 1974.

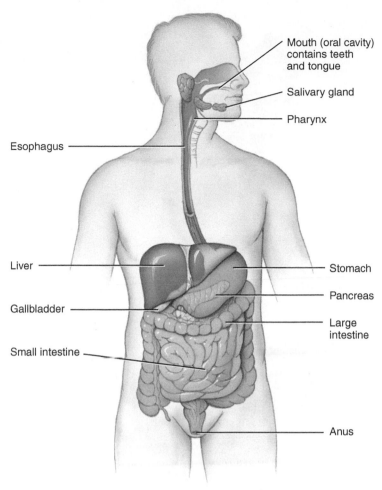

Mouth (oral cavity)
contains teeth
and tongue

Salivary gland

Pharynx

Esophagus

Liver

Gallbladder

Small intestine

Stomach

Pancreas

Large
intestine

Anus

Right lateral view of head and neck and anterior view of trunk

FIGURE 5.1 Gross Anatomy of
the Digestive System

EXERCISE 5.1   Gross Anatomy of the Digestive System

Digestive organs are not confined to one specific body cavity. As shown in Figure 5.1, digestive organs span the head, neck, thoracic cavity, abdominal cavity, and even the pelvic cavity. The length of the GI tract is advantageous to humans, since an extraordinary amount of surface area is required for digestion, water uptake, and nutrient uptake.

During this exercise, you will identify major digestive organs on a model of the human digestive system. Alternately, your instructor may ask you to identify these organs on a virtual cadaver dissection, a dissected fetal pig, or a dissected human cadaver. Label the listed structures on Figure 5.2.

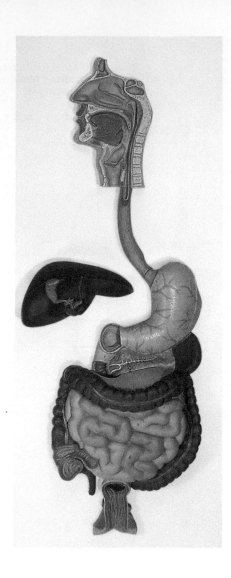

FIGURE 5.2 **Identification of Digestive Organs**

Organs of the Gastrointestinal Tract

1. Mouth
2. Pharynx
3. Esophagus
4. Stomach
5. Small intestine
   a. Duodenum
   b. Jejunum
   c. Ileum

6. Large intestine
   a. Ascending colon
   b. Transverse colon
   c. Descending colon
   d. Sigmoid colon
7. Rectum
8. Anus

Accessory Organs of the Digestive System

1. Liver
2. Gallbladder
3. Pancreas

Which organ is labeled A below? Nutrient molecules are absorbed in this region of the GI tract.

_____

_____

Which organ is labeled B below? This organ is responsible for bile production.

_____

_____

Which organ is labeled C below? This organ stores excess bile.

_____

_____

Which organ is labeled D below? This organ is connected to the esophagus and the duodenum.

_____

_____

Which organ is labeled E below? In the body, the head of this organ resides within the loop of the duodenum.

_____

_____

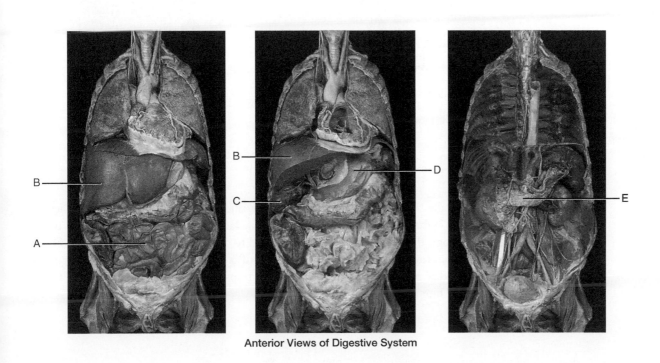

Anterior Views of Digestive System

CHEMICAL DIGESTION VERSUS MECHANICAL DIGESTION

Food is digested both mechanically and chemically as it passes through the GI tract. **Mechanical digestion** refers to processes that physically tear food into smaller chunks. For example, teeth take part in mechanical digestion as they rip food into smaller pieces. **Chemical digestion** involves enzymes that digest food at the molecular level. An **enzyme** is simply a protein that speeds up a chemical reaction in the body. **Digestive enzymes** are responsible for breaking large nutrient molecules (example: proteins) into smaller subunits (example: amino acids).

Figure 5.3 illustrates the basic steps involved in chemical digestion. First, a digestive enzyme must physically bind to an ingested food molecule; this **substrate** (food molecule) will be altered during the reaction. Next, the digestive enzyme uses water to break chemical bonds in the substrate. Finally, the **products** of this reaction are released by the enzyme. Enzymes are not destroyed by this process, so they can be reused to digest many nutrient molecules. The three-dimensional shape of each enzyme gives it **specificity**; this means each enzyme can only bind to one type of substrate and speed up one specific reaction in the body.

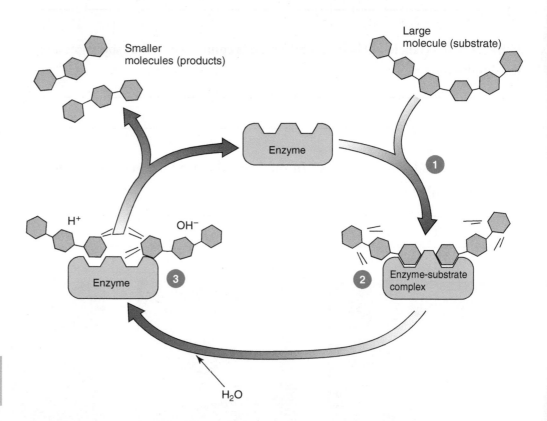

FIGURE 5.3 Activity of a Digestive Enzyme

Typically the name of an enzyme gives us clues about its role in the body. Lactase, for instance, is the enzyme that chemically digests lactose (milk sugar) in the small intestine. In biology, the suffix –*ase* is commonly used to denote the names of enzymes. Other enzyme names (example: trypsin) end with the suffix –*in*, which reminds us that enzymes are composed of protein. Hemoglobin, albumin, trypsin, and insulin are all proteins found inside of the human body.

Enzymes require specific pH and temperature conditions in order to function properly. If enzymes are exposed to less desirable conditions, then they may **denature** (lose their unique three-dimensional shape) and lose the ability to speed up reactions. Although temperature stays relatively constant throughout the GI tract, a variety of pH levels are encountered by food during digestion. Pure water has a **neutral** pH of 7; within the body, saliva and blood exhibit pH levels that are relatively close to neutral. Since gastric juices have a pH well below 7, **acidic** conditions are found in the stomach. **Basic** conditions are found within the small intestine, since the pH is higher than 7. Figure 5.4 summarizes the pH scale, which ranges from 0–14. Since pH values are mathematical logarithms (p) that abbreviate the concentration of hydrogen ions (H^+), no units are associated with pH values.

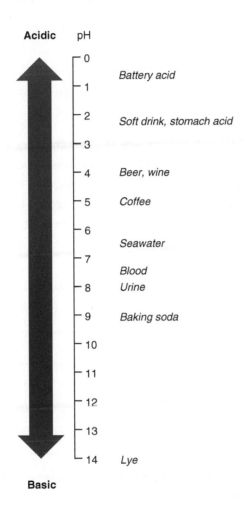

FIGURE 5.4 **The pH Scale**

In the following experiments, you will digest foods *in vitro*—outside of a living organism—with the help of digestive enzymes. At the same time, you will examine enzyme specificity, the effects of pH on enzyme activity, and the effects of temperature on enzyme activity. Since chemical digestion occurs at the molecular level, this process can be difficult to visualize with the naked eye. For this reason, we will use chemical indicators to confirm whether or not digestion is taking place.

Lactose is a sugar found in dairy products, such as milk, cheese, and ice cream. The enzyme lactase chemically digests lactose (a disaccharide) into glucose and galactose (monosaccharides) within the small intestine. Following digestion and absorption, these monosaccharides must travel through the bloodstream to reach tissue cells. Lactose digestion is summarized by the following equation:

$$\text{Lactose} + H_2O \xrightarrow{\text{lactase}} \text{glucose} + \text{galactose}$$

Sucrose, maltose, and lactose are all disaccharides that contain glucose. (Maltose is composed of two glucose subunits, while sucrose is composed of a glucose subunit and a fructose subunit.) Can an enzyme really tell the difference between these sugars? To determine the specificity of lactase, we will test its activity on sucrose, maltose, and lactose. Glucose test strips will be used to determine whether or not digestion occurred in each test tube.

Formulate a hypothesis to test regarding the ability of lactase to digest sucrose, maltose, and lactose.

_____

_____

Procedure

1. Collect four test tubes and rinse them thoroughly with distilled water prior to use.

2. Label the test tubes 1–4 and place them in a test tube rack. *Be sure to place labels near the top of each tube.*

3. Use plastic transfer pipettes to fill each test tube with the contents specified below. **Use a separate transfer pipette with each solution to prevent cross-contamination. (See photos on top of next page.)**
   a. Tube 1 = 1 mL milk
   b. Tube 2 = 1 mL milk
   c. Tube 3 = 1 mL of 1% sucrose solution
   d. Tube 4 = 1 mL of 1% maltose solution

1 mL

0.5 mL

4. Determine the initial amount of glucose present in each tube with a glucose test strip. Record the results in Table 5.1.

"PRECISION" Glucose Test Strip
100 test strips
PROCEDURES:
1. Remove one glucose test strip from the amber bag.
2. Reseal the bag as soon as possible.
3. Do not touch test pad with fingers.
4. Completely immerse test pad into specimen for 1-2 seconds.
5. Run the test strip along the edge of the container to remove excess liquid.
6. Read after two minutes for urine test, or after three minutes for water solutions of glucose.
7. Match color of test pad to color scale.

concentration mg/dL

0   100   300   1000   3000

PRECISION LABS, INC.
1-800-733-0266

5. Add the solutions specified below to each test tube. **Use a separate transfer pipette with each solution to prevent cross-contamination.**
   a. Tube 1 = 1 mL of distilled water
   b. Tube 2 = 1 mL of 0.5% lactase supplement
   c. Tube 3 = 1 mL of 0.5% lactase supplement
   d. Tube 4 = 1 mL of 0.5% lactase supplement

6. Thoroughly mix the contents of each test tube.

7. Place the test tubes into a 37°C water bath for 5 minutes.

8. Remove the test tubes from the water bath and place them in a test tube rack.

9. Use glucose test strips to determine the final amount of glucose present in each test tube. Record the results in Table 5.1.

10. Dispose of your solutions in the hazardous waste container provided by your instructor. **Do not pour any solutions down the sink without prior approval from your instructor.**

11. Thoroughly rinse the test tubes after use.

TABLE 5.1   **Digestion of Sugars by Lactase**

Tube	Contents	Initial Concentration of Glucose (mg/dL)	Final Concentration of Glucose (mg/dL)
1 – negative control	1 mL milk 1 mL distilled water		
2 – lactose	1 mL milk 1 mL of 0.5% lactase supplement		
3 – sucrose	1 mL of 1% sucrose solution 1 mL of 0.5% lactase supplement		
4 – maltose	1 mL of 1% maltose solution 1 mL of 0.5% lactase supplement		

Review Questions for Step by Step 5.2

Based on your results, can lactase digest a variety of substrates, or just one specific substrate? If necessary, modify your initial hypothesis based on the results of this experiment.

_____

_____

Lactase is the name of the enzyme that digests lactose. What is the likely name for an enzyme that digests sucrose? _____ Maltose?
_____

Which digestive disorder is caused by lactase deficiency? What are possible treatment options for this disorder?

_____

_____

Proteins are abundant in foods such as meat, fish, dairy products, eggs, and beans. Dietary proteins are chemically digested by several enzymes in the stomach and the small intestine. Within the small intestine, for instance, an enzyme named trypsin digests proteins into **peptides** (short chains of amino acids). These peptides must be further digested into amino acids prior to absorption. The activity of trypsin is summarized by the following equation:

$$\text{proteins} + \text{water} \xrightarrow{\text{trypsin}} \text{peptides}$$

During this exercise, trypsin will be used to chemically digest milk proteins. Casein is the protein that gives milk a white color, so its digestion will convert milk into a transparent fluid. Figure 5.5 illustrates the effects of casein digestion on the appearance of milk. During this experiment, you will also determine the optimum pH for trypsin activity.

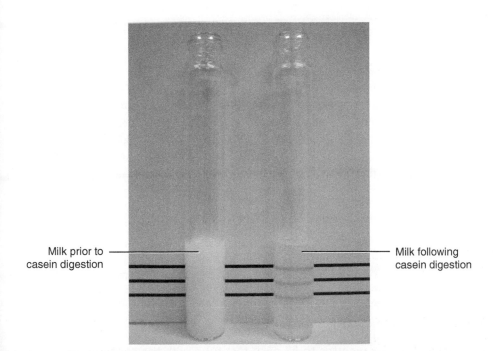

Milk prior to casein digestion

Milk following casein digestion

FIGURE 5.5   **Effect of Casein Digestion on the Color of Milk**

Biuret reagent will be used to confirm whether milk proteins were digested by trypsin. Figure 5.6 shows the possible results for the Biuret test. Originally, Biuret reagent is blue in color. In the absence of proteins and peptides, Biuret reagent remains blue. A pink color indicates peptides are present, while a purple color indicates proteins are present.

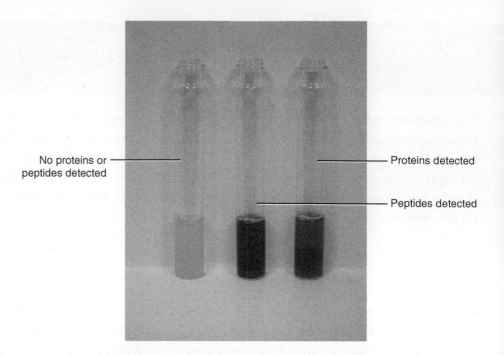

No proteins or peptides detected

Proteins detected

Peptides detected

FIGURE 5.6 Results of Biuret Test

Using your knowledge of the digestive system, formulate a hypothesis to test regarding the optimum pH for trypsin activity.

_____

_____

CAUTION Biuret reagent contains sodium hydroxide, which is a caustic base. Use care when handling Biuret reagent to avoid irritation of the skin and eyes. In addition to wearing the proper safety gear—gloves and safety goggles—use this reagent in a well-ventilated area. If Biuret reagent comes in direct contact with your skin or eyes, flush the area with water for at least 15 minutes. Notify your lab instructor of the situation to determine whether medical attention is necessary.

Procedure

1. Collect five test tubes and rinse them thoroughly with distilled water prior to use.

2. Label the test tubes 1–5 and place them in a test tube rack. *Be sure to place labels near the top of each tube.*

3. Use plastic transfer pipettes to fill each test tube with the contents specified by Table 5.2. **Use a separate transfer pipette with each solution to prevent cross-contamination.**

4. Thoroughly mix the contents of each test tube.

5. Place the test tubes into a 37°C water bath for 15 minutes.

6. Remove the test tubes from the water bath and place them in a test tube rack.

7. Observe the color and opacity / transparency of each milk solution. Record the results in Table 5.2.

8. Add 1 mL of Biuret reagent to each tube with a plastic transfer pipette. Thoroughly mix each test tube to distribute the Biuret reagent.

9. Record the final color of each test tube in Table 5.2.

10. Dispose of your solutions in the hazardous waste container provided by your instructor. **Do not pour any solutions down the sink without prior approval from your instructor.**

11. Thoroughly rinse the test tubes after use.

TABLE 5.2   Digestion of Milk Proteins by Trypsin

Tube	Contents	Appearance of Milk after Incubation	Final Color of Biuret Reagent	Results (+ or − for peptides)
1 – negative control	1 mL of pH 7 milk 1 mL of distilled water			
2 – pH 5	1 mL of pH 5 milk 1 mL of 0.5% trypsin solution			
3 – pH 6	1 mL of pH 6 milk 1 mL of 0.5% trypsin solution			
4 – pH 7	1 mL of pH 7 milk 1 mL of 0.5% trypsin solution			
5 – pH 8	1 mL of pH 8 milk 1 mL of 0.5% trypsin solution			

Review Questions for Exercise 5.3

Based on your results, what is the optimum pH for trypsin activity? If necessary, modify your initial hypothesis based on the result of this experiment.

_____

_____

What would happen to trypsin if it was placed into an acidic environment, such as the stomach? Why?

_____

_____

In nature, many plants store excess glucose in the form of starch. Rice, potatoes, wheat, corn, and peas are all rich in starch for this very reason. Starch is also added to certain processed foods, such as pudding and gravy mixes, as a thickening agent. Within the mouth and small intestine, the enzyme amylase digests starch into disaccharides and trisaccharides. This reaction is summarized by the following equation:

$$\text{Starch} + \text{H}_2\text{O} \xrightarrow{\quad\text{amylase}\quad} \text{disaccharides and trisaccharides}$$

In the following exercise, you will use amylase to digest starch in prepackaged pudding. At the same time, you will determine the optimum temperature for amylase activity. Starch digestion will be monitored using a chemical indicator: iodine reagent. Originally, iodine reagent is amber in color. In the absence of starch, iodine reagent retains its original color. If starch is present, however, a purple or black color develops inside of the test tube. Figure 5.7 shows the possible results for the iodine test.

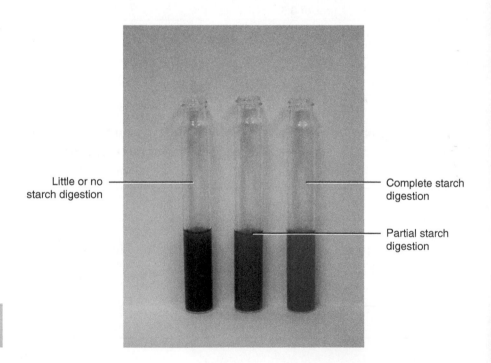

Little or no starch digestion

Complete starch digestion

Partial starch digestion

FIGURE 5.7   **Results of Iodine Test**

Formulate a hypothesis to test regarding the optimum temperature for amylase activity.

_____

_____

CAUTION Use care when handling iodine reagent to avoid clothing stains and potential irritation of the skin and eyes. In addition to wearing the proper safety gear — gloves and safety goggles — use this reagent in a well-ventilated area. If iodine reagent comes in direct contact with your skin or eyes, flush the area with water for at least 15 minutes. Notify your lab instructor of the situation to determine whether medical attention is necessary.

Procedure

1. Collect five test tubes and rinse them thoroughly with distilled water prior to use.

2. Label the test tubes 1–5 and place them in a test tube rack. *Be sure to place labels near the top of each tube.*

3. Use plastic transfer pipettes to fill each test tube with the contents specified by Table 5.3. **Use a separate transfer pipette with each solution to prevent cross-contamination.**

4. Place each test tube into the water bath (or ice bath) specified by Table 5.3. **Use a metal test tube holder or heat-resistant gloves to transfer Tube 5 into the 80°C water bath.**

5. Let each test tube equilibrate in the water bath for 5 minutes.

6. Add 1 mL of 5% vanilla pudding solution to each test tube. If possible, perform this step without removing the tubes from their temperature environments.

7. Thoroughly mix the contents of each test tube. After mixing each tube, quickly return it to the water bath.

8. Let the test tubes incubate in the water baths for an additional 10 minutes.

9. Following the incubation period, add 5 drops of iodine reagent to each test tube. If possible, perform this step without removing the tubes from their temperature environments.

10. Remove your tubes from the water baths, mix them thoroughly, and place them in a test tube rack. **Use a metal test tube holder or heat-resistant gloves to remove Tube 5 from the 80°C water bath.**

11. Examine the color of each test tube and record the results in Table 5.3.

12. Dispose of your solutions in the hazardous waste container provided by your instructor. **Do not pour any solutions down the sink without prior approval from your instructor.**

13. Thoroughly rinse the test tubes after use.

TABLE 5.3   Digestion of Starch by Amylase

Tube	Contents	Color after Iodine Addition	Did Starch Digestion Occur?
1 – negative control (≈37°C)	1 mL of distilled water		
2 – ice bath (≈4°C)	1 mL of 1% amylase solution		
3 – room temperature (≈22°C)	1 mL of 1% amylase solution		
4 – body temperature (≈37°C)	1 mL of 1% amylase solution		
5 – 80°C	1 mL of 1% amylase solution		

Review Questions for Exercise 5.4

Based on your results, what is the optimal temperature for amylase activity? If necessary, modify your initial hypothesis based on the result of this experiment.

_____

_____

Do you think every enzyme in your body has the same optimal temperature? Why or why not?

_____

_____

EXERCISE 5.5   Effects of Bile on Lipase Activity

Lipids are considered **hydrophobic** because they are repelled by water molecules. As a result, oil and water tend to separate from one another in products such as salad dressing. Since enzymes reside in watery body fluids, how do they encounter lipids during the digestive process? It takes the help of **emulsifier** molecules to suspend fats and oils into water. Soaps and detergents are actually emulsifiers that we use in everyday life. These emulsifiers break lipid globules into smaller droplets, which ultimately become suspended in water. Once this suspension is formed, it can be whisked away from dirty items such as clothes, dishes, and skin. **Bile** is the emulsifier that is produced within the human body. Although bile is created in the liver, excess bile is actually stored by the gallbladder. Ducts transfer bile from these organs to the small intestine, where it emulsifies fats and oils.

In the following exercise, you will digest cream lipids using an enzyme called lipase. At the same time, you will determine whether bile is necessary for lipid digestion. The activity of lipase is summarized by the following equation:

$$\text{Lipids} + \text{water} \xrightarrow{\text{Lipase}} \text{fatty acids and monoglycerides}$$

Fatty acids are one product of lipid digestion, as shown in the equation above. Just like any other acid, fatty acids will lower the pH of the solution once digestion takes place. For this reason, we can use a pH indicator called phenol red to monitor lipid digestion. Phenol red is pink in basic solutions, orange in neutral solutions, and yellow in acidic conditions. Figure 5.8 shows the colors produced by the phenol red indicator and the pH range associated with each color.

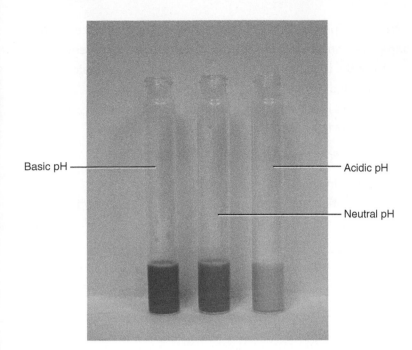

Basic pH ———— ———— Acidic pH

———— Neutral pH

FIGURE 5.8 Interpreting the Color of Phenol Red

Formulate a hypothesis to test regarding the effects of bile on the rate (speed) of lipid digestion.

_____

_____

CAUTION Use care when handling phenol red to avoid potential irritation of the skin and eyes. In addition to wearing the proper safety gear — gloves and safety goggles — use this reagent in a well-ventilated area. If phenol red comes in direct contact with your skin or eyes, flush the area with water for at least 15 minutes. Notify your lab instructor of the situation to determine whether medical attention is necessary.

Procedure

1. Collect three test tubes and rinse them thoroughly with distilled water prior to use.

2. Label the test tubes 1–3 and place them in a test tube rack. _Be sure to place labels near the top of each tube._

3. Fill each test tube with the contents specified by Table 5.4. **Use a separate transfer pipette with each solution to prevent cross-contamination.**

4. Add one drop of 0.1M NaOH (sodium hydroxide) to each tube.

5. Mix the contents of each tube thoroughly, and record the initial colors observed in Table 5.4.

6. Using plastic transfer pipettes, add the solutions listed below to each test tube. **Use a separate transfer pipette with each solution to prevent cross-contamination.** Note the time of addition in the spaces provided below.

    a. Tube 1 = 1 mL distilled water   Time of Addition: _____

    b. Tube 2 = 1 mL of 0.5% lipase solution   Time of Addition: _____

    c. Tube 3 = 1 mL of 0.5% lipase solution   Time of Addition: _____

7. Thoroughly mix the contents of each test tube.

8. Incubate your test tubes in a 37°C water bath for 30 minutes. If a color change occurs during the incubation period, then note the time of this event *and* the color change observed in the spaces provided below.

    a. Test Tube 1 _____

    b. Test Tube 2 _____

    c. Test Tube 3 _____

9. Remove your test tubes from the water bath, and place them in a test tube rack.

10. Record the final color of each test tube and the time required for digestion (if applicable) in Table 5.4.

11. Dispose of your solutions in the hazardous waste container provided by your instructor. **Do not pour any solutions down the sink without prior approval from your instructor.**

12. Thoroughly rinse the test tubes after use.

TABLE 5.4   **Digestion of Cream by Lipase**

Tube	Contents	Initial Color	Final Color	Time Required for Digestion (if applicable)
1 – negative control	1 mL cream 1 mL of 1% bile salt solution 1 mL of 0.4% phenol red solution			
2 – lipase without emulsifier	1 mL cream 1 mL of distilled water 1 mL of 0.4% phenol red solution			
3 – lipase with emulsifier	1 mL cream 1 mL of 1% bile salt solution 1 mL of 0.4% phenol red solution			

Review Questions for Exercise 5.5

Based on your results, is bile necessary for lipid digestion? Revise your initial hypothesis based on the results of this experiment, if necessary.

_____

_____

Can a person still produce bile if his or her gallbladder is removed? Why or why not?

_____

_____

EXERCISE 5.6 Disorders of the Digestive System

In 1996, the National Center for Health Statistics (http://www.cdc.gov/nchs) reported that 60 to 70 million people in the United States were living with digestive disorders. These individuals typically visit **gastroenterologists**, who specialize in the diagnosis and treatment of both acute (short-term) and chronic (long-term) digestive disorders. (The word _enteric_ is commonly used in medical terminology when referring to the intestines.) Lactose intolerance, celiac disease, irritable bowel syndrome, and colon cancer are just a few examples of the digestive disorders that may afflict a person's health and general well-being.

During this exercise, you will use your knowledge of the digestive system to diagnose two patients. Isabel and Maria are seeking medical attention for different digestive disorders. Use the symptoms and test results provided to make a logical diagnosis for each patient.

Isabel

Isabel was convinced her abdominal pain was merely a sign of indigestion. For the past few weeks, she felt a burning sensation in her belly as she tossed and turned restlessly throughout the night. The pain usually subsided during the daytime, except for times when her stomach was empty and growling. Following a particularly fitful night of sleep, Isabel's husband urged her to make an appointment with a doctor. Isabel's doctor performed a physical examination and collected blood for additional testing. Based on her symptoms, the physician also recommended endoscopy to view the lining of her gastrointestinal tract. At first, Isabel was hesitant about this procedure, since she was not familiar with it. The doctor explained that an endoscope is simply a thin, pliable tube with a small light source and camera on the end. Isabel agreed to this procedure in hopes of receiving an accurate diagnosis. During endoscopy, an open sore was detected in the lining of Isabel's stomach; a picture from Isabel's endoscopy video is shown in Figure 5.9. Isabel's doctor prescribed antibiotics to treat her disorder. Following the full course of antibiotic therapy, Isabel's pain permanently subsided.

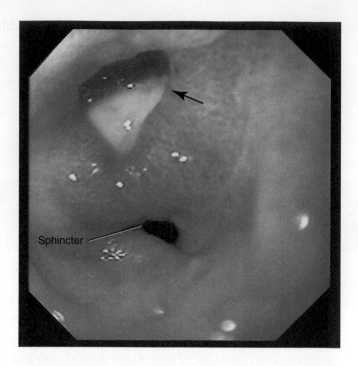

Sphincter

FIGURE 5.9   Isabel's Endoscopy
Results

What digestive disorder is suggested by Isabel's endoscopy results? _____

Antibiotic therapy successfully treated Isabel's disorder. What does this suggest as the
cause of Isabel's stomach sore? _____

Maria

On her way home from work one night, Maria felt an intense pain in the upper
right portion of her abdomen. Over the course of the next few hours, the pain
steadily intensified until she just couldn't stand it any longer. Maria's son rushed
her to the emergency room, where a physical examination was performed. Based
on the location of her pain, the physician decided to perform an ultrasound on her
abdominal organs. As shown in Figure 5.10, the ultrasound detected a hard mass
inside of Maria's gallbladder.

Which digestive disorder is suggested by Maria's ultrasound results? _____

What is a common treatment for this disorder? _____

Is the gallbladder considered a vital organ in the human body? Why or why not?

_____

_____

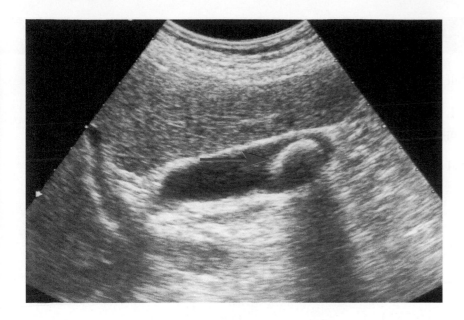

FIGURE 5.10 Ultrasound of Maria's Gallbladder

A CLOSER LOOK AT DIGESTIVE DISORDERS

If you would like to learn more about the symptoms, diagnosis, and treatment of a specific digestive disorder, you can access the National Digestive Diseases Information Clearinghouse (NDDIC) at http://digestive.niddk.nih.gov. The mission of NDDIC is "to increase knowledge and understanding about digestive diseases among people with these conditions and their families, health care professionals, and the general public." NDDIC is affiliated with the National Institute of Health (NIH), which is a branch of the United States Government.

1. What does the suffix –*ase* denote in biology?_____ _____

2. Teeth physically rip food into smaller pieces with the mouth. This process is an example of _____ digestion.

3. Label organs a–c on the picture below.

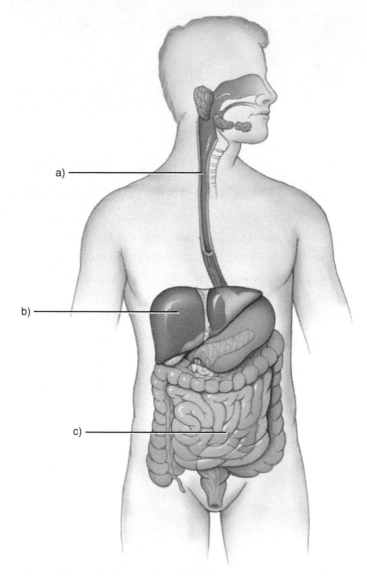

Right lateral view of head and neck and anterior view of trunk

4. What is the optimal temperature for digestive enzymes in the human body?
   a. 0°C
   b. 22°C
   c. 37°C
   d. 80°C

5. Which organ produces bile in the body?_____Which organ stores excess bile?_____

6. The small intestine has a pH of 8. Is this pH level considered acidic, basic, or neutral?_____

7. Label the substrate, enzyme, and product on the equation below.

$$\text{Proteins + water} \xrightarrow{\text{trypsin}} \text{peptides}$$

8. Label the pharynx and a salivary gland on the image below.

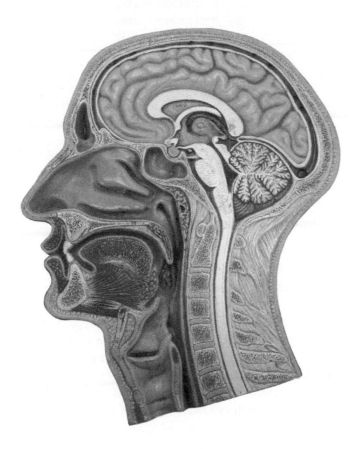

9. Lactose intolerance is caused by a deficiency of the enzyme _____.

10. Bile salts _____ lipids into smaller droplets.

LAB 6:
The Cardiovascular System

By the End of This Lab, You Should Be Able to Answer the Following Questions:

- What role does blood play in gas exchange, immune defense, and blood clotting? Which component(s) of blood is (are) involved in each of these tasks?

- In general, how can you tell whether a magnified blood smear appears normal or abnormal?

- What does the suffix *–emia* refer to in medical terminology?

- How many chambers and valves are found inside of the human heart?

- What is cardiac auscultation? What piece of equipment is used to auscultate the heart?

- What is the difference between systolic and diastolic blood pressure? How can these values be determined using a stethoscope and a sphygmomanometer?

- How do arteries and veins differ in structure, function, and microscopic appearance?

- Why is atherosclerosis considered a risk factor for strokes and heart attacks?

INTRODUCTION

With the exception of a few cell types, such as blood cells and eggs—or sperm, if you are male—most cells are confined to one specific location in your body. Like a person who is confined to bed rest, these cells cannot move to collect nutrients, discard wastes, or even communicate with one another. For this reason, cells rely on a transportation route known as the cardiovascular system for survival. As implied by its name, the cardiovascular system consists of the heart (*cardio-*), the blood, and the vessels (*-vascular*) that carry blood to every inch of your body. Figure 6.1 illustrates the general layout of the human cardiovascular system. Unless every component of this system is working properly, crucial substances (examples: oxygen, water, hormones, nutrients, and wastes) won't reach their target destinations quickly enough to sustain the life of each cell, and ultimately, the life of the entire body.

BLOOD COMPOSITION

Did you know that roughly 5 liters of blood are flowing through your body at this very moment? Think about how this volume of fluid relates to the size and weight of a 2-liter bottle of soda. For some people, it is hard to imagine 5 liters of liquid fitting inside of the heart, which is roughly the size of two clenched fists, and its adjoining blood vessels. Amazingly, though, the female body can withstand a 50% increase in blood volume over the course of a full-term pregnancy.* As discussed in Lab 3 (Observing Body Tissues),

---

*Source: The Merck Manuals Online Medical Library http://www.merck.com/mmhe/sec22/ch257/ch257d.html accessed 5-29-10.

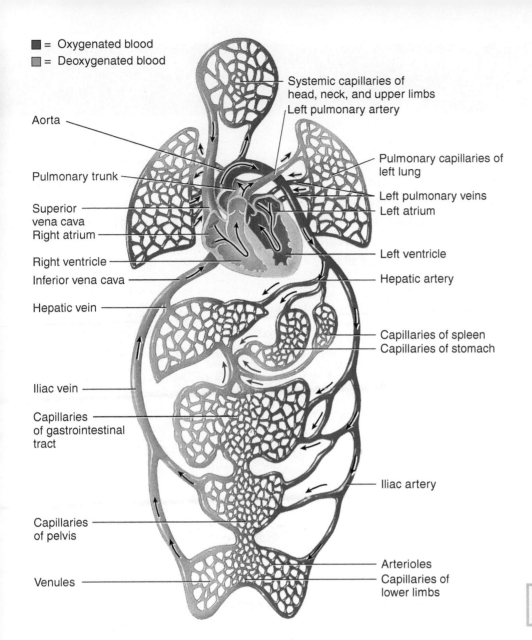

= Oxygenated blood
= Deoxygenated blood

Systemic capillaries of head, neck, and upper limbs

Left pulmonary artery

Aorta

Pulmonary trunk

Superior vena cava

Right atrium

Right ventricle

Inferior vena cava

Hepatic vein

Iliac vein

Capillaries of gastrointestinal tract

Capillaries of pelvis

Venules

Pulmonary capillaries of left lung

Left pulmonary veins

Left atrium

Left ventricle

Hepatic artery

Capillaries of spleen

Capillaries of stomach

Iliac artery

Arterioles

Capillaries of lower limbs

FIGURE 6.1   The Cardiovascular System

blood is one type of fluid connective tissue. **Plasma** is the fluid matrix that separates red blood cells, white blood cells, and platelets from one another. In addition to giving blood its fluidity, plasma also transports a number of dissolved substances—hormones, nutrients, wastes, electrolytes (salt ions), gases, and various proteins—from one area in the body to another.

The formed elements of blood (red blood cells, white blood cells, and platelets) are all derived from **hematopoietic** stem cells that reside in the red bone marrow. (In biology, the prefix *hemo-* is commonly used when referring to blood.) **Erythrocytes**, or red (*erythro-*) blood cells (*-cytes*), transport oxygen from the lungs to every cell in the body. Although they are smaller in size than white blood cells, erythrocytes are normally the most abundant cell type in blood. **Leukocytes**, or white (*leuko-*) blood cells, protect the body against infectious disease. Since leukocytes are transparent, they are usually stained purple on microscope slides to increase their contrast and visibility. Platelets, or **thrombocytes**, are involved in the blood-clotting cascade. (The prefix *thrombo-* refers to a blood clot.) Platelets are actually cell fragments, and as a result, they are smaller than either red blood cells or white blood cells. When a blood vessel gets ripped or torn, platelets interact with clotting proteins to create a blood clot. Blood clots temporarily seal gaps in damaged blood vessels to prevent excessive blood loss.

EXERCISE 6.1 Characteristics of a Normal Blood Smear

During this exercise, you will examine a prepared microscope slide of a normal blood smear. As you observe this slide, be sure to note the shape *and* abundance of red blood cells, white blood cells, and platelets. This information will help you diagnose blood disorders in the following exercise. Compare your focused microscope slide to the photomicrograph provided below, and answer the review questions that accompany this slide.

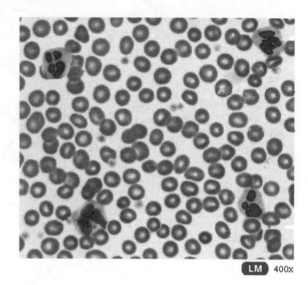

FIGURE 6.2
Normal Blood Smear

LM 400x

Label a red blood cell, a white blood cell, and a platelet on Figure 6.2.

Which cell type is most abundant in a normal blood sample? _____

How would you describe the shape of red blood cells in a normal blood smear?

_____

_____

EXERCISE 6.2 Blood Disorders

Monica, Malcolm, and Andrew are seeking medical attention for different blood disorders. Use the symptoms and test results provided to make a logical diagnosis for each patient: anemia, leukemia, or sickle cell anemia. The names *anemia*, *leukemia*, and *sickle cell anemia* actually give us clues about the composition and/or microscopic appearance of blood samples from patients with these disorders.

- The term *sickle* refers to a curved object.
- The prefix *leuko-* refers to white blood cells.
- The prefix *an-* means without or lacking.
- The suffix *–emia* refers to a blood disorder.

Monica

Monica adopted a strict vegan lifestyle during her second year of college. Although she intended to incorporate lots of fruits, vegetables, and legumes into her diet, Monica's hectic schedule—and dwindling bank account—often led to meals of crackers, popcorn, or noodles. One month after changing her diet, she couldn't believe how baggy her clothes had become. Who knew you could lose weight by eating nothing but carbohydrates? By the end of the fall semester, though, Monica's dress size was the last thing on her mind. For three consecutive days, she couldn't muster up the strength to get dressed or go to school, let alone crack open her textbooks. Even Monica's roommate Ingrid was alarmed by how pale her complexion looked. When Monica experienced shortness of breath on her way from the couch to the bathroom, she knew it was time to see a doctor.

After listening to Monica's symptoms, her physician performed a thorough physical examination. Afterward, a blood sample was drawn for laboratory analysis. Monica's blood smear is shown in Figure 6.3. Her complete blood count (CBC) revealed an abnormally low number of red blood cells, and low iron levels were detected in her serum sample.

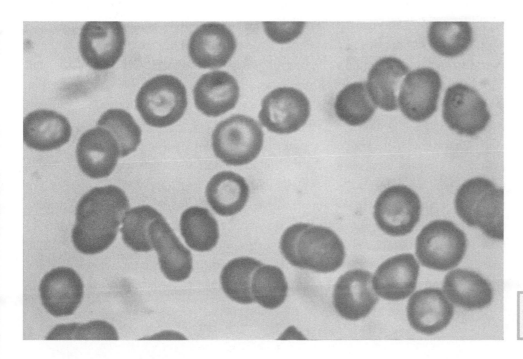

FIGURE 6.3 Monica's Blood Smear

Which blood disorder is suggested by Monica's symptoms and test results?

_____

Monica's blood tests revealed low levels of iron in her serum. What treatment options exist to correct this issue?

_____

_____

Why does Monica's complexion looked pale as a result of this blood disorder?

_____

_____

Why is iron essential for blood health?

_____

_____

Malcolm

From a very young age, Malcolm loved to play outside with his neighborhood friends. In fact, on most nights, his parents had to chase him down and drag him inside at bedtime. Shortly after Malcolm's ninth birthday, though, his parents noticed a change in his energy level, hobbies, and general demeanor. Instead of playing outdoors after school, Malcolm would lie on the couch, groan about body aches, and insist he was too tired to eat dinner. Convinced that his son was in the middle of a growth spurt, Malcolm's dad gently rubbed his back one night and explained it was common to feel growing pains.

The next morning before school, Malcolm tried to convince his mom that his teeth didn't need another brushing. After all, he had barely touched his supper the night before. At her wit's end, Malcolm's mom grabbed his arm, dragged him into the bathroom, and watched as he begrudgingly brushed his teeth. As Malcolm spit out the toothpaste, though, his mom was shocked to see more blood in the sink than toothpaste foam. When she asked why his gums were bleeding, Malcolm just shrugged and said it happened all the time.

That same day, Malcolm's parents drove him to the pediatrician's office. After checking his vital signs, the doctor examined Malcolm's gums, sore limbs, and swollen lymph nodes. Dr. Ross was particularly struck by Malcolm's pallor, not to mention the bruises that seemed to pop up for no apparent reason. Given the nature of Malcolm's symptoms, a blood sample was drawn for laboratory analysis; Malcolm's blood smear is shown in Figure 6.4. His complete blood count (CBC) revealed abnormally high numbers of white blood cells. In contrast, his platelet count and red blood cell count were extremely low.

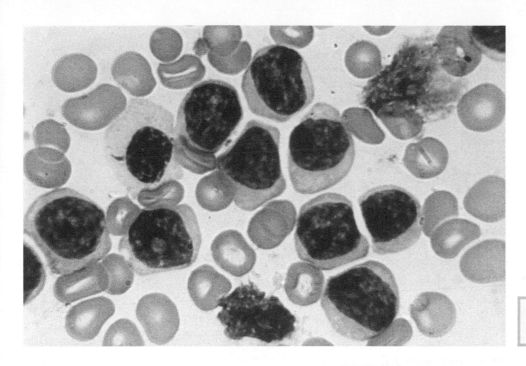

FIGURE 6.4   Malcolm's Blood Smear

Which blood disorder is suggested by Malcolm's symptoms and test results?

_____

Why does Malcolm's complexion appear pale as a result of this disorder?

_____

_____

Why does Malcolm bruise and bleed easily as a result of this disorder?

_____

_____

Andrew

At the moment Andrew was conceived, he inherited a blood disorder from his mom and his dad. Ironically, neither one of Andrew's parents actually lives with the disorder that he inherited. Genetic testing revealed the cause of Andrew's blood disorder: DNA mutations (similar to spelling errors) in both of his hemoglobin genes. Since hemoglobin fills the interiors of red blood cells, any change in its three-dimensional shape may affect (1) its ability to bind oxygen and (2) the shape of the cells that house it. As shown in Figure 6.5, Andrew's red blood cells tend to take on crescent shapes, particularly when oxygen levels plummet within his body. Due to their jagged shape, these cells have difficulty squeezing through narrow blood vessels. When they clump together and obstruct blood flow, Andrew feels severe pain in the affected region(s) of his body.

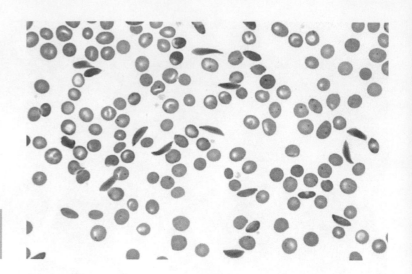

FIGURE 6.5 Andrew's Blood Smear

Which blood disorder is suggested by Andrew's symptoms and test results?

_____

What treatment options exist for Andrew's blood disease?

_____

_____

Can you think of another blood disorder that is caused by DNA mutations?

_____

A CLOSER LOOK AT BLOOD DISORDERS

If you would like to learn more about the symptoms, diagnosis, and treatment options for a specific blood disorder, you can access National Heart Lung and Blood Institute's website at http://www.nhlbi.nih.gov.

THE HEART

Did you know that the human heart beats roughly 100,000 times every day? Although this number is extraordinary when we think about it, many of us take our hearts for granted, especially if they have always functioned properly. The heart essentially serves as a muscular pump, since it creates the pressure required to push blood through adjoining arteries. Figure 6.6 illustrates an anterior (front) view of a heart that was dissected along the frontal plane. This view mimics the orientation of a patient's heart as seen from the perspective of a physician. As shown in Figure 6.6, the human heart contains four chambers, four valves, and a **septum** (wall) that separates the left and right chambers from one another. (To avoid confusion, the terms *left* and *right* always refer to the *patient's* perspective of his or her body.) The upper chambers of the heart are called **atria**, and the lower chambers of the heart are called **ventricles**. **Heart valves** ensure one-way blood flow from one chamber to another.

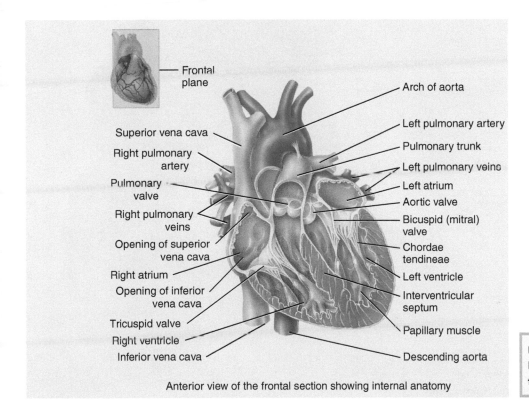

Frontal plane

Superior vena cava

Right pulmonary artery

Pulmonary valve

Right pulmonary veins

Opening of superior vena cava

Right atrium

Opening of inferior vena cava

Tricuspid valve

Right ventricle

Inferior vena cava

Arch of aorta

Left pulmonary artery

Pulmonary trunk

Left pulmonary veins

Left atrium

Aortic valve

Bicuspid (mitral) valve

Chordae tendineae

Left ventricle

Interventricular septum

Papillary muscle

Descending aorta

Anterior view of the frontal section showing internal anatomy

FIGURE 6.6   **Anterior View of a Human Heart Dissected along the Frontal Plane**

During this exercise, you will identify key structures on a dissected sheep heart, a model of the human heart, or a virtual dissection of the human heart. Label each structure listed below on Figure 6.7.

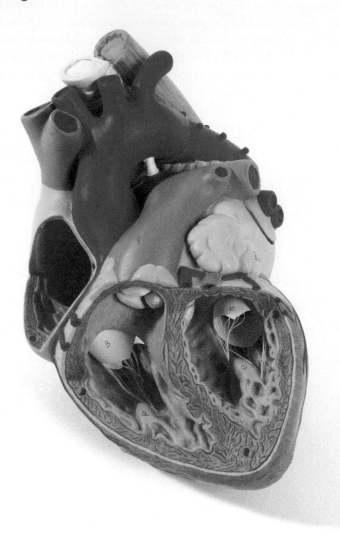

FIGURE 6.7   Identifying Structures in the Human Heart

Heart Chambers

- Left and right atria
- Left and right ventricles

Valves

- Tricuspid valve
- Bicuspid (mitral) valve
- Pulmonary valve
- Aortic valve

Adjoining Blood Vessels

- Aorta
- Pulmonary trunk
- Superior vena cava
- Inferior vena cava (not shown in Figure 6.7)

Miscellaneous Structures

- Interventricular septum

Review Questions for Exercise 6.3

Which chamber of the heart is labeled A below? When this chamber contracts, blood is pushed into the aorta.

_____

Which chamber of the heart is labeled B below? The superior and inferior venae cavae are connected to this chamber.

_____

Is the pulmonary trunk labeled C or D below? How can you tell?

_____

_____

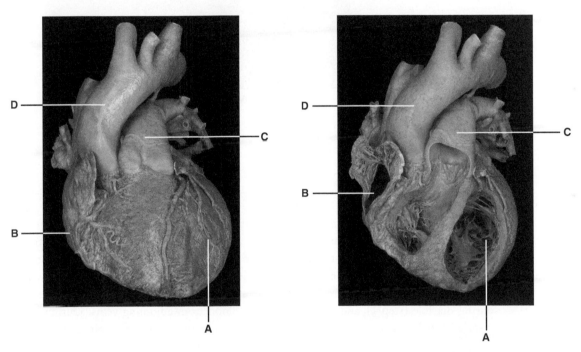

Left Anterolateral View of the Human Heart

Which heart valve is labeled A below? This valve is located between the right atrium and the right ventricle. _____

Which heart valve is labeled B below? This valve is located between the left atrium and the left ventricle. _____

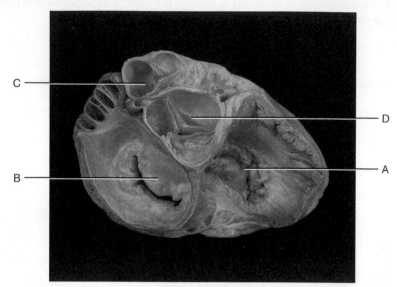

Superior View of Human Heart Valves

THE CARDIAC CYCLE

Sometimes people assume that the entire heart contracts simultaneously. As illustrated in Figure 6.8, though, each heartbeat actually consists of three distinct steps. Collectively, these steps are referred to as the **cardiac cycle**. The term **systole** refers to contraction of heart chambers, while **diastole** refers to relaxation of heart chambers. (In Greek, the words *systole* and *diastole* mean contraction and expansion, respectively.) The entire heart relaxes during diastole (step 1), and the atria fill with blood from adjoining veins. During atrial systole (step 2), blood is pushed out of the contracting atria and into adjoining ventricles. During ventricular systole (step 3), blood is pushed out of the contracting ventricles and into adjoining arteries. Blood will not surge through the aorta or pulmonary trunk again until another cardiac cycle (heartbeat) is successfully completed.

Health professionals use stethoscopes to hear internal body sounds, such as the sounds produced by a beating heart. The practice of listening to body sounds is called **auscultation** within the medical field. **Cardiac auscultation** is performed to assess cardiovascular health and, in certain cases, to diagnose cardiovascular disorders. Normally, the heart produces a "lub-dub" sound during each heartbeat. The "lub" sound is produced as the atrioventricular valves snap shut; the "dub" sound is produced as the aortic and pulmonary valves snap shut. If a valve does not shut properly, then blood may regurgitate (flow backwards) and create additional sounds (heart murmurs). For instance, a "shhhh" noise may be heard between the "lub" and "dub" sounds. It is important to note that heart murmurs are relatively common in the population, and they do not always indicate the presence of a cardiovascular disorder.

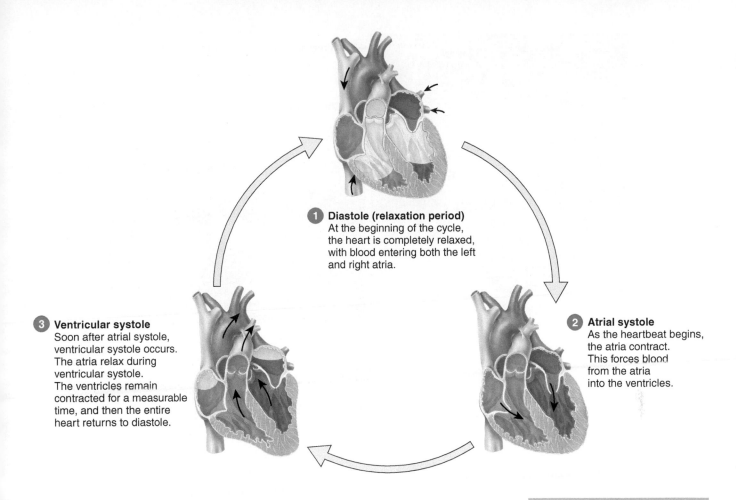

1 Diastole (relaxation period)
At the beginning of the cycle, the heart is completely relaxed, with blood entering both the left and right atria.

3 Ventricular systole
Soon after atrial systole, ventricular systole occurs. The atria relax during ventricular systole. The ventricles remain contracted for a measurable time, and then the entire heart returns to diastole.

2 Atrial systole
As the heartbeat begins, the atria contract. This forces blood from the atria into the ventricles.

FIGURE 6.8 **The Cardiac Cycle**

STEP BY STEP 6.4   Listening to Your Heartbeat

Visualizing THE LAB

During this exercise, you will auscultate your own heart with a stethoscope. The parts of a typical stethoscope are labeled on Figure 6.9. Before the class performs this exercise, your instructor may play audio files of normal and abnormal heart sounds. If desired, you may also access audio links to cardiac auscultation online. You can access a virtual auscultation tool called *Diagnostics 101*, which is provided for students by Welch Allyn, at http://www.welchallyn.com/wafor/students/diagnosis 101/index.html.

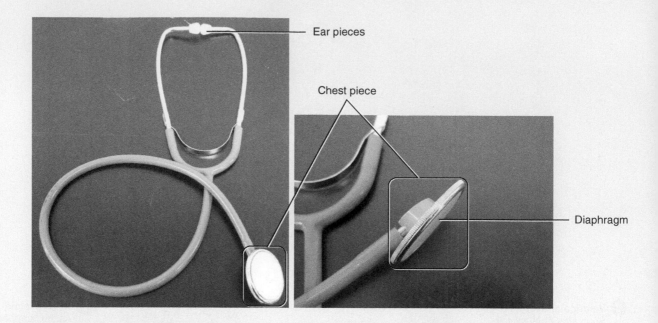

Ear pieces

Chest piece

Diaphragm

FIGURE 6.9 **Parts of the Stethoscope**

Procedure

Note: Perform this entire procedure while sitting down.

1. Use an alcohol swab to clean the earpieces of the stethoscope.

2

2. Tilt the stethoscope anteriorly (toward your face) while securing the earpieces inside of your ear canals. Adjust the earpieces, as necessary, to minimize background noises.

3

4

3. Warm the stethoscope's chest piece with your hands for 30 seconds.

4. Place the diaphragm (the flat end of the chest piece) slightly above your left nipple. Move the stethoscope around slowly until you hear "lub-dub" sounds. If you are having difficulty hearing your heartbeat, place the chest piece directly on your skin.

4 ▶

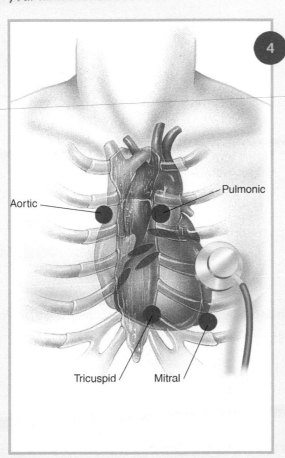

Aortic

Pulmonic

Tricuspid   Mitral

5. Count the number of heartbeats produced during a 30-second interval, and multiply this number by two to determine your resting heart rate. *Reminder: Each "lub-dub" represents one heartbeat.*

My Resting Heart Rate:_____ beats per minute (bpm)

HEART RATE

Throughout the course of a typical day, your heart rate varies as you move from one activity—or one emotional state—to another. Over the course of a lifetime, your resting heart rate may also change due to certain medical issues, the aging process, or a combination of both. **Heart rate** is defined as the number of beats per minute (bpm) produced by the heart at a given time. Typically, a resting heart rate of 60–100 bpm is considered normal for adults.* Heart rate can either be measured directly, as done in Step-by-Step 6.4, or indirectly by determining your **pulse rate**. Each pulse, or pressure wave, felt inside of an artery is a direct result of the previous heartbeat.

EXERCISE 6.5   Determining Pulse Rate

During this exercise, you will work in pairs to measure pulse rate. Perform each step in a sitting position, unless otherwise indicated.

1. Locate your partner's pulse by pressing your index finger and middle finger firmly on the radial artery. As shown in Figure 6.10, the radial artery is located on the inner wrist below the thumb.

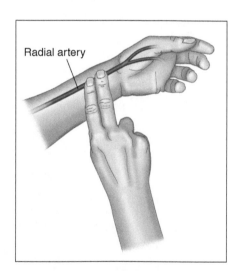

FIGURE 6.10   **Location of the Radial Artery**

2. Count the number of pulses you feel during a 30-second period. Multiply this number by two to determine your partner's resting pulse rate.

3. Switch roles with your partner and repeat steps 1–2.

   My Resting Heart Rate (Pulse Rate) _____ **bpm**

---

* Source: http://www.mayoclinic.com/health/heart-rate/an01906.

4. Compare the heart rate values you obtained through cardiac auscultation and pulse rate determination. How do these values compare to one another?

_____

_____

5. Have your partner perform moderate exercise, such as jumping jacks, for 3 minutes.

6. Determine your partner's pulse rate immediately after exercise.

7. Switch roles with your partner and repeat steps 5–6.

My Heart Rate (Pulse Rate) Following Moderate Exercise _____ **bpm**

What effect does moderate exercise have on your heart rate? Why is heart rate altered by physical activity?

_____

_____

BLOOD VESSELS

How many miles of blood vessels are coiled inside of your body? Well, the number is so gigantic that we need a reference for the sake of comparison. The earth's circumference is roughly 25,000 miles at its widest point: the equator. If every single artery, capillary, and vein in your body were attached end to end, this tube would wrap around the earth twice . . . with about 10,000 miles to spare.* Even though oxygenated blood leaves the heart in a single artery, millions of branches develop along the way. These branches resemble the highways, main streets, and side streets that diverge from a single interstate. As a result of this extensive branching, every cell in your body can interact with blood simultaneously.

Blood vessels differ from one another based on function, location, and structure. By definition, an **artery** carries blood away from the heart and toward capillary beds. The largest artery in the body—the **aorta**—picks up oxygenated blood from the left ventricle of the heart. As blood travels through the aorta, it dumps into smaller arteries along the way. Each one of these branches connects to a specific organ or region of the body. **Capillaries** are incredibly narrow, so blood cells must travel through them in single file. It is here, in capillary beds, that materials are swapped between blood and tissue cells. Blood releases oxygen, nutrients, and hormones (when applicable) to surrounding tissue cells. At the same time, wastes and hormones (when applicable) diffuse into blood for elimination and distribution, respectively. Finally, **veins** carry blood away from capillary beds and back toward the heart. Throughout the body, millions of small blood vessels converge into two large veins: the inferior and superior venae cavae. Before blood reenters the aorta, though, it must travel through a circuit that connects the lungs to the heart. After gas exchange occurs in the lungs, oxygenated blood returns to the left side of the heart and prepares for another trip around the body.

---

* http://www.nia.nih.gov/HealthInformation/Publications/AgingHeartsandArteries/chapter04.html

Figures 6.11a and b depict the major arteries and veins in the human body. Often times, the name of a blood vessel specifies its location in the body. For instance, subclavian arteries are located underneath (*sub-*) the left and right collar bones (*-clavian*). By the same token, subclavian veins also reside in this general vicinity. With that said, keep in mind that

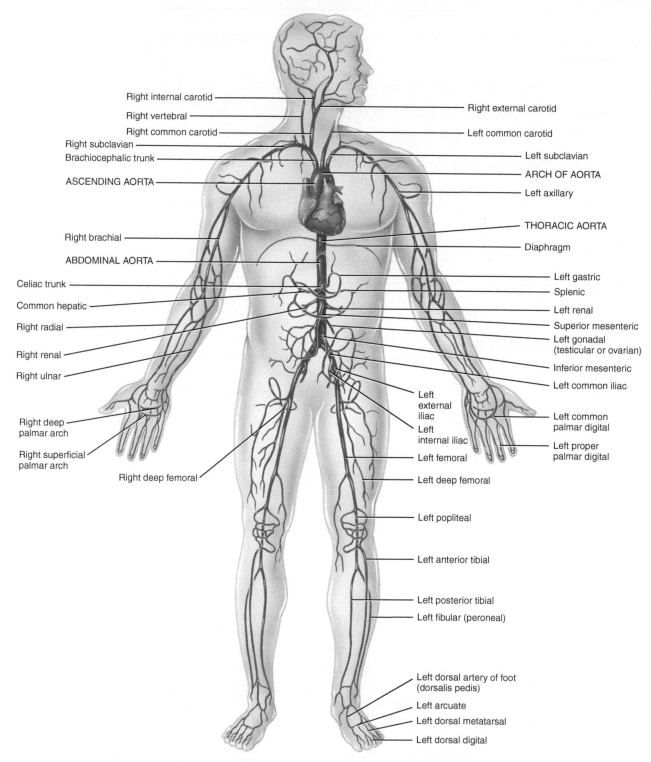

(a) Overall anterior view of the principal branches of the aorta

FIGURE 6.11 Major Arteries and Veins of the Human Body

most arteries lie **deep** in veins in the body. In other words, arteries tend to be located farther away from the skin than veins. Since veins are **superficial** to arteries, many veins can be seen directly through the skin.

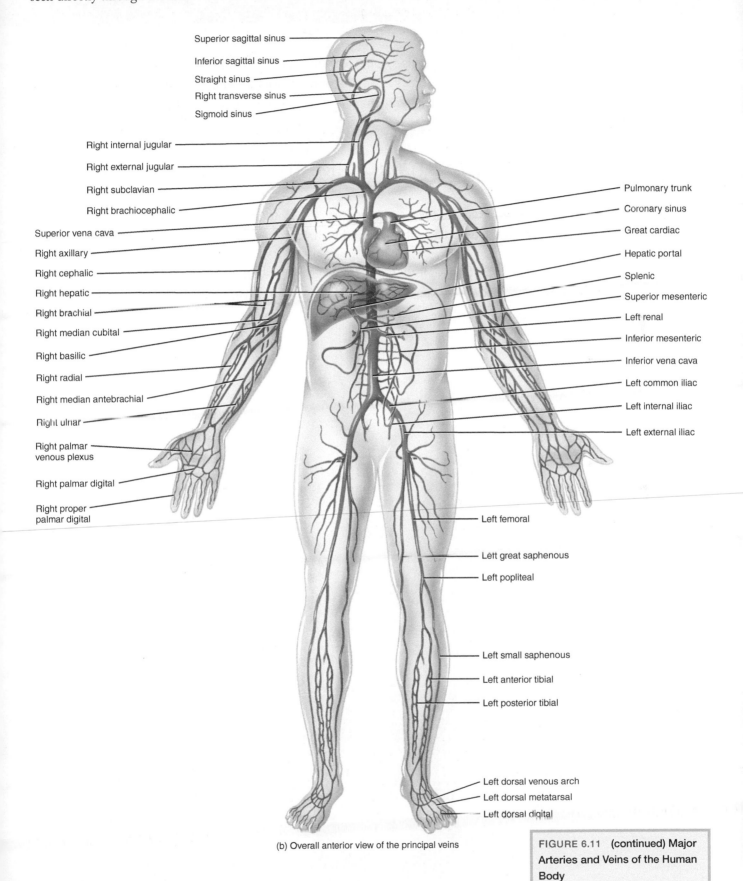

Superior sagittal sinus
Inferior sagittal sinus
Straight sinus
Right transverse sinus
Sigmoid sinus

Right internal jugular
Right external jugular
Right subclavian
Right brachiocephalic
Superior vena cava
Right axillary
Right cephalic
Right hepatic
Right brachial
Right median cubital
Right basilic
Right radial
Right median antebrachial
Right ulnar
Right palmar venous plexus
Right palmar digital
Right proper palmar digital

Pulmonary trunk
Coronary sinus
Great cardiac
Hepatic portal
Splenic
Superior mesenteric
Left renal
Inferior mesenteric
Inferior vena cava
Left common iliac
Left internal iliac
Left external iliac

Left femoral
Left great saphenous
Left popliteal

Left small saphenous
Left anterior tibial
Left posterior tibial

Left dorsal venous arch
Left dorsal metatarsal
Left dorsal digital

(b) Overall anterior view of the principal veins

FIGURE 6.11 (continued) Major Arteries and Veins of the Human Body

6-18

During this exercise, you will trace the path of one blood cell as it travels through the cardiovascular system. Refer to Figures 6.6 and 6.10, as needed, to review the location of each heart chamber, heart valve, and blood vessel.

THE SYSTEMIC CIRCUIT

Arrange the following blood vessels in sequence on the flowchart. The goal is to trace blood flow from the left ventricle to the right atrium.

- Renal vein
- Aorta
- Renal artery

- Renal capillary
- Inferior vena cava

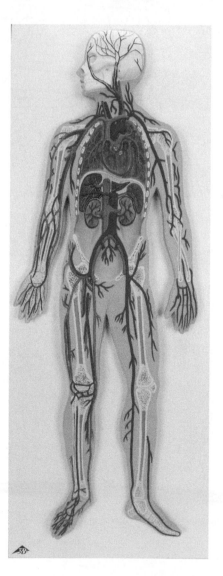

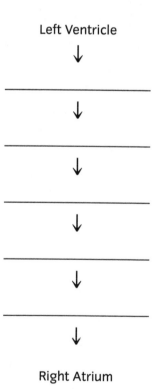

Left Ventricle

↓

_____

↓

_____

↓

_____

↓

_____

↓

_____

↓

Right Atrium

THE HEART AND PULMONARY CIRCUIT

Arrange the valves, blood vessels, and heart chambers listed below in sequence on the flowchart. The goal is to trace blood flow from the venae cavae to the aorta.

- Right atrium
- Right ventricle
- Left atrium
- Left ventricle
- Aortic valve
- Tricuspid valve
- Bicuspid (mitral) valve
- Pulmonary valve
- Pulmonary trunk

Inferior Vena Cava and Superior Vena Cava

↓

_____

↓

_____

↓

_____

↓

_____

↓

_____

↓

Pulmonary Arteries

↓

Pulmonary Capillary Beds

↓

Pulmonary Veins

↓

_____

↓

_____

↓

_____

↓

Aorta

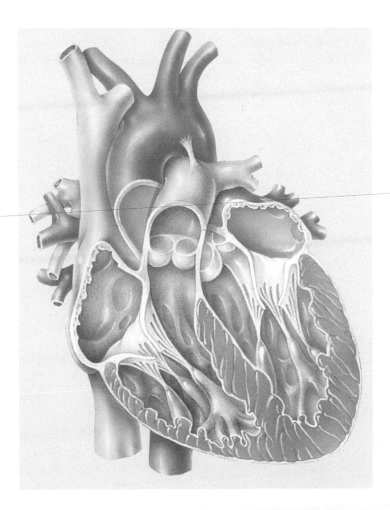

Review Questions for Exercise 6.6

Do all arteries carry oxygenated blood? Explain your answer. (*Hint*: How does deoxygenated blood reach the lungs?)

_____

_____

Do all veins carry deoxygenated blood? Explain your answer. (*Hint*: How does oxygenated blood reach the heart?)

_____

_____

COMPOSITION OF BLOOD VESSELS

Arteries, capillaries, and veins differ based on more than just their functions and relative locations. Their construction also differs in three fundamental ways: tissue composition, wall thickness, and the diameter of their **lumens** (hollow interiors). Figure 6.12 shows the tissues that construct the walls of arteries, capillaries, and veins. The lumen of each blood vessel is surrounded **endothelium**; this tissue comes in direct contact with blood, since it is the innermost (*endo-*) tissue layer (*-thelium*). Endothelium is

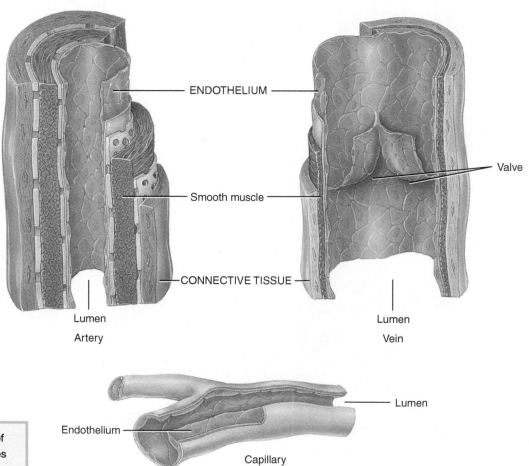

FIGURE 6.12 **Composition of Arteries, Veins, and Capillaries**

composed of *simple squamous epithelium*, which is the thinnest tissue type in the body. Capillaries have incredibly thin walls, since they are composed primarily of endothelium. In addition to the endothelial layer, the walls of arteries and veins also contain layers of smooth muscle and connective tissue. The relative thickness of these layers – not to mention the relative thickness of the wall itself - is greater in arteries than veins. As a result, the lumen diameter is smaller in arteries relative to veins.

EXERCISE 6.7   Identifying Arteries and Veins on a Microscope Slide

During this exercise, you will observe cross sections of arteries and veins on a prepared microscope slide. Use cues such as wall thickness and lumen diameter to distinguish these blood vessels. Once you have identified an artery and vein on the microscope slide, label them accordingly on Figure 6.13.

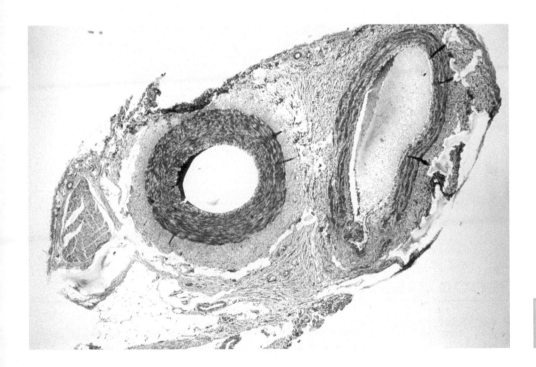

FIGURE 6.13   **Cross Section of an Artery and Vein**

A CLOSER LOOK AT ATHEROSCLEROSIS

Atherosclerosis (or arteriosclerosis) is a vascular disease that affects the integrity of arterial walls. When fatty **plaque** accumulates inside of an artery, its walls harden (-*sclerosis*) and lose their elasticity. Since the walls are no longer flexible, they cannot expand to counteract high blood pressures. As a result, there is an increased risk that the blood vessel will eventually rupture. Since plaque deposits also thicken arterial walls, the lumen inevitably narrows in size; this phenomenon is shown in Figure 6.14. As a result, atherosclerosis increases the risk of a complete blockage, either directly (example: plaque deposits obstructing the entire lumen) or indirectly (example: blood clots interacting with plaque deposits). For this reason, atherosclerosis is a risk factor for other cardiovascular diseases, such as strokes and heart attacks.

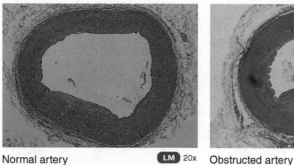
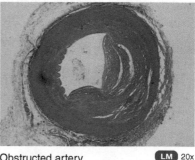

Normal artery   LM 20x   Obstructed artery   LM 20x

FIGURE 6.14 **Cross Section of a Normal Artery and an Atherosclerotic Artery**

BLOOD PRESSURE

We all know that blood pressure is an important facet of our health, but what exactly do those two numbers mean? **Blood pressure** is literally the force that blood exerts on the walls (blood vessels) surrounding it. Your blood pressure rises dramatically every time your ventricles contract. When the left ventricle contracts, for instance, its walls tightly clench together. With nowhere else to go, the blood that once filled this chamber squeezes through a one-way valve. Every time it contracts, the left ventricle generates enough blood pressure to (1) open the flaps of the aortic valve *and* (2) thrust blood towards distant capillary beds, including those located above the heart. Blood pressure peaks as blood exits the heart, but it decreases as blood migrates through the body.

TABLE 6.1   Blood Pressure Interpretation Chart

Top Number (systolic) in mm Hg		Bottom Number (diastolic) in mm Hg	Blood Pressure Classification
Below 120	and	Below 80	Normal
120–139	or	80–89	Prehypertension
140–159	or	90–99	Stage 1 Hypertension
160 or higher	or	100 or higher	Stage 2 Hypertension

Source: *The Seventh Report of the Joint National Committee on Prevention Detection, Evaluation, and Treatment of High Blood Pressure*, 2003. This report can be accessed online at http://www.nhlbi.nih.gov/guidelines/hypertension/express.pdf

Blood pressure fluctuates as the heart progresses through each cardiac cycle. **Systolic blood pressure** is the force blood exerts on arteries as the heart (or more specifically, the ventricles) contracts. **Diastolic blood pressure** is the force blood exerts on arteries while the heart relaxes. Across most of the world, blood pressure is measured and reported in units called millimeters of mercury (mmHg). A blood pressure reading of 120/80mmHg tells us the highest (systolic) and lowest (diastolic) blood pressures encountered by an artery *at a specific point in time*. Both blood pressure values are of interest to medical professionals who are assessing your cardiovascular health. Table 6.1 provides information on how to interpret your blood pressure readings. Elevated (*hyper-*) blood pressure is commonly called **hypertension**, while low (*hypo-*) blood pressure is commonly called **hypotension**. Chronic hypertension puts unnecessary stress, or mechanical force, on your blood vessels. Over time, this can increase your susceptibility to other cardiovascular diseases.

Typically, health professionals measure blood pressure at the brachial artery. Figure 6.15 shows the general location of the brachial artery in the arm. Two pieces of equipment are traditionally used to measure blood pressure: a stethoscope and a sphygmomanometer. The parts of a typical sphygmomanometer are labeled on Figure 6.16. (Refer to Figure 6.9, as needed, to review the parts of the stethoscope.) A **sphygmomanometer** is composed of an inflatable arm cuff, a manually operated inflation bulb, an air valve that controls pressure in the arm cuff, and a manometer that measures and displays pressure in mmHg.

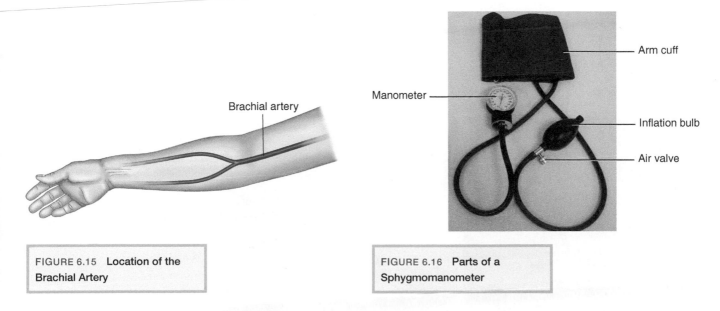

FIGURE 6.15   **Location of the Brachial Artery**

FIGURE 6.16   **Parts of a Sphygmomanometer**

During this exercise, you will work in pairs to measure resting blood pressure. Perform each step while sitting, unless otherwise stated.

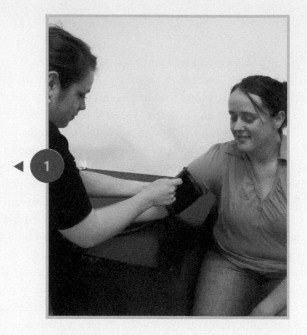

1. Secure the cuff of the sphygmomanometer snugly on your partner's upper arm.

2. Locate your partner's pulse at the brachial artery, and place the stethoscope's diaphragm directly over this region. **Note**: The arm should be supported so that the brachial artery is at the same level as the heart.

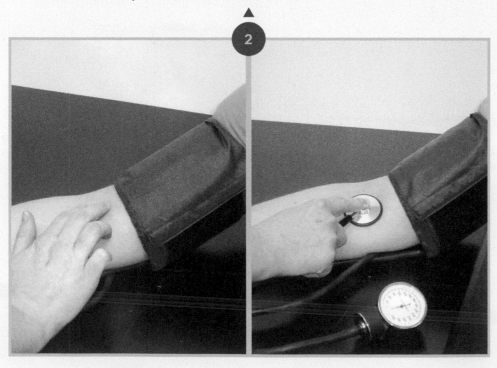

3. Inflate the blood pressure cuff up to 180 mmHg. At this pressure, the brachial artery completely closes, so blood stops flowing into the forearm.

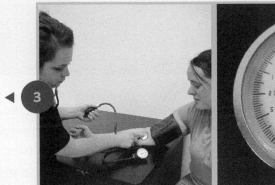 

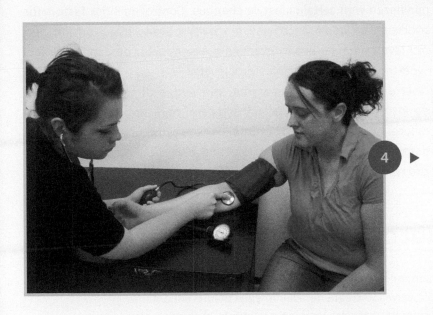

4. Use the air valve to slowly release pressure from the cuff. Listen intently through the stethoscope to detect any noises produced in the brachial artery.

5. Record the blood pressure at which the first sounds (Korotkoff sounds) are heard; this is your lab partner's systolic blood pressure. At this pressure, blood is able to flow through the artery during systole, but the artery closes up during diastole.

6. Continue listening until sounds cease to be heard; this is your lab partner's diastolic blood pressure. At this point, blood can freely flow through the artery again.

7. Completely deflate the cuff and remove it from your partner's arm.

8. Switch roles with your lab partner and repeat this procedure.

My resting blood pressure: _____ / _____ mmHg

According to the Centers for Disease Control and Prevention (CDC, http://www.cdc.gov/HeartDisease/facts.htm), cardiovascular disease (CVD) is the primary cause of death for men *and* women in the United States. Strokes, or cerebrovascular disease, rank as the third leading cause of death (http://www.cdc.gov/nchs/fastats/stroke.htm). Although some risk factors for cardiovascular disease are not controllable—such as family history, age, and gender—other risk factors can be regulated and/or minimized with certain lifestyle changes. Controllable risk factors for cardiovascular disease include tobacco use, inactivity, obesity, hypertension, and high cholesterol levels. Regular health exams are essential for detecting risk factors such as hypertension and high cholesterol levels.

During this exercise, you will use your knowledge of the cardiovascular system to diagnose three patients. Taylor, Syd, and Sara are seeking medical attention for different cardiovascular disorders. Use the symptoms and test results provided to determine which disorder is affecting each patient: cyanotic heart disease, cerebrovascular disease, or a myocardial infarction.

TERMINOLOGY USED IN THIS EXERCISE

Magnetic Resonance Imaging (MRI)—medical imaging technique that uses magnets and radio waves to create detailed images of internal organs

Ultrasound Imaging—medical imaging technique that uses high-frequency sound waves to produce images of internal organs

Electrocardiography (ECG or EKG)—test that measures the electrical activity of the heart

Angiography—medical imaging technique that uses dye and X-rays to visualize blood vessels

Echocardiography—test that uses sound waves to create an image of the heart

Hemorrhage—interrupted blood flow that is caused by a ruptured blood vessel

Ischemia—interrupted blood flow that is caused by blockage of a blood vessel

Infarct—localized region of cells that are dead or dying due to loss of blood supply

Coronary—pertaining to the heart

Cyanosis—bluish skin coloration that results from low oxygen levels

Taylor

While Taylor grilled streaks on the patio one night, he felt an intense headache brewing on the side of his head. Given its intensity and sudden onset, it felt unlike any other headache he had experienced before. As he rubbed his temple with one hand, Taylor asked his girlfriend Lindsay to grab the ibuprofen from the bathroom cabinet.

The next thing he knew, Lindsay was shrieking as he lay prostrate on the hard concrete patio. Feeling overwhelmed, Taylor tried to tell Lindsay that he didn't know why he was on the ground, that his head was throbbing, and that the left side of his body felt useless and numb. Lindsay looked increasingly panicked as slurred syllables rolled out of his mouth. With the inadvertent help of a neighbor— who had called the police after Lindsay started shrieking—Taylor reached the emergency room in record time. The attending physician immediately ordered an MRI of Taylor's brain. As shown in Figure 6.17, the MRI detected a blocked blood vessel in Taylor's cerebrum.

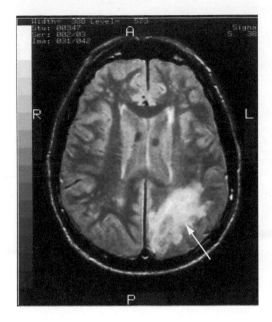

FIGURE 6.17 Taylor's MRI

Which vascular disease is suggested by Taylor's symptoms and test results?

_____

Was this disease caused by ischemia or hemorrhaging?

_____

What treatment options exist for Taylor's condition?

_____

_____

Syd

After 12 years of loyalty to the same aircraft company, Syd never thought he'd see the day when a pink slip found its way to his desk. As he tossed and turned in bed, searching for the right way to tell his wife, he couldn't stop his mind from racing obsessively. Why didn't I save more money when I had the chance? How could they do this to me after 12 years? Why didn't I take that other job 3 years ago? As his rage spun out of control, Syd ignored the painful indigestion that was spreading through his chest. Just what I need, he thought: another problem. As Syd rolled over, he instinctively grabbed his chest as a sharp pain spread across it. Unbeknownst to Syd, his wife couldn't sleep, either, due to his incessant tossing and turning. When she drowsily asked her husband what on earth was going on, all Syd could say was, "My chest."

Twenty minutes later, Syd arrived at the county hospital. An ECG detected abnormal electrical activity in his heart. Coronary angiography also detected a blockage in one of Syd's coronary arteries. Syd's coronary angiogram is shown in Figure 6.18.

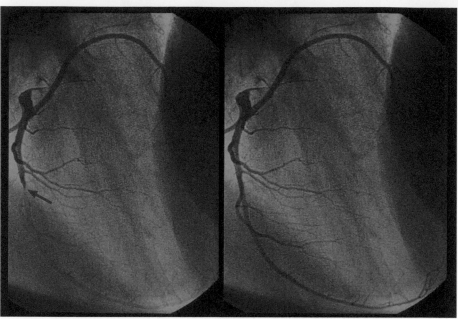

FIGURE 6.18 Syd's Coronary Angiogram

Syd's Coronary Angiogram   Normal Coronary Angiogram

Which cardiovascular disease is suggested by Syd's symptoms and test results?

_____

What treatment options exist for Syd's condition?

_____

_____

Do you think stress may have played a role in the progression of Syd's disease? How does stress affect physiological factors, such as blood pressure and heart rate?

_____

_____

Sara

When Sara was born, it was apparent to everyone in the delivery room—including her parents—that something wasn't right. She was breathing rapidly for no apparent reason; her lips and nail beds also displayed an odd, bluish tint. Sara was transferred to the neonatal intensive care unit (NICU), where an oxygen monitor confirmed she had extremely low blood oxygen levels. Abnormal heart sounds were also detected by the attending physician, who decided to order an echocardiogram. As shown in Figure 6.19, the echocardiogram revealed a septal defect in Sara's heart. This hole allowed blood to mix from the left and right sides of the heart.

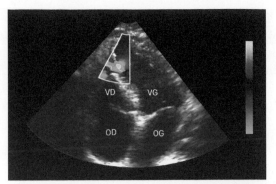

FIGURE 6.19 Echocardiogram of Sara's Heart

VD = right ventricle   VG = left ventricle
OD = right atrium   OG = left atrium

Which cardiovascular disorder is suggested by Sara's symptoms and test results?

_____

Why do Sara's lips and nail beds display a bluish tint?

_____

_____

What treatment options exist for Sara's condition?

_____

_____

A CLOSER LOOK AT CARDIOVASCULAR DISORDERS

If you would like to learn more about the symptoms, diagnosis, and treatment of a specific cardiovascular disorder, you can access the National Heart Lung and Blood Institute's website at http://www.nhlbi.nih.gov.

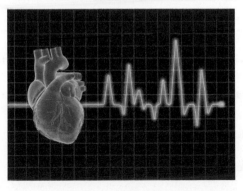

To learn more about the prevention, treatment, and warning signs for strokes or cardiovascular disease, you can also visit the American Stroke Association's website at http://www.strokeassociation.org or the American Heart Association's website at http://www.heart.org.

1. Label the left ventricle, right atrium, aorta, and tricuspid valve on the diagram below.

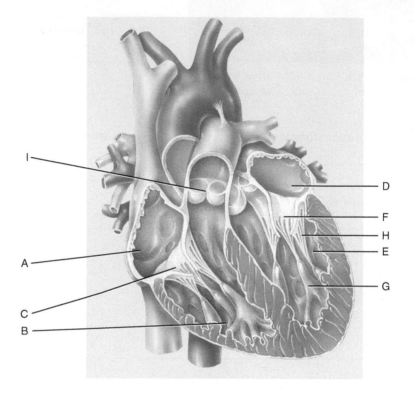

2. What is the difference between systolic and diastolic blood pressure? If your blood pressure is 125/75 mmHg, which number represents your diastolic blood pressure?

   _____

   _____

3. What structures in the heart produce the sounds of a heartbeat? _____

4. _____ is/are generally caused by blockage of a coronary artery.
   a. Strokes   c. Hypertension
   b. Heart attacks   d. Anemia

5. What is the difference between hypertension and hypotension?

   _____

   _____

   _____

6. Label a red blood cell, a white blood cell, and a platelet on the picture below. What is the function of each of these cells (or cell fragments)?

_____

_____

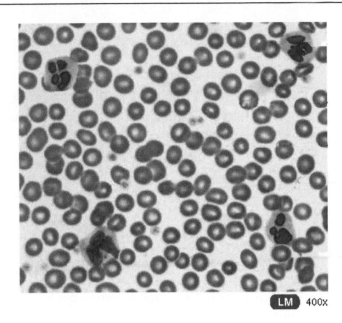

LM 400x

7. The suffix _____ is used when referring to a blood disorder.
   a. −algia
   b. −it is
   c. −emia
   d. −osis

8. What is cardiac auscultation? What piece of equipment is used to auscultate heart?

_____

_____

9. Label the artery and vein on the image below.

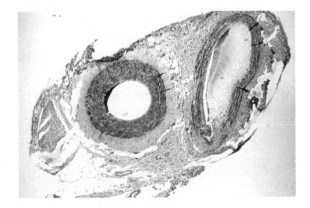

10. What is atherosclerosis? Why is atherosclerosis considered a risk factor for strokes and heart attacks?

_____

_____

LAB 7:
The Respiratory System

By the End of This Lab, You Should Be Able to Answer the Following Questions:

- Why is oxygen required by the body for survival?

- How do the left and right lungs differ from one another?

- What type of epithelium is found within alveoli? Why is this cell shape and cell arrangement necessary for gas exchange?

- What is pulmonary auscultation? What piece of equipment is used to auscultate the lungs?

- Why is carbon dioxide considered a toxic waste product?

- What is spirometry? Why is spirometry performed on patients with certain respiratory disorders?

INTRODUCTION

Most people know that oxygen is crucial for survival, but why is this so? The answer lies in a small energy molecule that cells "spend" to fuel their daily activities: adenosine triphosphate, or ATP. Believe it or not, your cells would perish unless the energy trapped in dietary nutrients (example: carbohydrates) was converted into molecules of ATP. Under normal conditions, ATP is produced through a series of reactions called aerobic cellular respiration. The term **aerobic** tells us this process requires oxygen. Gas exchange, or **respiration**, must occur between (1) tissue cells and blood, as well as (2) blood and inhaled air to sustain this critical process. The following equation summarizes cellular respiration:

$$C_6H_{12}O_6 \text{ (glucose)} + 6\ O_2 \rightarrow 36\text{-}38\ ATP + 6\ CO_2 + 6\ H_2O$$

At the level of the human body, one glucose molecule holds a trivial amount of energy. At the level of a cell, though, one glucose molecule carries enough energy to complete a variety of tasks. The idea of a cell "spending" glucose on tasks instead of ATP is analogous to sticking a fifty dollar bill into a vending machine. Not only would this be a waste of money, since the vending machine can't give back suitable change, but the machine itself is not even configured to acknowledge this type of currency. Similarly, cells are wired to use ATP exclusively in their day-to-day activities: hormone production, muscle contraction, cell division, and so on.

Carbon dioxide accumulates in body fluids as a result of cellular respiration. Carbon dioxide is toxic to cells because it forms carbonic acid in aqueous solutions, such as cytoplasm and blood plasma. The following equation summarizes this event:

$$CO_2 + H_2O \rightleftarrows \underset{\text{(carbonic acid)}}{H_2CO_3} \rightleftarrows \underset{\text{(hydrogen ion)}}{H^+} + \underset{\text{(bicarbonate ion)}}{HCO_3^-}$$

To maintain appropriate pH levels, the body must eliminate carbon dioxide in a timely manner. The respiratory system is responsible for bringing oxygen into the body, as well as removing carbon dioxide from the body. Since blood transports gases to *and* from the lungs, gas exchange requires an intimate association between the respiratory system and the cardiovascular system. Figure 7.1 illustrates the anatomy of the respiratory system.

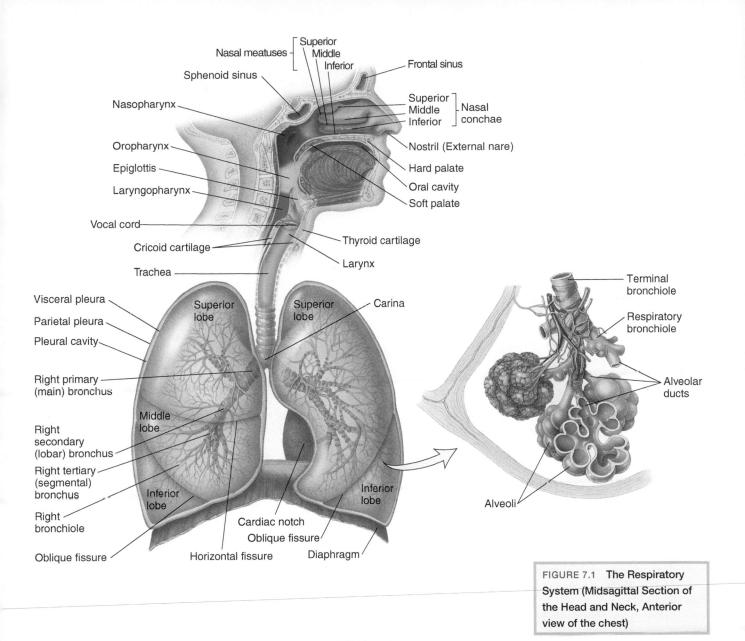

Nasal meatuses — Superior / Middle / Inferior
Sphenoid sinus
Frontal sinus
Nasopharynx
Superior / Middle / Inferior — Nasal conchae
Oropharynx
Nostril (External nare)
Epiglottis
Hard palate
Laryngopharynx
Oral cavity
Soft palate
Vocal cord
Thyroid cartilage
Cricoid cartilage
Larynx
Trachea
Visceral pleura
Superior lobe
Superior lobe
Carina
Parietal pleura
Pleural cavity
Terminal bronchiole
Respiratory bronchiole
Right primary (main) bronchus
Alveolar ducts
Middle lobe
Right secondary (lobar) bronchus
Right tertiary (segmental) bronchus
Right bronchiole
Inferior lobe
Inferior lobe
Alveoli
Cardiac notch
Oblique fissure
Oblique fissure
Horizontal fissure
Diaphragm

FIGURE 7.1 **The Respiratory System (Midsagittal Section of the Head and Neck, Anterior view of the chest)**

EXERCISE 7.1   Gross Anatomy of the Respiratory System

The respiratory system is unique with respect to its location in the human body; no other organ system is confined exclusively to the head, neck, and thoracic cavity. As inhaled air swirls through the respiratory tract—extending from the mouth and nose to tiny air sacs (**alveoli**) within in the lungs—it is filtered, moistened, and heated along the way.

 During this exercise, you will identify key structures on an anatomical model of the respiratory system. Alternately, your instructor may ask you to identify these structures on a dissected fetal pig, a virtual cadaver dissection, or a dissected human cadaver. Label the following structures on Figures 7.2–7.4.

1. Paranasal Sinuses (Only Frontal Sinus and Sphenoid Sinus are Shown on figure 7.2)

2. Nostrils (External Nares)

3. Nasal Conchae (Turbinate Bones)

4. Pharynx

   a. Nasopharynx

   b. Oropharynx

   c. Laryngopharynx

5. Epiglottis

6. Larynx

7. Trachea

8. Lungs

   a. Superior Lobes

   b. Middle Lobe (Right Lung Only)

   c. Inferior Lobes

   d. Cardiac Notch

9. Diaphragm Muscle

10. Intercostal Muscles

11. Bronchi

    a. Primary (Main) Bronchi

    b. Secondary (Lobar) Bronchi

    c. Tertiary (Segmental) Bronchi

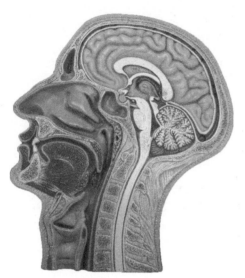

Label structures 1–7 on Figure 7.2

FIGURE 7.2 **Midsagittal View of the Head and Neck**

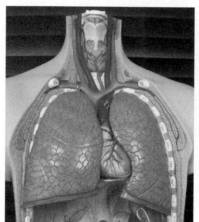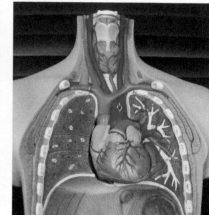

Label structures 6–10 on Figure 7.3

FIGURE 7.3 **Anterior Views of the Respiratory System**

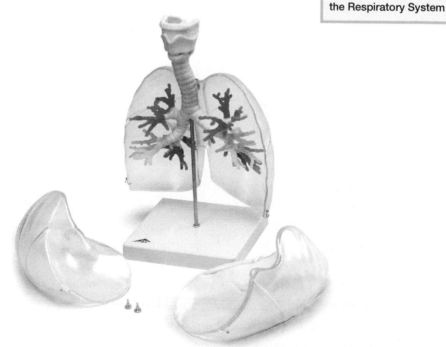

FIGURE 7.4 **Bronchial Tree with Larynx**

Label structures 6, 7, and 11 on Figure 7.4

Review Questions for Exercise 7.1

Label the left and right lungs on the image below. How can you tell them apart?

_____

_____

Label the diaphragm muscle, the trachea, and bronchi on the image below.

Is the trachea anterior **to (situated in front of) or** posterior **to (situated behind) the esophagus?** _____

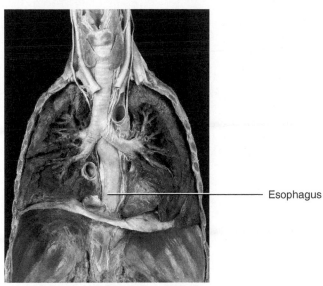

—————— Esophagus

Anterior View of the Respiratory System

Label the frontal sinus, the nasal conchae, and the epiglottis on the image below.

What is the function of the epiglottis? _____

_____

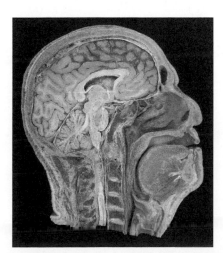

Median (Midsagittal) Views of the Head and Neck

Label the bronchi, trachea, larynx, and epiglottis on the image below.

Rings of hyaline cartilage are found within the larynx and the trachea. What function do you think these rings of cartilage serve? _____

_____

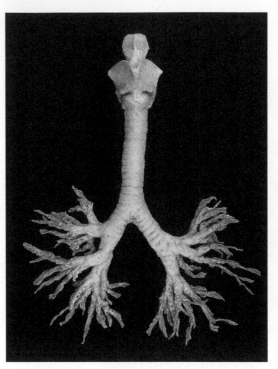

Anterior View of the Larynx and Bronchial Tree

INTERNAL AND EXTERNAL RESPIRATION

Bronchioles terminate in microscopic air sacs called **alveoli**, which are covered by pulmonary capillaries. (The term **pulmonary** is commonly used in the medical field when referring to the lungs.) The thin walls of alveoli allow gas exchange to occur between inhaled air and blood within the pulmonary capillaries. This process is called **external respiration**, since gas exchange is occurring between blood and air from the external environment. Figure 7.5 illustrates the directions in which oxygen and carbon dioxide move during this process. Carbon dioxide diffuses out of blood and into alveoli, while oxygen diffuses out of alveoli and into the bloodstream.

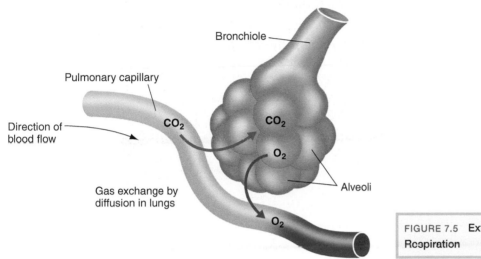

FIGURE 7.5 External Respiration

Following external respiration, blood returns to the heart and gets pumped throughout the entire body. When blood enters tissue capillaries, a second round of gas exchange occurs. This process is called **internal respiration**, since gas exchange is occurring between cells *and* fluids that reside inside of the body. Figure 7.6 illustrates the directions in which oxygen and carbon dioxide move during this process. Oxygen diffuses out of blood and toward tissue cells, while carbon dioxide diffuses away from tissue cells and into the bloodstream.

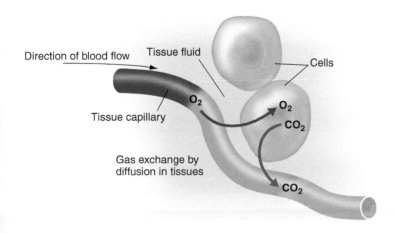

FIGURE 7.6 Internal Respiration

To understand how an organ functions, it is helpful to view its structure at the microscopic level. During this exercise, you will examine prepared microscope slides of the trachea and normal lung tissue. Be sure to note the characteristics of healthy alveoli during this exercise; this knowledge will help you identify alveolar damage during Exercise 7.6 (Disorders of the Respiratory System). Compare your focused microscope slides to the photomicrographs provided below, and answer the review questions that accompany each slide.

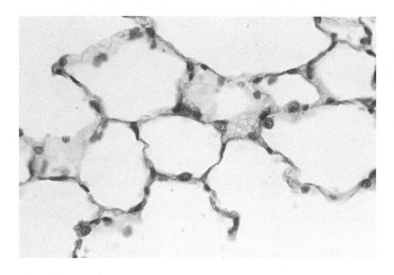

FIGURE 7.7   Histology of Lung Tissue

Label several alveoli on Figure 7.7.

Are the epithelial cells that construct alveoli squamous, cuboidal, or columnar in shape?

_____

Are the epithelial cells that construct alveoli simple, stratified, or pseudostratified in arrangement? _____

Why do you think this specific cell shape *and* cell arrangement is necessary for gas exchange?

_____

_____

Hundreds of millions of alveoli are found inside of your lungs. Why do you think so many alveoli are required for normal respiration?

_____

_____

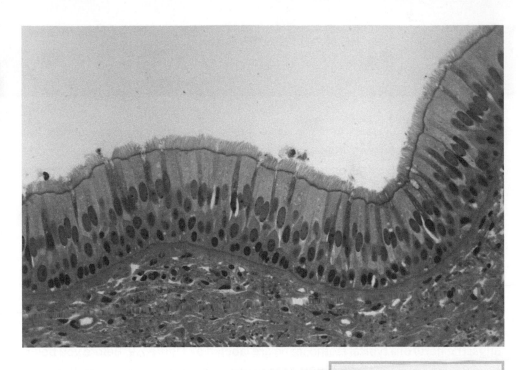

FIGURE 7.8 Histology of the Trachea

Are the epithelial cells that line the trachea (as shown in Figure 7.8) squamous, cuboidal, or columnar in shape? _____

Are the epithelial cells that line the trachea simple, stratified, or pseudostratified in arrangement? _____

The surface of a ciliated epithelial cell is covered with small, hair-like projections. When cilia beat together in unison, objects (or fluids) in the outside environment can be moved in a specific direction. Impaired ciliary function may result in frequent respiratory infections; this is the case for many individuals who suffer from Primary Ciliary Dyskinesia (PCD). If you would like to watch a video that shows normal and abnormal cilia in action, you can access the PCD Foundation's website at *http://pcdfoundation.org/video/ciliavideo.htm.*

Label cilia on the epithelial cells shown above. What substances do these hair-like projections move in the respiratory tract?

_____

_____

_____

Health professionals use stethoscopes to hear internal body sounds, such as the breath sounds produced within a patient's lungs. The practice of listening to body sounds is called **auscultation** within the medical field. **Pulmonary auscultation** is commonly performed to assess respiratory health and, in some cases, to diagnose respiratory disorders. Normal breath sounds are created as air travels through the respiratory tract, given its turbulent nature and velocity. Abnormal breath sounds are produced by patients with certain respiratory disorders, such as asthma, pneumonia, bronchitis, or emphysema. **Crackles/rales** (a crackling sound), **stridor** (a harsh, high-pitched sound), **wheezing** (a whistling sound), and **rhonchi** (a snoring sound) are examples of abnormal breath sounds.

During this exercise, you will use a stethoscope to auscultate your partner's lungs; the parts of a typical stethoscope are labeled on Figure 7.9. Before you begin this exercise, your instructor may choose to play audio files of normal and abnormal breath sounds. Prior to lab, you can also access audio files of lung sounds from a reputable website: *The Merck Manual's Online Medical Library* (http://www.merck.com/mmpe/index.html). Simply search for "auscultation" and click on "Approach to the Patient with Pulmonary Symptoms: Auscultation".

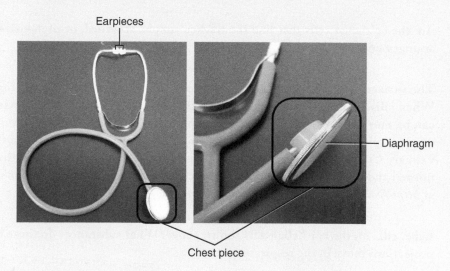

FIGURE 7.9 **Parts of the Stethoscope**

Earpieces

Diaphragm

Chest piece

Procedure

1. Use an alcohol swab to clean the earpieces on your stethoscope.

2. Tilt the stethoscope anteriorly (toward your face) while securing the earpieces inside of your ear canals. Adjust the earpieces, as necessary, to minimize background noises.

3. Warm the stethoscope's diaphragm (the flat side of the chestpiece) in your hands for 30 seconds.

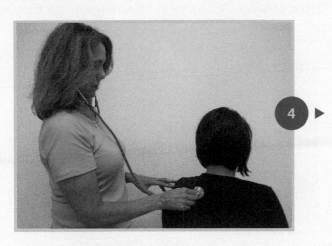

4. Place the diaphragm on your partner's back at the first auscultation site (L1), which is illustrated on the diagram below. Slowly move the stethoscope around until you hear breath sounds, and listen to one full respiratory cycle at this site. (A respiratory cycle consists of one inhale followed by one exhale.)

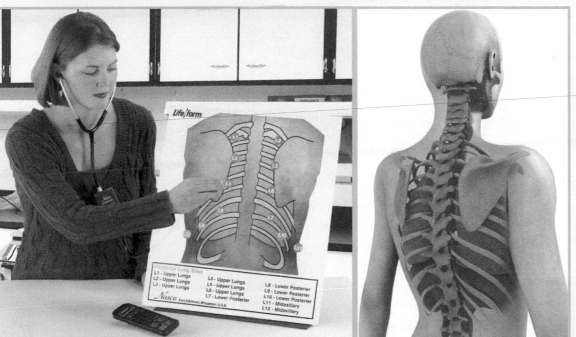

5. Move the stethoscope to the second auscultation site (L2), and listen to one full respiratory cycle.

6. Compare the sounds you heard at the first two auscultation sites; these sites mirror one another on the left and right sides of the body.

7. Repeat steps 4–6 until you have listened to each auscultation site on your partner's back.

8. Switch roles with your partner and repeat this procedure.

A CLOSER LOOK AT PULMONARY AUSCULTATION

If you would like to learn more about the pulmonary disorders that result in crackles (rales), stridor, wheezing, and rhonchi, you can access the *Pulmonary Disorders* link on *The Merck Manual's Online Medical Dictionary* at http://www.merck.com/mmpe/index.html.

Visualizing THE LAB

STEP BY STEP 7.4 Visualizing the Effects of Dissolved Carbon Dioxide on pH

Every time you exhale air, carbon dioxide gas is released from your body. During this exercise, you will monitor the effects of dissolved carbon dioxide on the pH of distilled water. Your own exhaled air will serve as the source for carbon dioxide, and a pH indicator called bromothymol blue will detect any pH changes that may occur. Figure 7.10 shows the various colors produced by bromothymol blue, as well as the pH range associated with each color. This pH indicator only displays a blue color when the pH level is 7.6 or higher. Between pH 6.0-7.6, bromothymol blue turns green in color. A yellow color indicates the solution has dipped below pH 6.0.

Formulate a hypothesis to test that describes how exhaled air will affect the pH of distilled water.

_____

_____

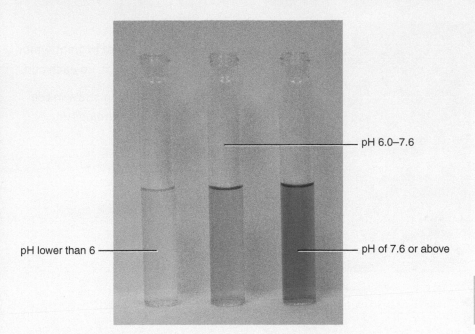

pH 6.0–7.6

pH lower than 6

pH of 7.6 or above

FIGURE 7.10 Interpreting the
Color of Bromothymol Blue

Procedure

1. Collect two drinking cups and a 100 mL graduated cylinder.

2. Label one cup *Control* and the other cup *Exhaled Air*.

3. Using the graduated cylinder,
   add 100 mL of distilled water
   to each cup.

◄ 3

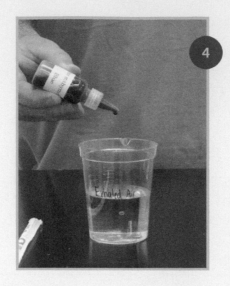

4. Add 20 drops of bromothymol blue pH indicator to each cup.

5. Place a drinking straw in the cup labeled *Exhaled Air*.

6. Cover the top of each drinking cup with a piece of plastic wrap.

7. Blow bubbles through the straw of the *Exhaled Air* cup for one minute.

8. Record your observations in Table 7.1.

9. Dispose of your solutions in the hazardous waste container provided by your instructor. **Do not pour any solutions down the sink without prior approval from your instructor.**

TABLE 7.1 Effects of Dissolved Carbon Dioxide on the pH of Distilled Water

Cup	Initial Color	Final Color
Control		
Exhaled Air		

Review Questions for Step by Step 7.4

What effect did exhaled air have on the pH of distilled water? Was your initial hypothesis supported or rejected? Revise your initial hypothesis, if necessary, based on the results from this experiment.

_____

_____

_____

Which molecule is produced when carbon dioxide combines with water?

_____

Do you think exhaled air would change the pH of a buffered solution? How could you test this hypothesis?

_____

_____

_____

What potential sources of error exist in this experiment? What improvements (if any) could be made to this experiment?_____

_____

_____

RESPIRATORY VOLUMES

Just as each heartbeat (aka cardiac cycle) consists of three distinct phases, each breath (aka respiratory cycle) actually consists of two distinct phases: **inspiration** (inhalation) and **expiration** (exhalation). Due to changes in the size of the thoracic cavity—and ultimately, changes in the size of the lungs—air rushes into the lungs during inspiration and out of the lungs during expiration. The number of breaths you take per minute is called your **respiratory rate**, or breathing rate. On average, a resting adult takes 12–18 breaths every minute. As you have undoubtedly noticed, your respiratory rate does vary in response to physical activity, stress, relaxation, and even changes to your respiratory health. The volume of air you inhale and exhale may fluctuate in response to these factors as well. Each red hill on Figure 7.11 represents one distinct breath taken by a person. Some of these breaths obviously differ based on volume and/or frequency, given the shapes and sizes of these hills.

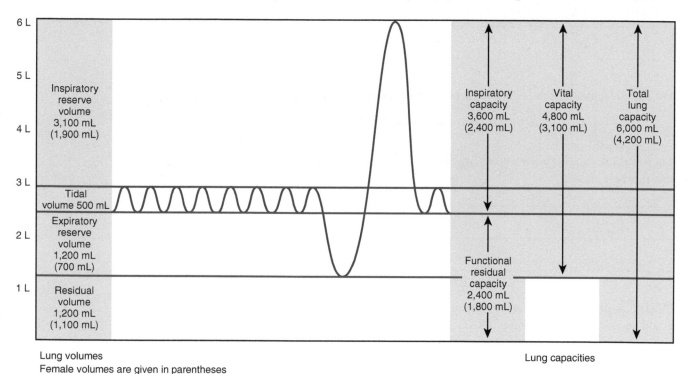

Lung volumes
Female volumes are given in parentheses

Lung capacities

FIGURE 7.11 **Average Respiratory Volumes for Males and Females.**

Spirometry tests are used by pulmonologists to assess lung function. In addition to measuring a patient's respiratory rate, spirometry measures the following lung function values, which are visually depicted on Figure 7.11:

- **Tidal volume** refers to the volume of air inspired or expired during one normal breath. As shown in Figure 7.11, tidal volume only utilizes approximately 10% of total lung capacity.

- The terms **inspiratory reserve volume** and **expiratory reserve volume** refer to the additional volume of air — *beyond the tidal volume* — that can be inspired or expired, respectively, during forceful breathing.

- After inspiring as deeply as possible, the maximum volume of air that can be expired is called the **vital capacity** of the lungs.

- Regardless of the intensity of expiration, approximately one liter of air will remain inside of the lungs; this is called the **residual volume** of the lungs.

Source: Medline Plus Medical Encyclopedia. Accessed 12-1-10 from http://www.nlm.nih.gov/medlineplus/ency/article/002341.htm.

During this exercise, you will perform spirometry and estimate your tidal volume, inspiratory reserve volume, expiratory reserve volume, and vital capacity. The parts of a typical dry spirometer are labeled on Figure 7.12.

Disposable mouthpiece

Inlet hole for exhaled air/
Attachment site for
disposable mouthpiece

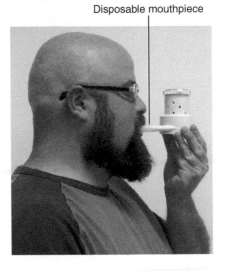

Air pressure gauge
Units = milliliters (mL) or cubic centimeters (cc)

FIGURE 7.12   **Dry Spirometer**

●●● SAFETY NOTE
Be sure to place a clean, disposable mouthpiece on the spirometer before inserting it into your mouth.

During this experiment, we are not following the standards and guidelines employed by accredited medical institutions. The results are not intended to diagnose respiratory disorders.

Procedure

Steps to Complete before Taking Each Measurement: Make sure the spirometer is set to zero, and take several normal breaths.

Measuring Your Tidal Volume (TV): Inhale normally, and then exhale normally into the mouthpiece. Record your tidal volume in Table 7.2.

Measuring Your Expiratory Reserve Volume (ERV): Inhale normally, exhale normally, and then exhale as deeply as possible into the spirometer. Record your expiratory reserve volume in Table 7.2.

Vital Capacity (VC): Inhale as much air as possible, and then exhale as much air as possible into the spirometer. Record your vital capacity in Table 7.2.

Measuring Your Inspiratory Reserve Volume (IRV): The inspiratory reserve volume cannot be measured directly with most dry spirometers. Use the following equation to calculate your IRV:

$$IRV = VC - (TV + ERV)$$

Record your inspiratory reserve volume in Table 7.2.

TABLE 7.2 Respiratory Volumes

Measurement	Volume (mL)
Tidal Volume	
Expiratory Reserve Volume	
Vital Capacity	
Inspiratory Reserve Volume	

Review Questions for Exercise 7.5

How do your respiratory volumes compare to the average volumes listed in Figure 7.11? If you are female, be sure to compare your results with the volumes in parentheses.

_____

_____

_____

To improve the accuracy of the results, what improvements could be made to this experiment?

_____

_____

_____

Name one respiratory disorder that typically affects a person's lung function (i.e., respiratory volumes). Why is this so?

_____

_____

_____

EXERCISE 7.6   Disorders of the Respiratory System

During this exercise, you will use your knowledge of the respiratory system to diagnose three patients. Fawn, Dave, and Michael are seeking medical attention for different respiratory disorders. Use the symptoms and test results provided to make a logical diagnosis for each patient: asthma, pneumothorax, or emphysema.

Terminology Used in this Exercise:

Emphysema—abnormal distension (expansion) of a body tissue, resulting in decreased elasticity and permanent tissue damage

Pneumothorax—accumulation of air in the pleural cavity surrounding a lung

The prefixes *pulmono-* and *pneumo-* are commonly used in medical terminology when referring to the lungs.

Computerized Axial Tomography (CAT Scan or CT Scan)—X-rays taken from many different angles to create thin, detailed images of internal body structures

Fawn

Fawn has been a heavy smoker since the age of 19. For years, she enjoyed smoking everywhere—at home and at work, in bars, at school—without feeling any sort of physical repercussion. Due to insistent pleas from her family, she did try to quit smoking at the ages of 32, 40, and 48, and most recently at the age of 55. For some reason, though, she just couldn't find a way to permanently kick this addiction. Given her shortness of breath and chronic coughing, Fawn's desire to quit was stronger than ever by the age of 56.

During a particularly severe bout of coughing and wheezing, Fawn's primary care physician referred her to a pulmonologist. Decreased breath sounds were picked up in Fawn's lungs during pulmonary auscultation. As a result, the pulmonologist elected to perform a pulmonary function test: spirometry. Fawn's vital capacity was abnormally low, even though her total lung capacity was abnormally high. As shown on a CT scan (Figure 7.13), vacant spaces had developed in Fawn's lungs where healthy alveoli had once resided. Figure 7.14 shows the structure of Fawn's alveoli as observed through a microscope.

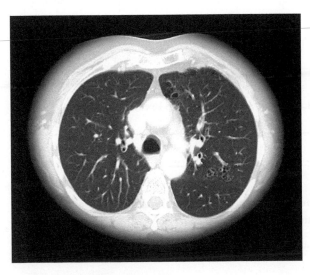

FIGURE 7.13   CT Scan of Fawn's Lungs

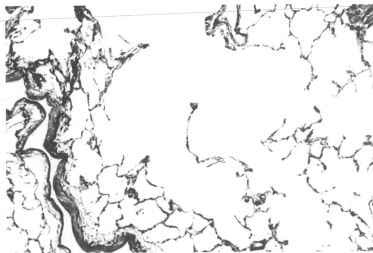

FIGURE 7.14   Microscopic Image of Fawn's Alveoli

Which respiratory disorder is suggested by Fawn's symptoms and test results?

_____

Why is Fawn's total lung capacity abnormally high, in spite of her shortness of breath and low vital capacity?

_____

_____

What treatment options (if any) exist for Fawn's condition?

_____

_____

Dave

Dave gravitated toward any sport that required speed and agility, particularly team sports such as soccer and basketball. By the age of 8, he was already considered a gifted athlete by his coaches, teammates, and competitors alike. Dave had never dealt with a serious health issue until he attended a rigorous basketball camp one summer. In the middle of sprints, Dave terrified himself—not to mention the other kids—as he dropped to his knees and gasped for air.

Feeling overwhelmed and anxious, Dave was rushed to the camp's on-site medical facility. His chest pains and incessant coughing made every second feel more like an hour. Without the aid of a stethoscope, the doctor could hear wheezing sounds as Dave struggled to inhale and exhale. Given his symptoms, it was apparent to the doctor that his airways were constricted. Dave was told his condition could either be caused by muscle spasms, inflammation, or a combination of both.

Which respiratory disorder is suggested by Dave's symptoms and test results?

_____

Why were wheezing sounds produced as Dave struggled to breathe?

_____

_____

What treatment options exist for Dave's condition, assuming it is caused by muscle spasms _and_ inflammation of his airways?

_____

_____

Michael

On his way home from work one night, Michael was involved in a serious, six-car pile-up on the freeway. As his chest slammed into the steering wheel, he felt a sharp pain radiate across the side of his chest. Before passing out, Michael feared that he would not survive the wreck, given his excruciating chest pain and his inability to breathe. At the hospital, X-rays revealed several fractured ribs, as well as evidence of a collapsed lung. A CT scan confirmed his right lung had indeed collapsed, as shown in Figure 7.15.

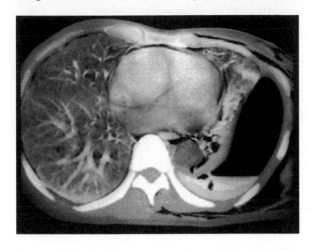

FIGURE 7.15 CT scan of Michael's Lungs

Which respiratory disorder is suggested by Michael's symptoms and test results?

_____

What treatment options exist for Michael's condition?

_____

_____

The CT scan in Figure 7.15 shows a transverse section, or horizontal slice, of Michael's thorax. Label Michael's heart, ribs, vertebra, right lung, and left lung on Figure 7.15. What landmarks help you distinguish the left and right sides of the body?

_____

_____

A CLOSER LOOK AT RESPIRATORY DISORDERS

The National Heart, Lung, and Blood Institute (NHLBI) is a division of the National Institutes of Health. NHLBI strives to "promote the prevention and treatment of heart, lung, and blood diseases and enhance the health of all individuals so that they can live longer and more fulfilling lives." If you would like to learn more about the symptoms, diagnosis, treatment, and/or prevention of a specific respiratory disorder, you can visit the National Heart Lung and Blood Institute's website at http://www.nhlbi.nih.gov.

1. Label the epiglottis, trachea, and left lung on the diagram below.

2. Structure 1 is called the _____
    a. Alveoli
    b. Nasal concha
    c. Frontal sinus
    d. Epiglottis

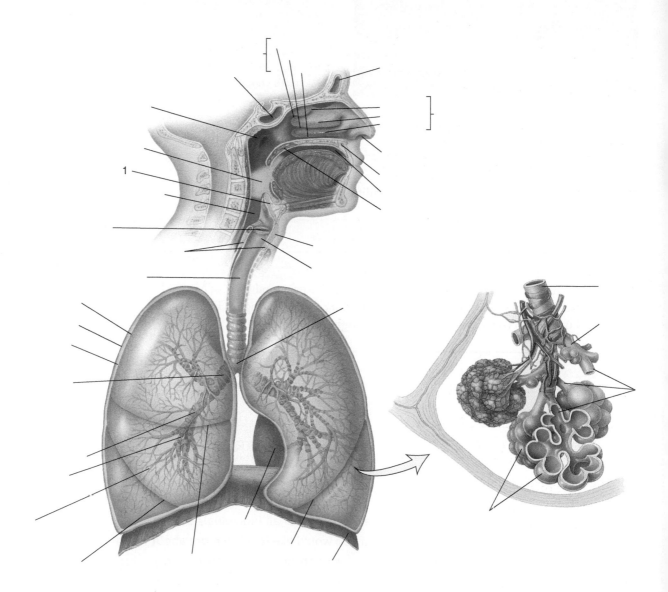

3. Arrange the following organs in sequence on the flowchart. The goal is to trace the path of inhaled air as it travels through the respiratory tract.

- **Alveoli**
- **Larynx**
- **Trachea**
- **Bronchioles**
- **Bronchi**
- **Pharynx**

Nasal Cavities
↓
_____
↓
_____
↓
_____
↓
_____
↓
_____
↓
_____

4. Why is carbon dioxide toxic to cells?

_____

_____

5. The walls of alveoli are composed of _____ epithelium, which is the thinnest tissue type in the body.
   a. stratified squamous
   b. simple cuboidal
   c. stratified columnar
   d. simple squamous

6. _____ is the medical term for a collapsed lung.
   a. Pneumothorax
   b. Pneumonia
   c. Emphysema
   d. Pulmonary fibrosis

7. What is pulmonary auscultation? What piece of equipment is used to auscultate the lungs?

_____

_____

8. _____ is labeled B on the spirogram below.
   a. Tidal volume
   b. Expiratory reserve volume
   c. Inspiratory reserve volume
   d. Vital capacity

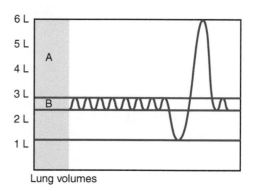

Lung volumes

9. Why is spirometry performed on patients? What is the difference between your vital capacity and the residual volume of your lungs?

_____

_____

_____

10. _____ is gas exchange between inhaled air and blood within the pulmonary capillaries.
    a. Cellular respiration
    b. External respiration
    c. Internal respiration
    d. All of the above

LAB 8:
The Urinary System

By the End of This Lab, You Should Be Able to Answer the Following Questions:

- How does metabolic waste differ from undigested waste?

- Which organ produces urea? Which organs filter urea out of the bloodstream?

- What differences exist between the male and female urethra?

- What type of epithelium is found within the urinary bladder?

- What is a glomerulus? How do dialysis membranes mimic the function of glomeruli?

- How are urine specimens physically assessed during urinalysis?

- Is it normal to find albumin, hemoglobin, or glucose in a patient's urine sample?

INTRODUCTION

When most of us hear the word *metabolism*, we think about the relationship between body weight, caloric intake, and physical activity levels. These factors certainly do affect your body's metabolism, but they are not the entire story. **Metabolism** refers to every single chemical reaction that occurs inside of your cells: hormone production, food digestion, DNA synthesis, ATP production, protein degradation…the list goes on and on. The root word *metabol* actually means *change* in Greek. This is fitting, since molecules are changed within our bodies as a result of different metabolic pathways.

Believe it or not, cooking shares similarities with the metabolic reactions that occur inside of the body. Let's say, for instance, you are cooking an omelette from scratch. The original ingredients—eggs, milk, vegetables, cheese, and meat—undergo physical changes (examples: chopping and mixing) as well as chemical changes (example: coagulation of proteins) to produce the final entrée. Certain parts of these raw ingredients—egg shells, vegetable peels, bones, and fat trimmings—are purposely removed, so this waste must be disposed of in one way or another. During a metabolic reaction, the original ingredients (substrate molecules) are combined or torn apart by enzymes, instead of a person, to create the final product. Waste materials also accumulate in cells as a result of metabolism. Certain types of **metabolic waste**, such as carbon dioxide and ammonia, are actually toxic to the cells that create them, so they must be disposed of in a timely fashion.

As discussed in Lab 7 (The Respiratory System), carbon dioxide is formed as glucose and is metabolized during cellular respiration. Carbon dioxide is toxic to cells because it forms carbonic acid in bodily fluids, such as plasma and cytoplasm. To maintain appropriate pH levels in the body, carbon dioxide must be eliminated in a rapid and consistent manner; this feat requires an intimate association between the cardiovascular and respiratory systems. Ammonia is toxic to cells because its basic nature *raises* the pH of bodily fluids. Ammonia (NH_3) is produced as certain amino acids get degraded within the liver. Since ammonia is such a toxic and volatile molecule, the liver combines it with carbon dioxide to create a less toxic waste product: urea (H_2N-CO-NH_2). The bloodstream carries urea from the liver to the kidneys, where it is filtered out of blood and placed into a fluid that ultimately becomes urine. Following its formation in the kidneys, urine is pushed through adjoining tubes (the ureters) to reach the urinary bladder. Eventually, urine will pass through another tube (the urethra) and get **voided**, or eliminated, from the body. Figure 8.1 shows the gross anatomy of the urinary system.

As the kidneys remove **nitrogenous** (nitrogen-containing) wastes from the bloodstream, they simultaneously juggle a slew of additional responsibilities. In response to certain hormones, such as antidiuretic hormone (ADH) and aldosterone, the kidneys also adjust the water/salt content of the blood to maintain homeostasis. Let's say excessive amounts of water and sodium are present in your blood. In response to hormones, the kidneys will **excrete** (separate and expel) the surplus quantity, along with urea and other nitrogenous wastes. Blood pressure is affected by many variables, including the volume of water in blood, as well as its concentration of sodium ions. For this reason, the kidneys play a vital role in blood pressure regulation.

If blood deviates from its normal pH of 7.4, then the kidneys adjust the levels of two additional substances: hydrogen ions (H^+) and buffer molecules. This action helps to prevent **acidosis** (an abnormal decrease in the pH of a body fluid) or **alkalosis** (an abnormal increase in the pH of a body fluid) from developing. If these responsibilities aren't complicated enough, the kidneys also produce a hormone called erythropoietin, which stimulates the production (*-poiesis*) of additional red blood cells (*erythro-*), when needed. Let's say you donate blood, lose a large quantity of blood due to injury, or even move to a city with a higher altitude. In each of these scenarios, your blood can no longer carry the same amount of oxygen as it once did. To solve this problem, the kidneys release erythropoietin, which stimulates red blood cell production in the red bone marrow. The ability of the kidneys to *simultaneously* perform these complex tasks helps us understand why—to date—no one has been able to engineer an implantable artificial kidney.

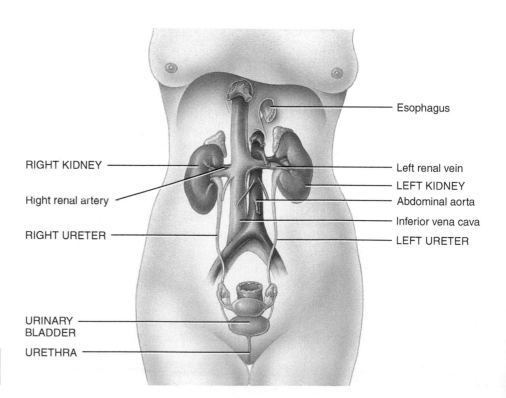

FIGURE 8.1 **Anterior View of the Urinary System**

During this exercise, you will identify organs on an anatomical model of the human urinary system. Alternately, your instructor may ask you to identify these organs on a virtual cadaver dissection, a dissected fetal pig, or a dissected human cadaver. You will also identify key structures on an anatomical model of the human kidney, a virtual kidney dissection, or a dissected sheep/human kidney. Label the structures listed below on Figures 8.2–8.4.

Terminology Used in This Exercise

Renal—pertaining to the kidneys
Cortex—outer tissues of an organ (renal cortex, adrenal cortex, cerebral cortex)
Medulla—inner tissues of an organ (renal medulla, medulla oblongata, medullary cavity of a bone)
Pelvis—a basin-shaped cavity (renal pelvis, bony pelvis)

1. Kidneys—label as *Right Kidney* and *Left Kidney*, respectively

    a. Renal Arteries—carry blood to the kidneys for filtration

    b. Renal Vein—carry filtered blood away from the kidneys

    c. Renal Capsule—tough membrane that surrounds the kidney

    d. Renal Cortex—outer tissues of the kidney

    e. Renal Medulla—inner tissues of the kidney (renal pyramids are located here)

    f. Renal Pelvis—collects and transports urine to the adjoining ureter

2. Ureters

3. Urinary Bladder

4. Urethra

    a. External Urethral Orifice

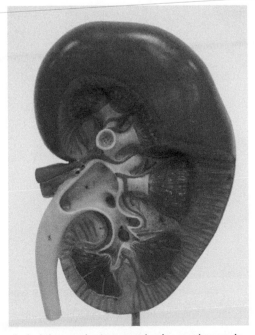

FIGURE 8.2 Sagittal Section of the Kidney

Label the renal artery, renal vein, renal capsule, renal cortex, renal medulla, renal pelvis, and ureter.

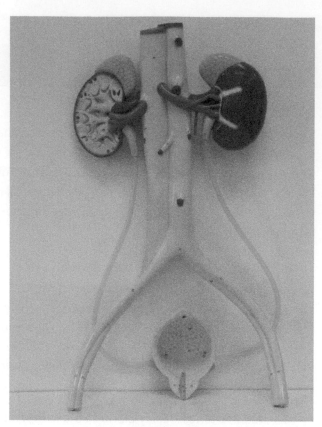

FIGURE 8.3   Model of Urinary
System (Anterior View)

Label the right/left kidney, right/left ureters, urinary bladder,
and urethra.

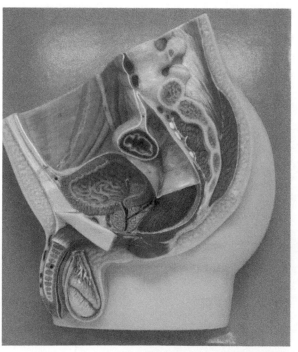 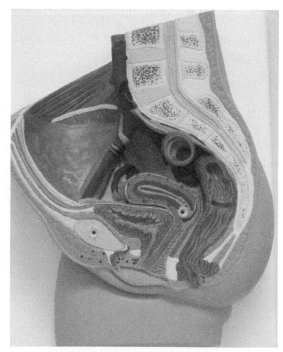

Label the urinary bladder, urethra, and external urethral orifice.

FIGURE 8.4   Sagittal Sections
of the Male and Female Urinary
Tracts

Label the left kidney, the left ureter, and the urinary bladder on the image below.

Is the urinary bladder located in the thoracic, abdominal, or pelvic cavity? Name another organ that is found in this body cavity.

_____

_____

How would you describe the location of the kidneys within the abdominal cavity? For instance, are the kidneys anterior (situated in front of) or posterior to (situated behind) the digestive organs? Are they medial (closer to the midline than . . .) or lateral to (farther from the midline than . . .) the spine?

_____

_____

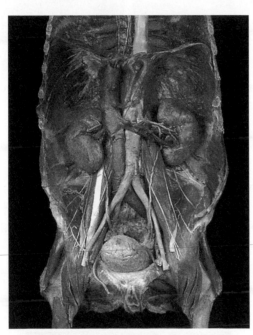

Anterior View of the Human Urinary System

Label the urinary bladder on each image at top of next page.

Note the location of the prostate gland relative to the male urinary bladder. Why does prostatitis, or inflammation (-*itis*) of the prostate, commonly lead to difficulty urinating?

_____

_____

Note the location of the uterus relative to the female urinary bladder. Why do women tend to urinate more frequently during pregnancy?

_____

_____

What differences exist between the male and female urethra?

_____

_____

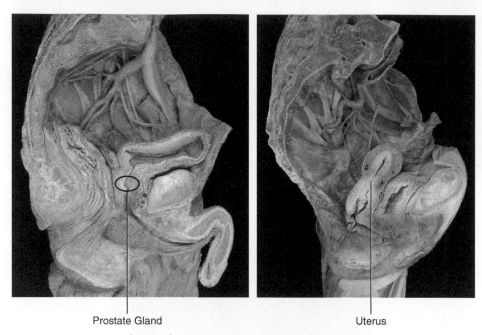

Prostate Gland   Uterus

Sagittal Sections of the Male and Female Pelvis

Label the renal artery and the ureter on the image below.

Label the renal cortex and renal medulla on the image below.

What type of muscle (skeletal, cardiac, or smooth) is found within the ureters? What role do you think this muscle plays in urine transport?

_____

_____

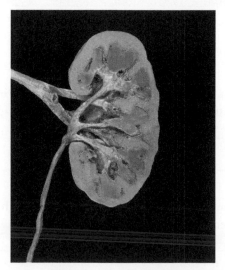

Dissected Human Kidney

URINE FORMATION

At first glance, the urinary system looks like one of the simplest organ systems in the human body. Appearances can be deceiving, though—especially when it comes to the humble, bean-shaped kidneys. Urine formation is *anything* but simple! Take a look at Figure 8.5, which illustrates the functional unit for urine formation: a microscopic structure called a **nephron**. (The term *nephros* actually means *kidney* in Greek. In medical terminology, the prefix *nephro-* is commonly used when referring to the kidneys.) If this nephron illustration looks intimidating, try to focus on its basic parts: a thin tubule intertwined with a capillary bed. Each one of your kidneys contains about a million nephrons, which continuously filter blood and create urine.

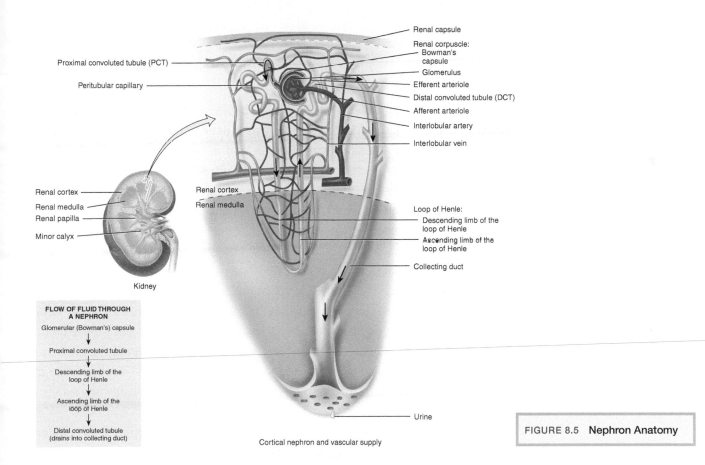

FIGURE 8.5 **Nephron Anatomy**

Cortical nephron and vascular supply

Amazingly, in a resting adult, more blood flows through the kidneys (units: mL/min) than through the brain, the heart, the lungs, or the organs of the GI tract.* As blood gets filtered through a ball of twisted capillaries—the glomerulus—the resulting filtrate enters a cavity inside of Bowman's capsule. **Glomerular filtration** removes molecules from your blood based on size, similar to a colander. Water, glucose, and other small molecules leave the bloodstream during this step. Cells and large plasma proteins are retained in the bloodstream, due to their larger sizes.

If every filtered substance ended up in urine, you would develop severe (if not fatal) dehydration and lose many essential nutrient molecules. Luckily, the next step in urine formation remedies this problem: tubular reabsorption. **Tubular reabsorption** transports glucose,

---

*Reference values for resting blood flow to organs of man. L. R. Williams and R. W. Leggett, *Clin. Phys. Physiol. Meas.*, 1989; 10: 187.

amino acids, water, and salts back to blood within the peritubular capillaries, which are situated around (*peri-*) the tubule carrying the filtrate. Before the filtrate is officially called urine, though, a process called **tubular secretion** must also occur. It turns out that certain waste molecules, such as drug metabolites, are too large to get filtered out of blood in glomeruli. As a result, these waste molecules must be actively secreted from blood. Tubular secretion allows these large waste molecules to enter the filtrate, which is housed inside of the nephron's tubule. Blood pH can also be regulated, as needed, through secretion of hydrogen ions and/or buffer molecules. Once glomerular filtration, tubular reabsorption, and tubular secretion are successfully completed, urine enters a collecting duct and travels to the renal pelvis.

Visualizing THE LAB

STEP BY STEP 8.2   Filtering Blood through a Semipermeable Membrane

Within glomeruli, molecules are either filtered out of blood *or* retained by blood based on one criterion: size. The term **dialysis** is used to describe any process that separates molecules based on size. Dialysis can occur **in vivo** (inside of a living organism) in an organ such as the kidney, or **in vitro** (in a setting outside a living organism) with the help of a semipermeable barrier. When dialysis is performed using a semipermeable membrane, the size of the membrane's pores is critical to the process. Ultimately, the pore size determines which blood constituents (examples: urea, glucose, and cells) are small enough to permeate the membrane and leave the bloodstream.

During this exercise, you will work in groups to filter sheep blood through dialysis tubing. Alternatively, dialysis may be performed on either (1) a synthetic blood sample or (2) blood from grocery-purchased chicken livers. Since these blood samples have already been **defibrinated**, they no longer contain fibrin: a fibrous protein involved in the clotting cascade. As a result, the blood is not capable of clotting. In terms of pore size, the dialysis tubing will mimic the walls of a healthy glomerulus.

Before you perform this experiment, formulate a hypothesis to test on the ability of albumin, urea, glucose, and sodium ions to permeate the dialysis membrane. Use the molecular weights provided below to help formulate your hypothesis. For reference, water has a molecular weight of 18 Daltons (Da).

Tested Substances

Albumin (the most abundant plasma protein in humans): molecular weight ≈ 65,000 Da

Glucose: molecular weight ≈ 180 Da

Urea: molecular weight ≈ 60 Da

Sodium ions: molecular weight ≈ 23 Da

Hydrogen ions: molecular weight ≈ 1 Da

Hypothesis:

_____

_____

Initially, the filtrate (solution submerging the dialysis bag) will consist solely of distilled water. If hydrogen ions enter the filtrate, will its pH increase or decrease? _____

> **SAFETY GUIDELINES**
> ### for Working with Defibrinated Sheep Blood or Chicken Liver Blood
>
> Occupational Safety and Health Administration (OSHA, *www.osha.gov*) Guidelines per *Occupational Safety and Health Standards → Bloodborne Pathogens*
>
> Defibrinated sheep blood is sterilized during the company's processing protocol. Grocery-purchased chicken livers are harvested from animals that were screened for the presence of pathogens per USDA guidelines. Blood from these two sources does not carry a risk of bloodborne pathogens and therefore does not meet OSHA's guidelines for the disposal of bloodborne pathogens.
>
> Wear gloves when handling defibrinated sheep blood or blood from grocery-purchased chicken livers. If splashing is possible, eye protection and a lab coat may also be worn. When you finish this experiment, the blood may be disposed of down the sink. The used dialysis tubing may be placed in a trash can.

Procedure

1. Collect one 100 mL graduated cylinder and two 250 mL beakers. Label one beaker *Blood* and the other beaker *Control*.

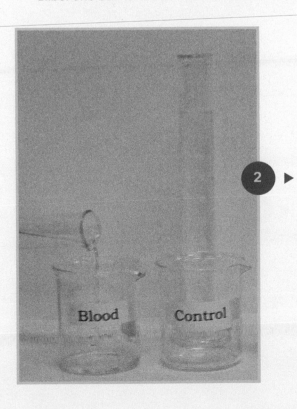

2. Using the graduated cylinder, add 100 mL of distilled water to each beaker.

3. Collect an 8-inch piece of dialysis tubing that has been presoaked in water.

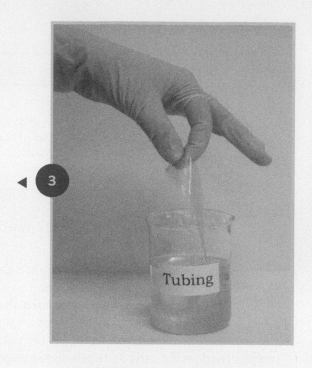

4. Tie a tight knot at one end of the dialysis tubing. Alternatively, you can seal this end with a tubing clip.

5. Using your index finger and thumb, rub the other end of the dialysis bag to create an opening for blood addition.

6. Transfer 10 mL of blood into the dialysis bag.
   Note: Be careful not to spill blood on the exterior of the bag!

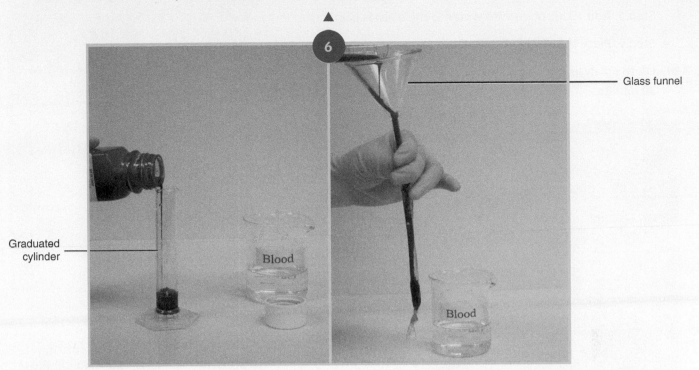

Glass funnel

Graduated cylinder

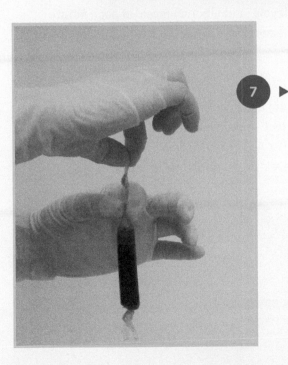

7. Close the open end of the dialysis bag with a tight knot. Alternatively, you can seal this end of the bag with a tubing clip.

8. Rinse the outside of the dialysis bag with distilled water. Be sure to rinse the knotted ends of the bag, as well, to remove any traces of blood.

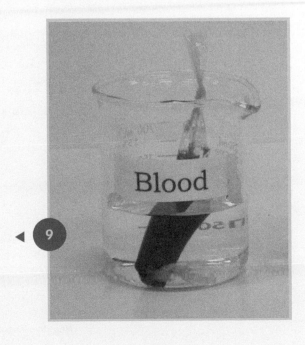

9. Place the dialysis bag into the beaker labeled *Blood*.

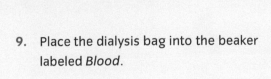

10. Modify steps 3–9 as indicated below to create the "Control" beaker.

   Step 6: Add 10 mL of *distilled water* to the dialysis bag.

   Step 9: Place the dialysis bag into the *Control* beaker.

11. Let the dialysis bags sit in their respective beakers for 30 minutes.

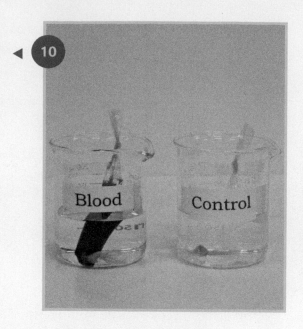

Collecting Results

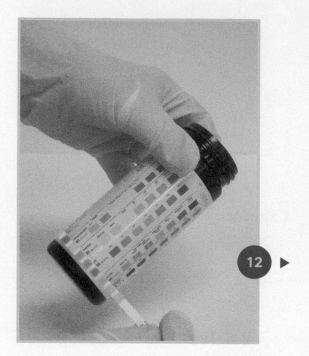

12. Read the directions on the urinalysis test strip bottle prior to use. Use a urinalysis test strip to determine the pH, glucose levels, and albumin levels of each filtrate (the liquid in each beaker). Record the results in Table 8.1.

13. Read the directions on the sodium chloride test strip bottle prior to use. Use a sodium chloride test strip to detect sodium ions in each filtrate. Alternatively, add 5 drops of 1% silver nitrate (1% $AgNO_3$) to each filtrate; sodium ions are present if a white precipitate develops. Record your results in Table 8.1.

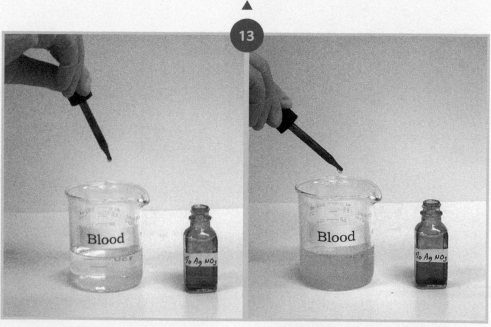

Negative Result for Silver Nitrate Test   Positive Result for Silver Nitrate Test

14. Read the directions on the UUN (urea nitrogen—urine) or BUN (urea nitrogen—blood) test strip bottle prior to use. Use a UUN or BUN test strip to detect urea in each filtrate. Record the results in Table 8.1.

TABLE 8.1 Filtration of Blood through Dialysis Tubing

	pH of filtrate	Was glucose detected in the filtrate?	Was albumin detected in the filtrate?	Were sodium ions detected in the filtrate?	Was urea detected in the filtrate?
Blood					
Control					

Review Questions for Step by Step 8.2

Was your initial hypothesis supported or rejected by the results of this experiment? If necessary, revise your hypothesis based on the results you collected.

_____

_____

How do your results compare with those obtained by other groups?

_____

_____

Plasma proteins are detected in urine as a result of certain glomerular diseases. What would this finding tell you about the structure/integrity of a patient's glomeruli?

_____

_____

To understand how an organ functions, it is helpful to view its structure at the microscopic level. During this exercise, you will examine prepared microscope slides of the kidney and the urinary bladder. Compare your focused microscope slides to the photomicrographs provided below, and answer the review questions that accompany each slide.

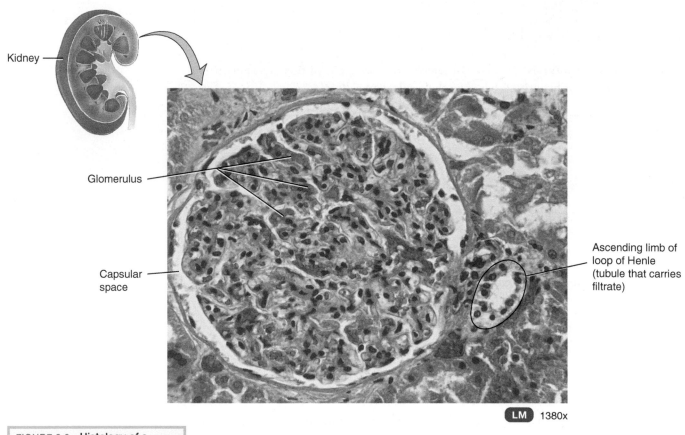

Kidney

Glomerulus

Capsular space

Ascending limb of loop of Henle (tubule that carries filtrate)

LM   1380x

FIGURE 8.6   **Histology of a Renal Corpuscle**

Are glomeruli composed of squamous, cuboidal, or columnar epithelial cells?

_____

Are glomeruli composed of simple, stratified, or pseudostratified epithelium?

_____

Why do you think this specific cell shape *and* cell arrangement is required for blood filtration?

_____

_____

The renal tubule labeled on Figure 8.6 is composed of _____.

a. simple squamous epithelium

b. simple cuboidal epithelium

c. stratified squamous epithelium

d. stratified cuboidal epithelium

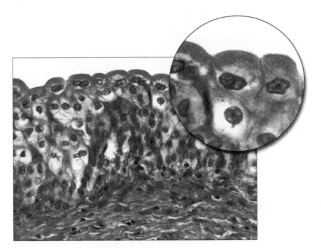
Relaxed State

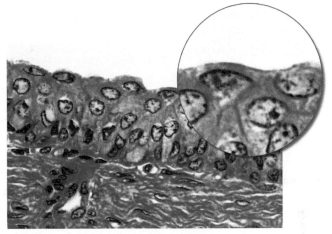
Distended (Stretched) State

FIGURE 8.7   Histology of the Urinary Bladder

Are the epithelial cells shown in Figure 8.7 squamous, cuboidal, columnar, or transitional in shape? _____

What role does this special type of epithelium play in the urinary bladder?

_____

_____

What type of muscle (skeletal, cardiac, or smooth) is found within the urinary bladder? What happens to the urinary bladder when this muscle contracts?

_____

_____

URINALYSIS

Urinalysis is a quick, simple, and noninvasive procedure that is routinely used to assess patient health. Urinalysis refers to the visual, chemical, and (in some cases) microscopic analysis of a patient's urine specimen. The color, odor, and clarity of the urine specimen are visually examined during urinalysis. Chemical analysis can detect abnormal molecules in a urine sample, such as glucose or plasma proteins. Table 8.2 gives an overview of abnormal urine constituents and the disorders they may indicate in patients. As we know from over-the-counter tests, urinalysis can also be used to detect pregnancy, ovulation, and even the use of illegal drugs.

TABLE 8.2 **Abnormal Urine Constituents**

Abnormal constituent	Description
Albumin	A normal constituent of plasma, it usually appears in only very small amounts in urine because it is too large to pass through capillary fenestrations(pores). The presence of excessive albumin in the urine—**albuminuria**—indicates an increase in the permeability of filtration membranes due to injury or disease, increased blood pressure, or irritation of kidney cells by substances, such as bacterial toxins, ether, or heavy metals.
Glucose	The presence of glucose in the urine is called **glucosuria** and usually indicates diabetes mellitus. Occasionally it may be caused by stress, which can cause excessive amounts of epinephrine to be secreted. Epinephrine stimulates the breakdown of glycogen and liberation of glucose from the liver.
Red blood cells (erythrocytes)	The presence of red blood cells in the urine is called **hematuria** and generally indicates a pathological condition. One cause is acute inflammation of the urinary organs as a result of disease or irritation from kidney stones. Other causes include tumors, trauma, and kidney disease, or possible contamination of the sample by menstrual blood.
Ketone bodies	High levels of ketone bodies in the urine, called **ketonuria**, may indicate diabetes mellitus, anorexia, starvation, or simply too little carbohydrate in the diet.
Bilirubin	When red blood cells are destroyed by macrophages, the globin portion of hemoglobin is split off and the heme is converted to biliverdin. Most of the biliverdin is converted to bilirubin, which gives bile its major pigmentation. An above-normal level of bilirubin in urine is called **bilirubinuria**.
Urobilinogen	The presence of urobilinogen (breakdown product of hemoglobin) in urine is called **urobilinogenuria**. Trace amounts are normal, but elevated urobilinogen may be due to hemolytic or pernicious anemia, infectious hepatitis, biliary obstruction, jaundice, cirrhosis, congestive heart failure, or infectious mononucleosis.
Casts	Casts are tiny masses of material that have hardened and assumed the shape of the lumen of the tubule in which they formed. They are then flushed out of the tubule when filtrate builds up behind them. Casts are identified by either the cells or substances that compose them or their appearance.
Microbes	The number and type of bacteria vary with specific infections in the urinary tract. One of the most common is *E. coli*. The most common fungus to appear in urine is the yeast *Candida albicans*, a cause of vaginitis. The most frequent protozoan seen is *Trichomonas vaginalis*, a cause of vaginitis in females and urethritis in males.

During this exercise, you will perform urinalysis in one of two ways, as specified by your instructor:

- Step-by-Step 8.4a: Performing Urinalysis on Your Own Urine Sample
- Step-by-Step 8.4b: Performing Urinalysis on Simulated Urine Samples

The results from Step-by-Step 8.4b will be used to complete Exercise 8.5: Diagnosing Disorders with Urinalysis.

STEP BY STEP 8.4A   Performing Urinalysis on Your Own Urine Sample

 SAFETY GUIDELINES
for Working with Human Urine

Occupational Safety and Health Administration (OSHA, *www.osha.gov*) Guidelines per *Occupational Safety and Health Standards → Bloodborne Pathogens → Standard 1910.1030*: Unless blood is visible in the specimen, urine is not regarded as blood or other potentially infectious material (OPIM).

Wear gloves and a lab coat when handling urine samples. If splashing is possible, eye protection may also be worn. When you finish urinalysis, the urine sample may be disposed of down the toilet. The used test strip and urine collection cup may be placed into a trash can. Alternatively, your instructor may ask you to dispose of the used specimen cup and test strip in a biohazard bag. Wash your hands thoroughly following urinalysis.

Note: This exercise is not being performed under the strict standards and guidelines of an accredited clinical laboratory. The urinalysis results from this exercise are not intended to diagnose health disorders.

Collecting Your Urine Specimen

1. Collect a sterile specimen cup and alcohol wipe before heading to the restroom.

2. Wash your hands thoroughly before entering a bathroom stall.

3. Clean your external genitalia with the alcohol wipe.

4. Begin urinating into the toilet, and start collecting urine in the specimen cup midstream. Once the fill line is reached on the specimen cup, finish urinating in the toilet.

5. Wash your hands thoroughly after you have collected the urine specimen.

Urinalysis Procedure

1. Note the physical characteristics of your urine sample (color, odor, and clarity) in Table 8.3.

2. Collect one urinalysis test strip and read the directions on the test strip bottle prior to use.

3. Dip the test strip into your urine specimen for the length of time specified by the manufacturer.

4. Compare the test strip to the interpretation chart found on the test strip bottle. Record your results in Table 8.3.

5. Clean your lab bench with disinfectant and wash your hands thoroughly.

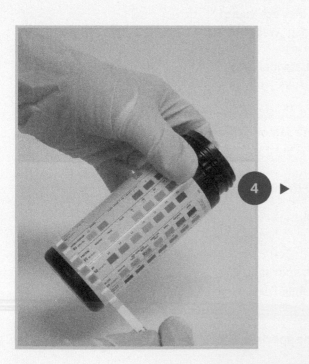

Note: This exercise can be completed in groups of four. Each group member is responsible for collecting and analyzing one simulated urine sample: Control Urine, Derek's Urine, Carrie's Urine, or Megan's Urine.

1. Collect a clean specimen cup (or a 100 mL beaker) and label it with the name of the patient you will test.

2. Using the graduated cylinder, transfer 80 mL of the simulated urine sample into the specimen cup.

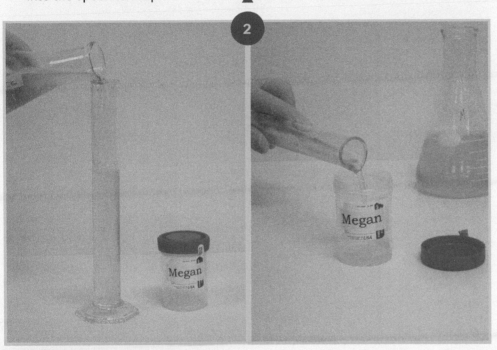

3. Record the physical characteristics of the urine sample (color, odor, and clarity) in Table 8.3.

4. Collect one urinalysis test strip, and read the directions on the test strip bottle prior to use.

5. Dip the test strip into the urine specimen for the length of time specified by the manufacturer.

6. Compare your test strip with the interpretation chart found on the test strip bottle. Record the results in Table 8.3.

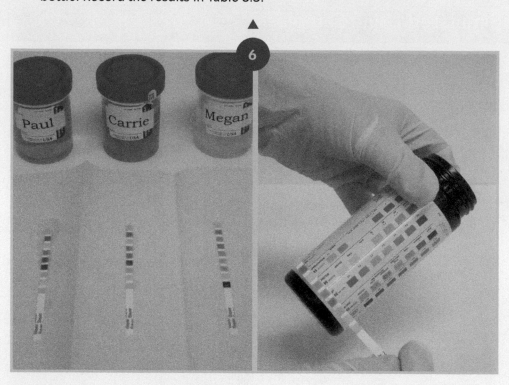

TABLE 8.3   Urinalysis Results

	Your Urine Specimen (8.4a)	Control Specimen (8.4b)	Derek's Urine Specimen (8.4b)	Carrie's Urine Specimen (8.4b)	Megan's Urine Specimen (8.4b)	Standard Values
Color						Pale yellow, yellow, amber
Odor						Little or no odor
Clarity						Clear or in some cases cloudy
Leukocytes (White Blood Cells)						Negative
Nitrites						Negative
Protein						Negative or trace amount
pH						4.6 to 8.0
Blood						Negative or trace amount
Ketones						Negative
Glucose						Negative

During this exercise, you will use your knowledge of the urinary system to diagnose three patients: Derek, Carrie, and Megan. In Step-by-Step 8.4b, you performed urinalysis on each patient's simulated urine sample. Compare these urinalysis results with the symptoms and test results provided below. Using this information, make a logical diagnosis for each patient: diabetes mellitus, kidney stones, or a urinary tract infection.

Terminology Used in This Exercise

Ultrasound Imaging (aka Sonography)—medical imaging technique that uses high-frequency sound waves to produce images of internal organs

Complete Blood Count (CBC)—a broad series of blood tests used to determine a patient's red blood cell count, white blood cell count, platelet count, etc.*

Comprehensive Metabolic Panel (CMP)—a broad panel of 14 blood tests used to evaluate kidney function, liver function, blood glucose levels, electrolyte levels, etc.*

Derek

For the past two days, Derek has felt an excruciating pain on the left side of his back. In the middle of the night, Derek's pain grew so unbearable that he asked his partner to call an ambulance. Upon arriving at the hospital, the attending physician asked Derek if he had noticed any other changes to his health. Hesitantly, Derek mentioned that his urine had been bloody for the past several days. Derek's physician ordered urinalysis and an abdominal ultrasound, as well as a complete blood count and a comprehensive metabolic panel. As shown in Derek's ultrasound (Figure 8.8), several dense masses were detected inside of his left kidney.

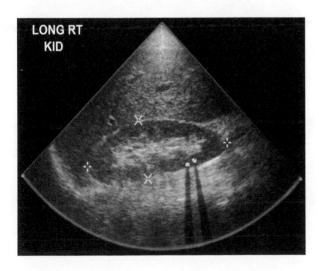

FIGURE 8.8   **Derek's Abdominal Ultrasound**

---

*If you would like to learn more about a specific laboratory test, you can access Lab Tests Online® at http://www.labtestsonline.org . This website, which is produced by the American Association for Clinical Chemistry (AACC), is a resource designed primarily for use by patients and caregivers.

Which disorder is suggested by Derek's symptoms and test results? _____

Red blood cells and hemoglobin were detected in Derek's urine. How do you think these substances ended up in his urine?

_____

_____

What treatment options exist for Derek's condition?

_____

_____

Carrie

Urination has become a painful event in Carrie's day-to-day life. Unfortunately, she also feels the urge to urinate more frequently than normal. During urinalysis, a medical technician noted turbidity (cloudiness) in her urine sample, as well as the presence of a pungent odor. During microscopic analysis, as shown in Figure 8.9, white blood cells and rod-shaped bacteria were detected in her urine specimen. As a follow-up, Carrie's urine was cultured for bacteria; the results confirmed that an abnormally high number of *E. coli* was present.

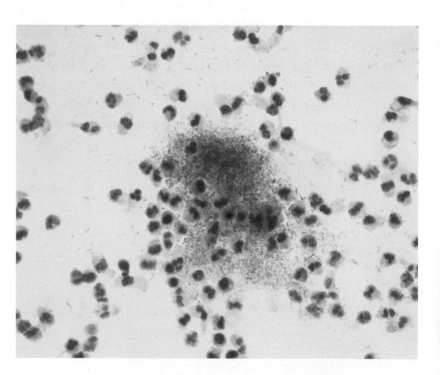

FIGURE 8.9 Microscopic Image of Carrie's Urine Sample

Which disorder is suggested by Carrie's symptoms and test results? _____

Why is this disorder is more prevalent in women versus men?

_____

_____

Why are white blood cells are present in Carrie's urine sample?

_____

_____

What treatment options exist for Carrie's disorder?

_____

_____

Megan

During the past several months, Megan's mom has grown increasingly concerned about her daughter's rapid weight loss and chronic fatigue. No matter how much food she eats, Megan constantly feels hungry. At the same time, Megan's thirst is impossible to quench, and she is urinating more frequently than ever. After listening to her symptoms, Megan's pediatrician ordered urinalysis, a complete blood count, and a comprehensive metabolic panel. The CMP revealed that Megan's blood glucose levels were abnormally high. Ketones were also detected in Megan's urine sample during urinalysis, and her urine exhibited a fruity smell.

Which disorder is suggested by Megan's symptoms and test results? _____

The CMP indicates that Megan's blood glucose levels are abnormally high. Urinalysis also confirmed the presence of glucose in her urine; apparently, Megan's nephrons cannot reabsorb all of the glucose floating around in her bloodstream. What does this suggest about the amount of glucose entering Megan's cells?

_____

_____

Why are ketones present in Megan's urine?

_____

_____

If Megan's pancreas is unable to produce insulin, what treatment options exist for her?

_____

_____

A CLOSER LOOK AT KIDNEY DISEASES AND UROLOGICAL DISORDERS

The National Institute of Diabetes and Digestive and Kidney Diseases (NIDDK, http://www.niddk.nih.gov) is a division of the National Institutes of Health. NIDDK "conducts and supports research on many of the most serious diseases affecting public health," including diabetes, obesity, urology, and renal disease. If you would like to learn more about a specific kidney disease or a urological disease, you can access the National Kidney and Urologic Diseases Information Clearinghouse at http://kidney.niddk.nih.gov/ . This website was established "to increase knowledge and understanding about diseases of the kidneys and urologic system among people with these conditions and their families, health care professionals, and the general public."

Lab Tests Online® is produced by the American Association for Clinical Chemistry (AACC). This resource "has been designed to help you, as a patient or family caregiver, to better understand the many clinical lab tests that are part of routine care as well as diagnosis and treatment of a broad range of conditions and diseases." Links to sample lab reports, reference ranges for laboratory tests, and the reliability of test results are also included on this site's homepage. If you would like to learn more about a specific laboratory test, you can access Lab Tests Online® at http://www.labtestsonline.org.

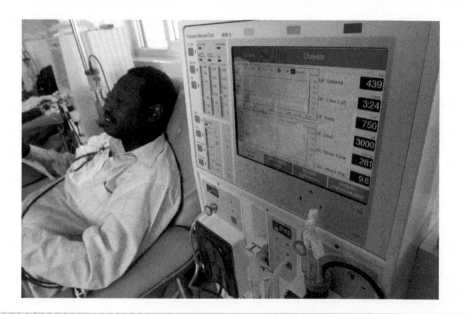

1. Urea is produced by the _____ and excreted by the _____.

2. What type of epithelium is found within the urinary bladder? _____

3. Which urinary organ is labeled D in the picture below? How does this tube differ in males and females?

    _____

    _____

    _____

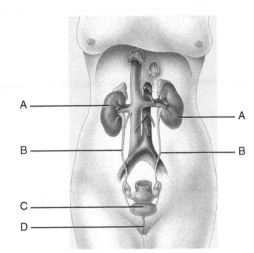

4. In a healthy patient, which blood component is too large to be filtered out of the glomerulus?
    a. Urea
    b. Water
    c. Glucose
    d. Albumin

5. A _____ is labeled A in the picture below. Blood is filtered within this ball of twisted capillaries.

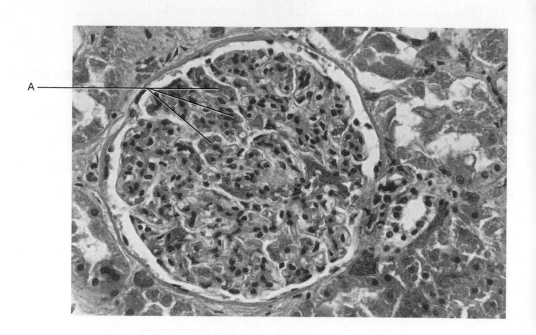

6. What is a nephron? What does the term *nephritis* mean?

_____

_____

_____

7. Dialysis selectively filters molecules based on _____.

8. Name two ways that urine is physically assessed during urinalysis.

_____

_____

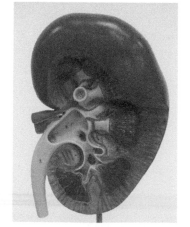

9. Label the renal capsule and the renal pelvis on the image below.

10. What does the presence of ketones in urine signify? Imagine that glucose and ketones are found in a patient's urine sample during urinalysis. What urological disorder is suggested by these results?

_____

_____

_____

LAB 9:
The Reproductive System

By the End of This Lab, You Should Be Able to Answer the Following Questions:

- In a female, which reproductive organ is homologous to the penis?

- Which organs produce gametes in the male and female reproductive systems?

- Where is the prostate gland located in relation to the urinary bladder? Where is the uterus located in relation to the urinary bladder?

- What similarities and differences exist between spermatogenesis and oogenesis?

- What is an ovarian follicle? How do primordial, primary, secondary, and mature follicles differ in appearance from one another on a microscope slide?

- How is sperm motility affected by chemicals such as caffeine, nicotine, and ethanol?

- What treatment options exist for viral STDs? Bacterial STDs?

INTRODUCTION

To maintain a presence on the planet, all living organisms must reproduce in one way or another. As humans, we may instinctively assume that complex organisms, such as ourselves, are better suited for survival and reproduction than simpler organisms, such as bacteria. Do you think this a valid assumption or a biased one? Since bacteria are unicellular, they certainly don't possess any tissues or organs dedicated to reproduction. Bacteria carry out **asexual reproduction**, whereby an organism (aka one bacterial cell) splits into two identical clones following DNA replication. Asexual reproduction may seem primitive to us, but there are definite advantages to this system. Since bacteria can reproduce in a matter of minutes, for instance, their numbers can grow exponentially within a matter of hours.

For humans, reproduction involves the fusion of **gametes** (sperm and eggs) produced by the male and female reproductive systems, respectively. **Sexual reproduction** increases our genetic diversity and thereby boosts the survival potential of our species as a whole. For example, an infectious disease that proves to be fatal in some (example: malaria) may not be fatal to others due to subtle genetic differences. On the other hand, though, it takes years—if not decades—for humans to pass from one generation to another. This is due, in part, to the fact that humans must complete various developmental stages—prenatal development, infancy, childhood, and adolescence—before reproduction is even possible.

Figures 9.1–9.2 show a side-by-side comparison of the male and female reproductive systems. Since so many striking differences are present, it is amazing that male and female genitalia arise from the same (*homo-*) original embryonic tissues. The term **homologous** is used when referring to different structures that arise from the same tissue origins during development. For instance, the **gonads** (testes and ovaries) are homologous organs that produce gametes and sex hormones. The penis and clitoris are homologous, erectile organs that become engorged with blood during sexual arousal. Tables 9.1–9.2 lists the homologous organs found in males and females, respectively, along with the function of each reproductive organ.

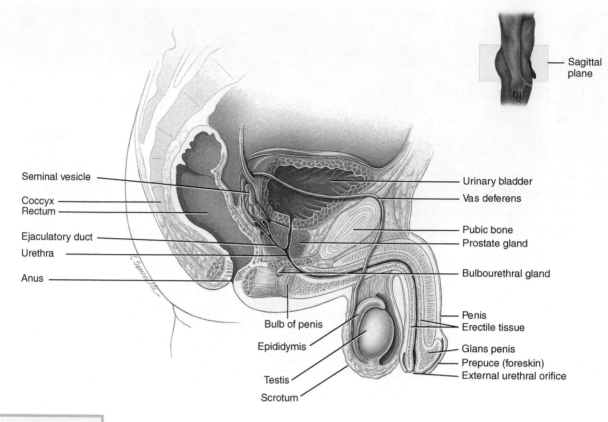

Sagittal plane

Seminal vesicle

Coccyx
Rectum

Ejaculatory duct

Urethra

Anus

Bulb of penis

Epididymis

Testis

Scrotum

Urinary bladder

Vas deferens

Pubic bone

Prostate gland

Bulbourethral gland

Penis
Erectile tissue

Glans penis
Prepuce (foreskin)
External urethral orifice

FIGURE 9.1 Sagittal Section of the Male Reproductive System

TABLE 9.1   Major Organs of the Male Reproductive System

Organ (Plural Form)	Function	Homologous Organs
Scrotum	Regulates the temperature of the testes	Labia Majora
Testis (Testes)	Sperm production Production of testosterone and related sex hormones	Ovaries
Epididymis (Epididymides)	Sperm storage and maturation	
Ductus Deferens (Ductus Deferentes) or Vas Deferens (Vasa Deferentia)	Transfers sperm from epididymis to ejaculatory duct prior to ejaculation	
Seminal Vesicle(s)	Contributes fluid to semen that is rich in fructose	
Ejaculatory Duct(s)	Passageway for sperm and seminal vesicle secretions	
Prostate Gland	Contributes a milky fluid to semen	Skene's Glands/Female Prostate
Urethra	Passageway for semen during ejaculation Passageway for urine during urination	
Bulbourethral Gland(s)/Cowper's Gland(s)	Secretes a clear, gelatinous fluid known as pre-ejaculate	Bartholin's Glands
Penis	Erectile, copulatory organ	Clitoris

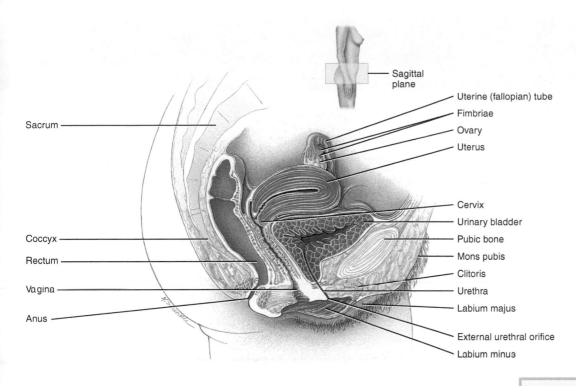

Sagittal plane

Sacrum

Coccyx

Rectum

Vagina

Anus

Uterine (fallopian) tube

Fimbriae

Ovary

Uterus

Cervix

Urinary bladder

Pubic bone

Mons pubis

Clitoris

Urethra

Labium majus

External urethral orifice

Labium minus

FIGURE 9.2 **Sagittal Section of the Female Reproductive System**

TABLE 9.2   **Major Organs of the Female Reproductive System**

Organ (Plural Form)	Function	Homologous Organs
Ovary (Ovaries)	Egg (ovum) production Production of estrogen, progesterone, and related sex hormones	Testes
Oviduct(s), Uterine Tube(s), or Fallopian Tube(s)	Transport ovulated eggs (ova) to uterus Typically, fertilization occurs in this organ	
Uterus	Site of gestation during pregnancy	
Vagina	Female copulatory organ Portion of birth canal Passageway for uterine secretions, such as menstrual fluid	
Clitoris	Erectile organ involved in sexual arousal	Penis
Labium Majus (Labia Majora)	Enclose and protect underlying organs	Scrotum
Labium Minus (Labia Minora)	Enclose and protect underlying organs Engorge with blood and swell during sexual arousal	
Skene's Glands or Female Prostate	Still unconfirmed—possibly involved in sexual stimulation and/or female ejaculation	Prostate Gland
Bartholin's Glands	Secrete a small amount of lubricating fluid as a result of sexual arousal	Bulbourethral Glands/Cowper's Glands

Although your sex was genetically determined at the moment of fertilization, the physical manifestation of your sex chromosomes—typically, XX for females and XY for males—relied upon the actions of hormones during development. As a result, it is possible for a person to develop physical sex characteristics (internal and/or external genitalia) that are associated with one gender, in spite of inheriting sex chromosomes that are associated with the opposite gender. Certain genetic syndromes are linked to sex reversal, such as XX male syndrome and XY female syndrome, while others are linked to the development of ambiguous genitalia, such as androgen insensitivity syndrome. If you would like to learn more about these genetic syndromes, you can visit the Genetics Home Reference at http://ghr.nlm.nih.gov/. This website is sponsored by the U.S. National Library of Medicine, which is part of the National Institutes of Health (NIH).

During this exercise, you will identify reproductive organs on anatomical models of the male and female reproductive systems. Alternatively, your instructor may ask you to identify these organs on a virtual cadaver dissection, a dissected fetal pig, or a dissected human cadaver. Label the listed organs on page 9-5 on Figures 9.3 and 9.4, respectively, and answer the review questions provided at the end of this exercise.

Terminology Used in This Exercise

Glans—the rounded head of an organ (examples: glans penis and glans clitoris)

Cervical—pertaining to the neck (example: cervical vertebrae) *or* a narrow, neck-like structure in the body (example: uterine cervix)

MALE REPRODUCTIVE ORGANS

1. Scrotum
2. Testis
3. Epididymis
4. Ductus (Vas) Deferens
5. Ejaculatory Duct
6. Urethra
7. Seminal vesicle
8. Prostate gland
9. Bulbourethral gland
10. Penis
11. Glans penis

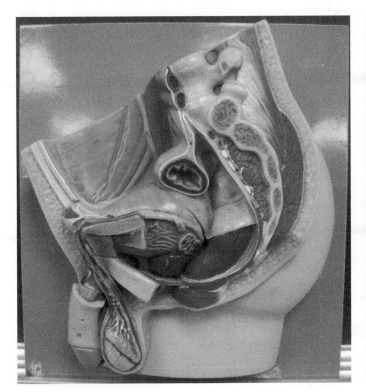 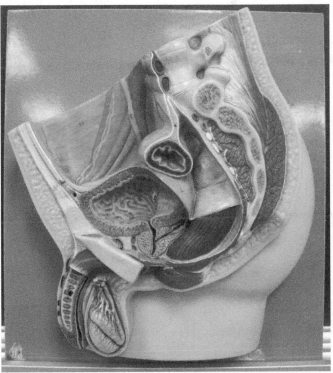

FIGURE 9.3 Identifying Organs of the Male Reproductive System

FEMALE REPRODUCTIVE ORGANS

1. Ovary

2. Oviduct (aka Fallopian Tube or Uterine Tube)

3. Uterus

4. Uterine Cervix

5. Vagina

6. Clitoris

7. Glans Clitoris

8. Labium Majus

9. Labium Minus

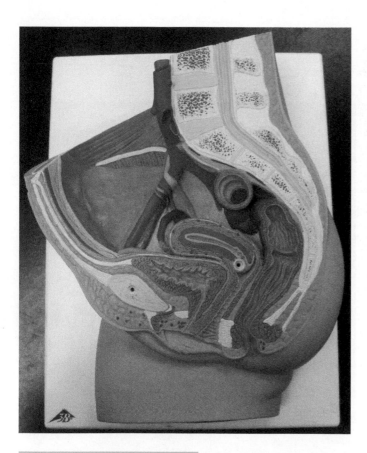 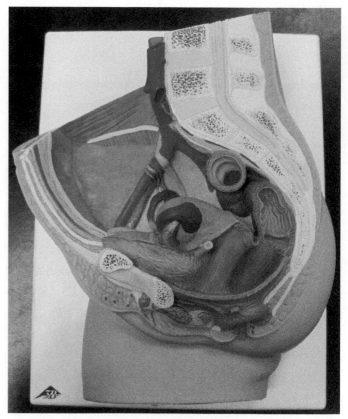

FIGURE 9.4 **Identifying Organs of the Female Reproductive System**

Review Questions for Exercise 9.1

Label the ovaries, oviducts, and uteri on the images provided below.

Which female reproductive organ is labeled A below? This organ is homologous to the penis. _____

How would you describe the location of the urethra in relation to the vagina? For example, is the urethra anterior (situated in front of) or posterior to (situated behind) the vagina?

_____

_____

How would you describe the location of the uterus in relation to the urinary bladder? For example, is the uterus superior (above) or inferior to (below) the urinary bladder? Why do women tend to urinate more frequently during pregnancy?

_____

_____

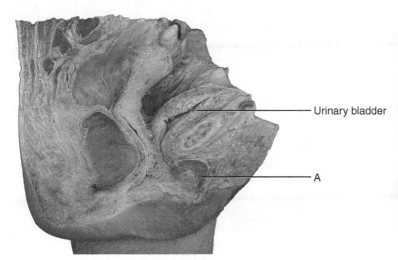

— Urinary bladder

— A

Sagittal Section of Female Pelvis, Medial View

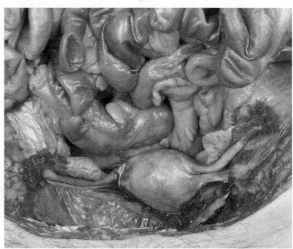

Anterior View of Female Reproductive Organs

Label the urinary bladder and the prostate gland on the images provided below.

How would you describe the location of the prostate gland in relation to the urinary bladder? For instance, is the prostate gland superior or inferior to the urinary bladder? Based on their relative locations, why does prostatitis—inflammation (-*itis*) of the prostate—commonly lead to difficulty urinating?

_____

_____

Name two ways in which the male and female urethra differ from one another.

_____

_____

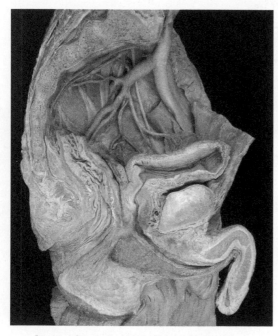

Sagittal Section of Male Pelvis, Medial View

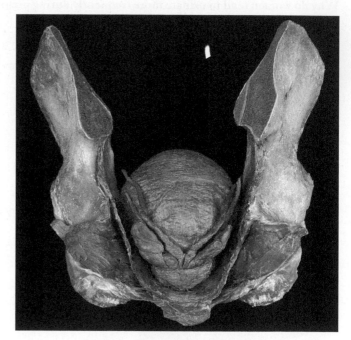

Male Pelvic Viscera, Posterior View

Label the testis and the epididymis on the image below.

What is the function of the epididymides?

_____

_____

Which ducts carry sperm away from the epididymides? During a vasectomy, these ducts are surgically altered to prevent sperm from entering the ejaculate.

_____

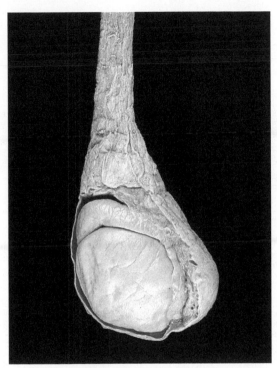

Testis Dissected, Lateral View

SPERMATOGENESIS AND OOGENESIS

With the exception of sperm and eggs—not to mention a few of their precursors in the gonads—human cells normally contain a total of 46 chromosomes each. The term *somatic cell* is used when referring to a body cell with 46 chromosomes (versus eggs and sperm). In Greek, the word *soma* actually means *body*. Years ago, in the midst of a fertilization event, two sets of chromosomes fused together in the egg that ultimately transformed into your first body cell (i.e., the zygote). You inherited one set of chromosomes from this egg; the other set was delivered by a penetrating sperm cell. Somatic cells, including your first body cell, are considered **diploid** (2n) because they carry two sets of chromosomes (*diplo-*) apiece.

How do we humans maintain a stable chromosome number as we progress from one generation to the next? The answer lies in the primary function of the testes and ovaries: **spermatogenesis** (sperm production) and **oogenesis** (egg production). A side-by-side comparison of spermatogenesis and oogenesis is shown in Figure 9.5. Although differences do exist between these two processes, **gametogenesis** (gamete production) involves the same basic mechanism in males and females: meiosis. **Meiosis** is a series of events that reduce the chromosome number in **spermatocytes** (immature sperm cells) and **oocytes** (immature egg cells). The word *meiosis* actually means *lessening* in Greek. By the end of meiosis, the number of chromosomes per cell has been cut in half from 46 to 23. Mature sperm and eggs are considered **haploid** (n), since they only carry one set of chromosomes (*haplo-*) a piece.

If you briefly glanced at Figure 9.5, you probably noticed that spermatogenesis and oogenesis share several similarities. For instance, they both start with diploid (2n) cells and

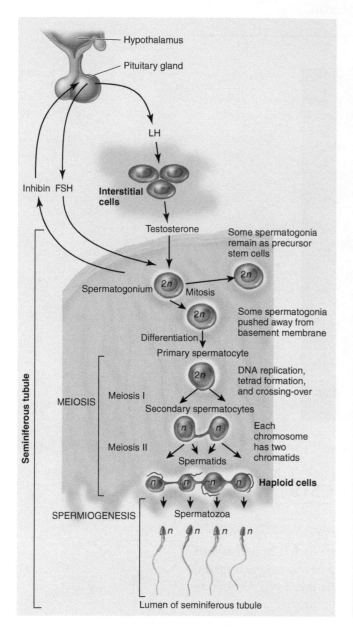
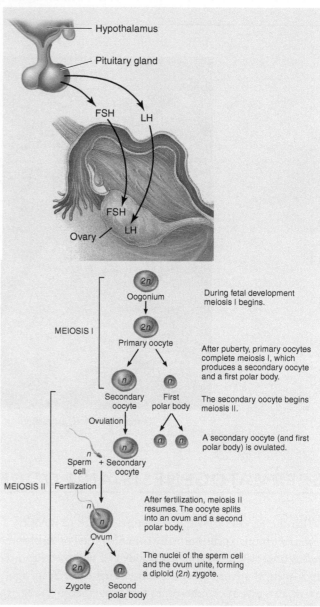

FIGURE 9.5 **Comparison of Spermatogenesis and Oogenesis**

end with production of haploid (n) gametes. The same **gonadotropic hormones**—follicle stimulating hormone (FSH) and luteinizing hormone (LH)—are also released in males and females to regulate (-*tropic*) the activities of the gonads. In reality, though, spermatogenesis and oogenesis also differ in several significant ways: their timing in regards to initiation and completion, the life span of associated stem cells (spermatogonia and oogonia), the number of functional gametes produced from a primary (aka parent) cell, the relative sizes of sperm and egg cells, and so on. Table 9.3 summarizes the major differences that exist between spermatogenesis and oogenesis.

TABLE 9.3 **Comparison of Spermatogenesis and Oogenesis**

Variable	Spermatogenesis	Oogenesis
Location	Testes	Ovaries
Time of Initiation	Process begins in males at puberty	Process begins in females during fetal development
Time of Completion	Each cycle takes ≈74 days to complete *,**	Oogenesis is not completed unless/until fertilization takes place
Number of Gametes Produced by Each Primary Cell	Four functional, haploid (n) sperm are produced from each primary spermatocyte	One functional, haploid (n) egg can be produced from each primary oocyte
Status of Stem Cells	The number of spermatogonia declines with age, but they are typically present in males until death	Oogonia are no longer present within the ovaries at birth
Time of Cessation	Not applicable—process gradually declines with age, but it typically continues until death ***	Menopause—average age ≈51 years old for females in the United States ****

EXERCISE 9.2   Histology of the Reproductive System

To understand how an organ functions, it is helpful to view its structure at the microscopic level. During this exercise, you will examine prepared microscope slides of the testis, penis, ovary, and oviduct. While viewing the testis and ovary, you will also identify maturing sperm and ovarian follicles at different stages of development. Compare your focused microscope slides to the photomicrographs provided in Figures 9.6–9.9, and answer the review questions that accompany each slide.

---

* Heller, C. G. Clermont, Y., Kinetics of the germinal epithelium in man, *Recent Prog Horm Res*. 1964; 20:545–571.
** The cycle of the seminiferous epithelium in humans: A need to revisit? Rupert P. Amann, *J Androl*, September 2008; 29:469–487.
*** J Androl, March–April 1987;8(2):64–68. Decrease in the number of human Ap and Ad spermatogonia and in the Ap/ Ad ratio with advancing age. New data on the spermatogonial stem cell. Nistal, M; Codesal, J; Paniagua, R; Santamaria, L.
**** National Institute on Aging. Accessed online 12-2-10 at http://www.nia.nih.gov/healthinformation/publications/menopause.htm.

HISTOLOGY OF THE TESTIS

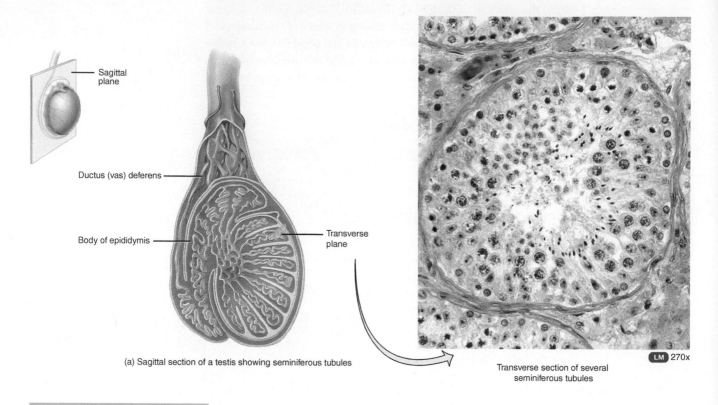

Sagittal plane

Ductus (vas) deferens

Body of epididymis

Transverse plane

(a) Sagittal section of a testis showing seminiferous tubules

LM 270x

Transverse section of several
seminiferous tubules

FIGURE 9.6   Cross Section of
the Testis

Spermatogenesis occurs within small, coiled tubes called seminiferous tubules that are packed inside of the testes. Once sperm are released into the lumen (open cavity) of this tubule, they get transported to the epididymis for storage.

Label a seminiferous tubule—and its lumen—on the photomicrograph above.

Spermatogonia are diploid (2n) stem cells that reside along the walls of seminiferous tubules. (In biology, the suffix *–gonium /-gonia* refers to a cell or structure that produces reproductive cells.) Label a spermatogonium on the photomicrograph above. In humans, how many chromosomes are normally found in each spermatogonium?

_____

Spermatids are immature sperm cells that get pushed toward the lumen of the seminiferous tubule. During this process, they develop tails and lose connections with one another. Label a spermatid on Figure 9.6. Since spermatids are haploid (n) cells, how many chromosomes are normally found in each one?

_____

How many sperm cells are produced from one primary spermatocyte?

_____

Assume a 25-year-old man produces 100 million sperm per day. If this man entered puberty at the age of 15, approximately how many sperm has he produced to date?

_____

HISTOLOGY OF THE PENIS

Label the urethra on the photomicrograph below. In males, what role does the urethra play in reproduction?

_____

_____

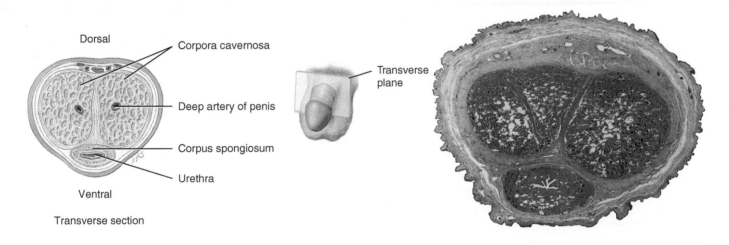

FIGURE 9.7   Cross Section of the Penis

The corpus cavernosa and corpus spongiosum are composed of spongy, erectile tissue. Label the corpus cavernosa and the corpus spongiosum on the photomicrograph above. How does this spongy tissue become physically erect during sexual arousal?

_____

_____

HISTOLOGY OF THE OVARY

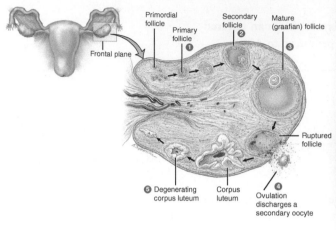

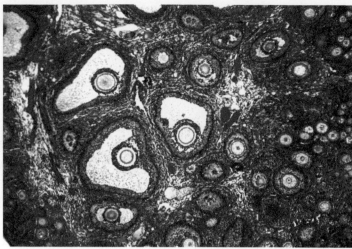

| FIGURE 9.8 | Cross Section of the Ovary |

Within the ovaries, each oocyte (immature egg cell) is surrounded by a follicle, or sac, of epithelial cells. Before the onset of puberty, ovaries primarily contain small, primordial follicles. A primordial follicle consists of a small oocyte that is surrounded by one layer of squamous epithelial cells. Label a primordial follicle on the photomicrograph above.

Once puberty is reached, primordial follicles may develop into primary follicles, secondary follicles, and mature follicles. Certain aspects of follicular development are driven by the follicle itself, while other aspects of this process are driven by follicle stimulating hormone (FSH). If multiple layers of epithelium develop around a follicle, it is now considered a secondary follicle. Label a secondary follicle—and the oocyte within this follicle—on the photomicrograph above.

The release of FSH may stimulate a secondary follicle to develop a fluid-filled cavity. At this point, the follicle is considered a mature (aka Graafian) follicle. As the fluid-filled cavity enlarges, the oocyte is pushed toward the edge of follicle. Label a mature (Graafian) follicle—and the oocyte within this follicle—on the photomicrograph above.

If a mature follicle responds to FSH signaling—meaning it gets recruited to be the dominant follicle—it will bulge at the perimeter of the ovary until luteinizing hormone (LH) stimulates ovulation. During ovulation, the dominant follicle bursts open and releases an oocyte from the ovary. In 2008, the first video footage of human ovulation was captured by Stephan Gordts and Ivo Brosens at the Leuven Institute for Fertility and Embryology in Belgium. You can access this video, which is titled *Human Ovulation Captured on Video*, on *NewScientist*'s website (http://www.newscientist.com). As shown in the video, how does an ovulated egg get transferred into the oviduct (fallopian tube)?

_____

_____

Assume a woman ovulates once every 28 days, and she winds up ovulating for a total of 35 years. In total, how many eggs will she ovulate throughout her reproductive years? _____

Do you think FSH or LH is detected in urine by ovulation tests? Explain your answer.

_____

_____

HISTOLOGY OF THE FALLOPIAN TUBE

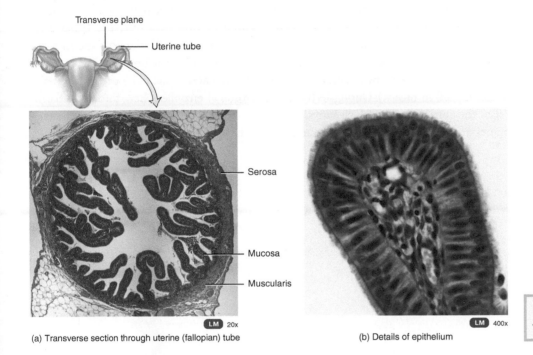

(a) Transverse section through uterine (fallopian) tube   LM 20x

(b) Details of epithelium   LM 400x

FIGURE 9.9   **Cross Section of the Fallopian Tube**

Label the lumen of the fallopian tube on the photomicrograph above.

The surface of a ciliated epithelial cell is covered with small, hair-like projections. When cilia beat together in unison, objects (or fluids) in the outside environment can be moved in a specific direction. Label a ciliated simple columnar epithelial cell on Figure 9.9b. What role do you think cilia play in the fallopian tubes?

_____

_____

_____

What type of muscle (smooth, skeletal, or cardiac) is found within the fallopian tubes? What happens to the egg inside of a fallopian tube when peristalsis occurs? Peristalsis (i.e., wave-like muscle contractions) also occurs throughout the digestive tract.

SPERM ANATOMY

Sperm are the only cells in humans that possess a unique appendage for motility: the flagellum. Given this fact, it makes sense that flagellated cells are produced exclusively within males. **Flagella** are long, tail-like appendages that beat to propel sperm through the female reproductive tract. This feat requires a considerable amount of fuel in the form of ATP molecules. This fuel (ATP) is produced by mitochondria residing in the middle piece, which is attached directly to the flagellum. Figure 9.10 shows the location of the middle piece in relation to the flagellum and a third structure: the head. The precious cargo carried by each sperm cell—chromosomes—is sequestered inside of a nucleus that sits within the head region. The tip of the head is also covered by a special organelle called the **acrosome**. (The prefix *acro-* refers to the top of an object.) During fertilization, acrosomal enzymes chew through a barrier called the zona pellucida, which surrounds the entire egg. Only then can the plasma membranes of the gametes fuse together, thus allowing the sperm to eject its nucleus into the egg.

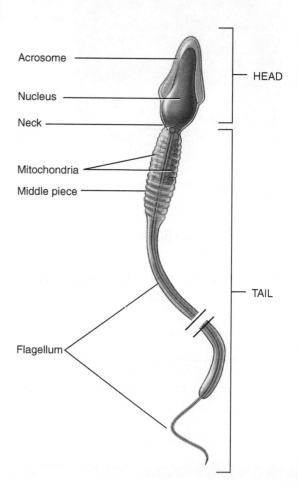

FIGURE 9.10 **Sperm Anatomy**

During this exercise, you will observe the activity of sperm that are living in bull semen. During Lab 2 (Observing Cells with Light Microscopy), you created wet mount preparations of hair, skin cells, and cheek cells. Using the same basic techniques, you will create a wet mount of bull semen for microscopic analysis. To create a wet mount, the specimen of interest—semen—will be suspended between a microscope slide and a thin cover slip. Since our specimen of interest is already suspended in liquid, it should not dry out during observation.

Translucent cells are difficult to observe unless they are stained prior to light microscopy, but staining techniques also tend to kill living cells. To increase the visibility of sperm cells, you will close the iris diaphragm on your light microscope prior to observation. Refer to Lab 2, as needed, to review the parts of the microscope and steps for proper usage.

Once your wet mount is focused under the microscope, you will also test the effects of a chemical on sperm motility. A list of potential test chemicals is provided below. Since heat will radiate away from the microscope's light source—and toward your living specimen—it is important to work quickly and efficiently during this experiment. Before you create a wet mount, gather the necessary materials for addition of your chemical of interest. This will allow you to add the chemical immediately after your initial observation.

Possible Test Substances

- 0.1%, 1%, 5%, or 10% Ethanol Solutions
- 1% Nicotine Solution
- 1% Caffeine Solution
- 10% Fructose Solution
- Buffers (Examples: pH 4, 7, or 10)
- Spermicidal Jelly, Cream, or Foam (may need to be liquefied with heat prior to addition)
- Bovine Ovarian Follicular Fluid

Before you perform this experiment, formulate a hypothesis to test regarding the likely effects of _____ on sperm motility.

_____

_____

The idea for this experiment was provided by Dr. Kent R. Thomas.

Procedure

1. Close the iris diaphragm on your microscope.

▲

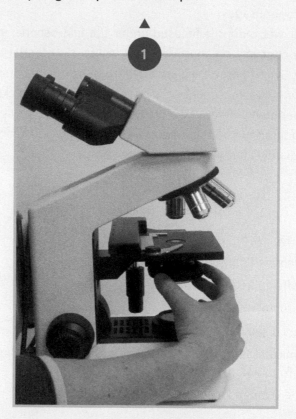

2. Collect a clean microscope slide, a cover slip, and the materials required for chemical addition: a disposable transfer pipette and your test substance.

3. Using a disposable transfer pipette, transfer one *small* drop of bull semen onto the center of your microscope slide.

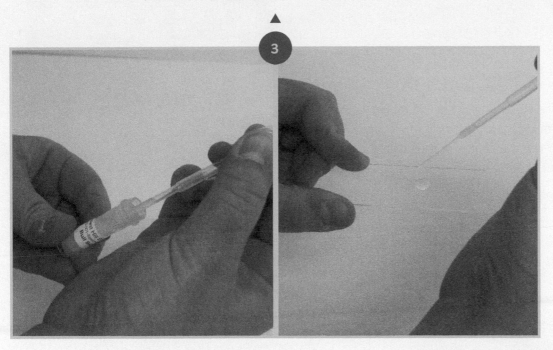

4. Starting at a 45° angle, gently lower the cover slip onto the drop of semen. *By the end of this process, the semen should be sand-wiched between the cover slip and the slide.*

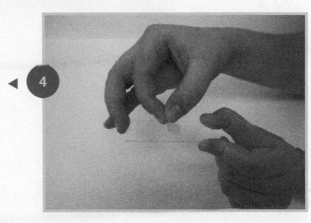

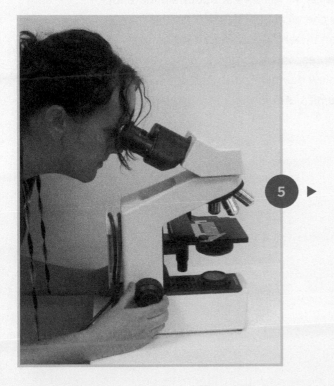

5. View your wet mount through a compound microscope; observe the shape and motility of the sperm cells under 400x magnification. (This requires the use of the 40x objective in conjunction with the 10x ocular lens.) Record your initial observations in Table 9.4 under *Sperm Activity Prior to Chemical Addition.*

6. While the wet mount is still focused on the microscope, add one *small* drop of your test substance to the edge of the cover slip.

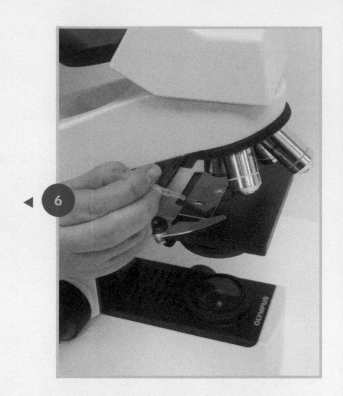

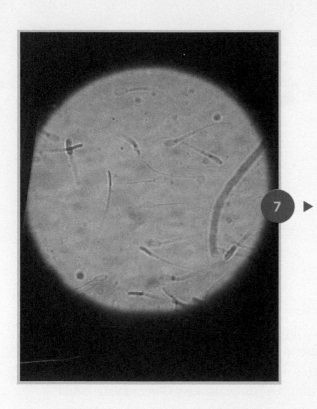

7. Continue observing the wet mount, and look for any obvious changes in sperm motility. (It will take about one minute for the chemical to diffuse across the slide.) Scan the entire wet mount for several minutes, and record your observations in Table 9.4 under *Sperm Activity Following Chemical Addition*.

8. Once the class results have been pooled, record this data in Table 9.4.

TABLE 9.4   **Effects of Chemicals on Sperm Motility**

Substance Tested	Sperm Activity Prior to Chemical Addition	Sperm Activity Following Chemical Addition	Possible Explanation for the Observed Results

Review Questions for Step by Step 9.3

Based on your observations, what effect (if any) did your test substance have on sperm motility? Is there a logical explanation for these results? If necessary, revise your initial hypothesis based on the results of this experiment.

_____

_____

_____

Did any other students test _____ (your test substance)? If so, how did your results compare with those obtained by other students?

_____

_____

_____

What are some potential sources of error in this experiment? What improvements (if any) could be made to this experiment?

_____

_____

_____

A CLOSER LOOK AT SEMEN ANALYSIS

During the previous experiment, you examined sperm motility in a **qualitative** (descriptive, subjective) manner. To obtain reliable, **quantitative** (numerical) results on sperm motility, most scientists and clinicians rely on computer-assisted semen analysis (CASA). These computerized assays not only provide great accuracy and reliability, but some can also complete full semen analyses in a matter of seconds. Factors such as sperm concentration (sperm count), sperm morphology (structure), total motility (% of motile sperm at a given time), and progressive (linear) vs. nonprogressive (curved) motility are evaluated by CASA and then compared to standard reference values.

In addition to its role in fertility testing, sperm motility is also of interest to scientists who study sperm competition and its role in biological evolution. **Sperm competition** relates to the ability of sperm *from two or more males* to fertilize the same egg(s) within a fertile female. Before publishing their findings in 2008, Jaclyn Nascimento and her colleagues at the University of California analyzed sperm motility in several primates, including humans, gorillas, and chimpanzees. Interestingly, the results from Nascimento's study revealed a correlation between sperm motility and the promiscuity of the tested species. These findings suggest that sperm may face selective pressures on speed (sperm competition) if a species behaves promiscuously.

Nascimento is featured in a short video on *NewScientist*'s website, where she summarizes her semen analysis techniques, results, and the conclusions drawn from her study. You can access this video online—titled *In Promiscuous Primates, Sperm Feel Need for Speed*—at http://www.newscientist.com.

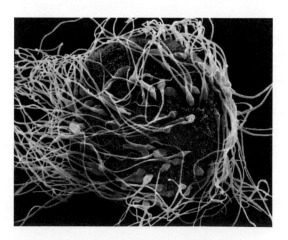

FIGURE 9.11   **Sperm Competion**

---

* The use of optical tweezers to study sperm competition and motility in primates. Jaclyn M. Nascimento, Linda Z. Shi, Stuart Meyers, Pascal Gagneux, Naida M. Loskutoff, Elliot L. Botvinick, and Michael W. Berns. *J R Soc Interface*, Mar 2008; 5:297–302.

SEXUALLY TRANSMITTED DISEASES

As discussed frequently in this day and age, **sexually transmitted diseases (STDs)** are primarily acquired through sexual contact (oral sex, vaginal sex, or anal sex) with an infected sexual partner. Certain STDs can also be transmitted through blood-to-blood contact, pregnancy, childbirth, and breastfeeding, or through use of infected tattoo needles/intravenous (IV) needles. As shown in Table 9.5, each STD is caused by a specific **pathogen**, or disease-causing agent. Depending on the STD, this pathogen may be bacterial, viral, parasitic, protozoan, or fungal in nature. Ultimately, the treatment guidelines for a particular STD depend on the nature of the pathogen involved in the infection. Bacterial STDs, for instance, are curable with antibiotic therapy. Viral STDs are not curable, but antiviral therapy may be prescribed to reduce a patient's viral load or to reduce the likelihood of recurrent outbreaks.

TABLE 9.5 **Sexually Transmitted Diseases**

Common Name	Scientific Name	Classification	Symptoms	Treatment
Chlamydia	*Chlamydia trachomatis*	Bacterial	Usually asymptomatic, may cause urethritis in males; leads to pelvic inflammatory disease in females	Antibiotics
Gonorrhea or "the clap"	*Neisseria gonorrhoeae*	Bacterial	Urethritis with excess pus discharge; may be asymptomatic in females, leading to sterility	Antibiotics
Syphilis	*Treponema pallidum*	Bacterial (spiral bacterium)	Primary stage results in a painless open sore or chancre; secondary stage is a rash , fever, and joint pain; tertiary stage results when organs begin to degenerate	Antibiotics in primary or secondary stage
Genital herpes	Type II herpes simplex virus (HSV)	Virus	Painful blisters on the external genitals of males and females, with possible internal blistering in females	Incurable, but outbreaks can be controlled with anti-inflammatory drugs
Genital warts	Human papillomavirus (HPV)	Virus	Cauliflower growths on the external genital area and internal growths in females; can also appear on or around the anus	Incurable; warts can be removed cryogenically
Trichomoniasis	Trichomonas vaginalis	Protozoan	Foul-smelling discharge and itching in females	Prescription drug metronidazole

Although some people do display visible signs of infection, many others are asymptomatic to themselves, as well as to their partners. Whether or not symptoms are present, infected individuals can transmit STDs to their sexual partners. Figure 9.12 estimates the number of new STD cases that arise each year in the United States. Given the prevalence of STDs across the nation, public health officials are educating communities on the transmission, detection, and prevention of STDs. For instance, Centers for Disease Control and Prevention (CDC) distributes information to the public through their website (http://www.cdc.gov/STD). Referrals to STD clinics are also available by calling 1-800-CDC-INFO.

There are several measures one can take to reduce the likelihood of acquiring or transmitting STDs. Abstaining from sexual activity is one logical way to prevent STD transmission. If two partners test negative for STDs and commit to a monogamous relationship, this greatly reduces—if not prevents—the transmission of asymptomatic STDs. Barrier techniques, such as latex condoms, greatly reduce the risk of STD transmission with proper use. The following is an excerpt from a report titled *Sexually Transmitted Diseases Treatment Guidelines, 2006* (http://www.cdc.gov/std/treatment/) released by the Centers for Disease Control (CDC):

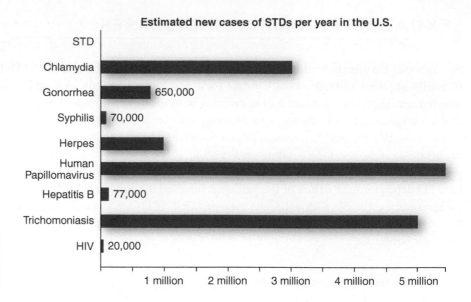

Estimated new cases of STDs per year in the U.S.

STD
- Chlamydia
- Gonorrhea — 650,000
- Syphilis — 70,000
- Herpes
- Human Papillomavirus
- Hepatitis B — 77,000
- Trichomoniasis
- HIV — 20,000

1 million | 2 million | 3 million | 4 million | 5 million

FIGURE 9.12 Estimated New Cases of STDs per Year in the United States

When used consistently and correctly, male latex condoms are highly effective in preventing the sexual transmission of HIV infection (i.e., HIV-negative partners in heterosexual serodiscordant relationships in which condoms were consistently used were 80% less likely to become HIV-infected compared with persons in similar relationships in which condoms were not used) and can reduce the risk for other STDs, including chlamydia, gonorrhea, and trichomoniasis.

With proper and consistent use, barrier techniques significantly reduce the risk of STD transmission. It is important to note, however, that condoms and other barrier techniques do not eliminate the possibility of STD transmission. On a final note, pre-exposure vaccines are available against hepatitis B and human papillomavirus (HPV), which are both viral STDs. For some individuals, these vaccinations are even covered by their health insurance plans.

Note: Graphic photographs of STD infections are included in the following exercise.

EXERCISE 9.4 Sexually Transmitted Diseases

During this exercise, you will use your knowledge of the reproductive system and STDs to diagnose four patients. Matt, Lydia, Buster, and Crystal are seeking medical attention for different sexually transmitted diseases. Use the symptoms and test results provided to make a logical diagnosis for each patient: syphilis, gonorrhea, genital herpes, or hepatitis B.

Matt

Matt felt a strange burning sensation as he urinated in the bathroom one morning. Perplexed by this sensation, he started to panic as he raced around to get ready for work. While driving to his office, Matt decided to search for clues about his painful symptom on the Internet. Before he got the chance to do so, though, he had to make another visit to the bathroom. Now, in addition to the burning sensation, one of his testicles felt swollen and painful. Nervous and confused, Matt called his physician and abruptly told his boss he had to leave for the remainder of the day.

After listening to Matt's symptoms, the doctor examined Matt's external genitalia. As shown in Figure 9.13, a small amount of green discharge was seeping from the tip of Matt's penis.

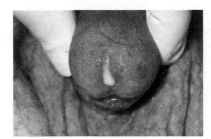

FIGURE 9.13 Matt's Discharge

To collect a culture for laboratory analysis, the doctor inserted a sterile swab into Matt's urethra. Matt's culture was transported to a clinical laboratory, where it was smeared on a microscope slide and subsequently stained. As shown in Figure 9.14, pairs of spherical bacteria were found in Matt's urethral culture during microscopic analysis.

Bacteria

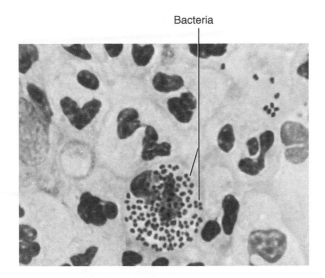

FIGURE 9.14 Photomicrograph of Matt's Urethral Culture

Which STD is suggested by Matt's symptoms and test results? _____

Matt's STD is bacterial in nature. What treatment options (if any) exist for Matt's infection?

_____

_____

Are any vaccines currently available against the pathogen that caused Matt's infection?

_____

Lydia

Lydia stayed home from work three days in a row due to nausea, vomiting, and abdominal pain. As shown in Figure 9.15, her skin and sclera (the whites of her eyes) had also developed a yellowish tint. Lydia suspected she had the flu, so she wasn't too worried about her symptoms. Nevertheless, she called her doctor's office and asked Monica—the physician assistant—if her symptoms truly merited an appointment. To her surprise, Monica urged her to come in immediately for evaluation and testing.

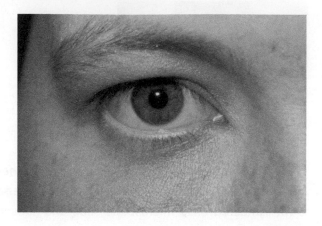

FIGURE 9.15 Lydia's Skin and Eye Discoloration

Since Lydia had developed jaundice (a yellow discoloration), her doctor was concerned about the health of her liver. During the physical examination, Lydia writhed in pain as the doctor palpated the right side of her abdominal cavity. Next, urine and blood samples were collected for laboratory analysis. As shown in Figure 9.16, Lydia's urine was abnormally dark in color. Her blood test was also positive for an antigen (foreign protein) belonging to a viral STD that targets the liver.

FIGURE 9.16 Lydia's Urine Specimen

Which STD is suggested by Lydia's symptoms and test results? _____

Why is the pathogen that caused Lydia's infection considered an STD, even though this virus targets cells within the liver?

_____

_____

Why would liver damage affect the color of Lydia's skin, sclera, and urine?

_____

_____

Lydia's STD is viral in nature. What treatment options (if any) exist for her infection?

_____

_____

Are any vaccines currently available against the pathogen that caused Lydia's STD?

_____

Buster

Roughly 5 weeks ago, Buster had his last sexual encounter. Given the fact that so much time had passed, he was shocked to find a prominent sore developing on the shaft of his penis (Figure 9.17). The sore did not feel painful, and it certainly did not appear to be spreading or growing in size. Just to be safe, though, Buster decided to visit a free health clinic and find out what was wrong.

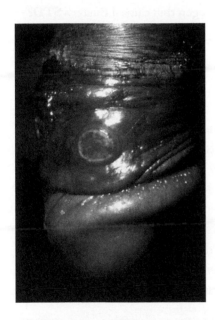

FIGURE 9.17 Buster's Chancre

During the physical exam, a chain of swollen lymph nodes was found along Buster's groin. Next, a culture was collected from the chancre (sore) on Buster's penis. A medical technician smeared Buster's culture onto a microscope slide and stained it prior to microscopy. As shown in Figure 9.18, Buster's culture contained a plethora of spiral-shaped bacteria. The doctor warned Buster that this STD—if left untreated—would lead to secondary and tertiary disease stages.

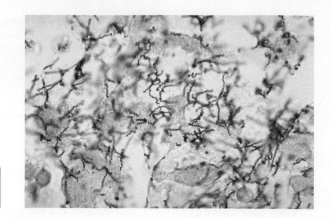

FIGURE 9.18 Microscopic Image of Buster's Culture

Which STD is suggested by Buster's symptoms and test results?

_____

Buster's STD is bacterial in nature. What treatment options (if any) exist for his infection?

_____

_____

Are any vaccines currently available against the pathogen that caused Buster's STD?

_____

Crystal

Because of her fever and achy muscles, Crystal simply assumed she was coming down with the flu. Needless to say, she was incredibly confused when—out of nowhere—her external genitalia started to feel itchy. Crystal used a hand mirror to examine her genitalia, and to her horror, she saw a large patch of red bumps forming along her labia. As shown in Figure 9.19, these bumps had transformed into painful blisters by the morning of her doctor's appointment. Following a physical examination, Crystal's doctor ordered blood tests to screen her for potential STDs. Several days later, the test results confirmed that DNA from a viral pathogen was present in Crystal's blood. Although this pathogen may cause recurrent outbreaks, the blisters from the initial infection would eventually scar over and heal.

FIGURE 9.19 Crystal's Blisters

Which STD is suggested by Crystal's symptoms and test results?

_____

Crystal's STD is viral in nature. What treatment options (if any) exist for her infection?

_____

_____

Are any vaccines currently available against the pathogen that caused Crystal's STD?

_____

A CLOSER LOOK AT STDS

If you would like to learn more about a specific STD—including its modes of transmission, prevalence, treatment guidelines, or ways to reduce its likelihood of transmission—you can access reputable information on the following websites:

CDC Division of STD Prevention: http://www.cdc.gov/std/default.htm

Find more information on STDs and HIV/AIDS in the Health Topics section on the World Health Organization (WHO) website: http://www.who.int

You can also call 1-800-CDC-INFO for additional information on STDs or referrals to STD clinics.

1. Label organs A–D on the diagram below.

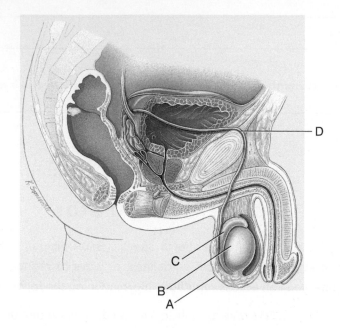

2. Sperm are produced inside of the _____, while eggs are produced inside of the _____. Why are these gonads considered homologous to one another?

   _____

   _____

3. How many chromosomes are normally found in sperm and eggs at the completion of spermatogenesis and oogenesis? _____ In contrast, how many chromosomes are normally found in somatic cells? _____

4. Which organ in the female reproductive system is homologous to the penis? What is the function of this organ?

   _____

   _____

5. Label organs A–D on the diagram below.

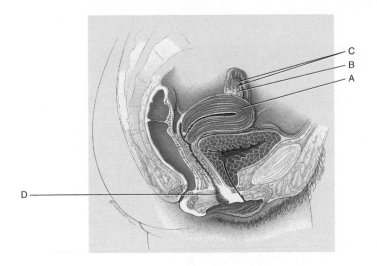

6. Describe three major differences that exist between spermatogenesis and oogenesis.

_____

_____

_____

_____

7. Label a mature (Graafian) follicle—and the oocyte within this follicle—on the picture below.

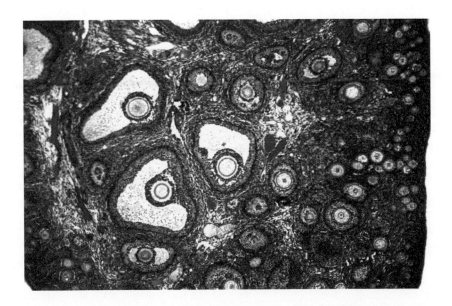

8. In general, what treatment options exist for viral STDs? Bacterial STDs?

_____

_____

9. Label a sperm flagellum on the picture below. What is the function of the flagellum?

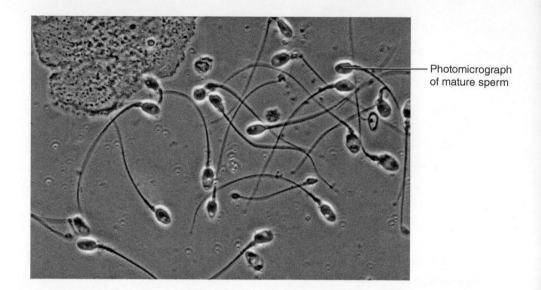

Photomicrograph of mature sperm

10. Name three ways to prevent or reduce the likelihood of STD transmission.

_____

_____

_____

LAB 10:
The Nervous System

By the End of This Lab, You Should Be Able to Answer the Following Questions:

- How are messages conducted through the axon of a neuron? On the other hand, how are messages relayed from one neuron to another?

- When viewing a microscope slide of nervous tissue, how do neurons differ in appearance from neuroglial cells?

- Where is the brain stem located in relation to the spinal cord? Where is the brain stem located in relation to the cerebrum, cerebellum, and diencephalon?

- What effects do stimulants have on reaction time, blood pressure, and heart rate?

- How do health professionals test the patellar reflex and the plantar reflex? Why do health professionals check reflex responses during physical exams?

- How does the structure of the brain change as a result of Alzheimer's disease?

INTRODUCTION

Rapid communication is an essential part of our everyday lives. Can you imagine living a week without Internet access, your cell phone, or face-to-face conversations? Take away nonverbal forms of communication, too—everything from ambulance sirens to stop lights—and havoc is likely to break loose! Communication allows us to solve problems, coordinate group activities, stay up to speed on current events, and avoid potential emergencies. Did you know that at this very moment, your body cells are swapping messages for the exact same reasons? Communication is so integral to survival, in fact, that two organ systems are devoted to this task: the endocrine system, which releases hormones (chemical messengers) into the bloodstream, and the nervous system, which is the focus of this lab period.

As shown in Figure 10.1, the framework of the nervous system is constructed out of the following organs: the brain, 12 pairs of cranial nerves, the spinal cord, and 31 pairs of spinal nerves. Since the brain and spinal cord are centrally located, they are collectively called the **central nervous system (CNS)**. Nerves extend from the brain (**cranial nerves**) and the spinal cord (**spinal nerves**) to innervate organs and limbs outside of the CNS. In reality, these nerves to split off into thousands of smaller branches along the way; this allows the nervous system to meticulously monitor the body from head to toe. Since nerves extend outside of the brain and spinal cord, they belong to the **peripheral nervous system (PNS)**. With the help of sensory organs, such as the skin and the eyes, the nervous system can monitor conditions in the external environment as well. Each day, your body encounters environmental changes—some internal, others external—that could potentially jeopardize your health. Fortunately, the nervous system is wired to rapidly identify risk factors, figure out whether a response is necessary, and implement responses, or solutions, in a timely manner.

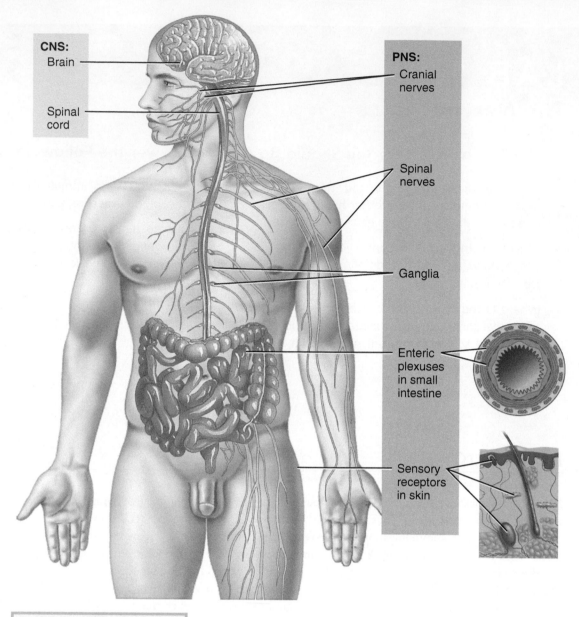

CNS:
Brain

Spinal
cord

PNS:
Cranial
nerves

Spinal
nerves

Ganglia

Enteric
plexuses
in small
intestine

Sensory
receptors
in skin

FIGURE 10.1   **Gross Anatomy of
the Nervous System**

NEURONS AND MESSAGE CONDUCTION

Each organ has a functional unit that carries out its basic activity, or function. When it comes to the nervous system, **neurons** (nerve cells) deserve this title, since they rapidly relay messages from one part of the body to another. (In medical terminology, the prefix *neuro-* is commonly used when referring to the nervous system.) The parts of a typical neuron are labeled on Figure 10.2. **Dendrites** are short tendrils that receive messages—chemicals called **neurotransmitters**—released by adjacent neurons. These messages are electrically conveyed to the **cell body**, which is sometimes called the **soma** of the cell. (In Greek, the word *soma* actually means *body*.) It is here within the cell body that the nucleus resides. Amazingly, the majority of cell bodies reside within the CNS, *regardless of a neuron's general location in the body.* How could a neuron reach the toe, for instance, if its cell body remains in the spinal cord? In some cases, the **axon**—a long, thin extension of the cell—is several feet long from one end to the other!

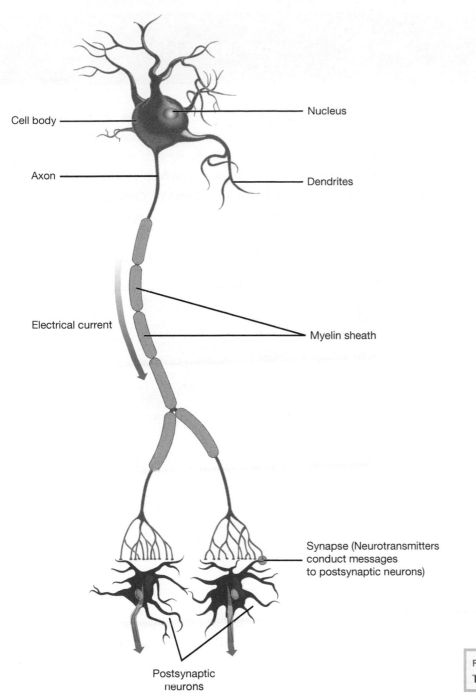

Cell body

Nucleus

Axon

Dendrites

Electrical current

Myelin sheath

Synapse (Neurotransmitters
conduct messages
to postsynaptic neurons)

Postsynaptic
neurons

FIGURE 10.2 **Structure of a
Typical Neuron**

Let's say a neuron gets bombarded with neurotransmitters that are stimulating in
nature, and as a result, it commits to relaying information upstream. This information must
reach neighboring cells that live near the end of its axon. How can messages travel through
the axon, which may be several feet long, and reach neighboring cells in a timely manner? This
is where **ions**—charged particles, such as sodium ions (Na^+)—and electrical activity come
into play. An **electric current** is generated when the axon starts to exchange ions with the
surrounding environment. Since electric currents flow at mind-boggling speeds, they travel
down the axon instantaneously, regardless of its length. Now, the message must reach neighboring cells, but there's a catch: electric currents cannot jump from one cell to another. This is
why neurotransmitters are integral to message conduction, even though they travel at slower
speeds than electric currents. As soon as neurotransmitters are released from the axon's terminal branches, they diffuse across short gaps (**synapses**) and toward neighboring cells.

To understand how neurons function, it is helpful to examine their structure at the microscopic level. During this exercise, you will observe prepared microscope slides of nervous tissue. As you view each slide, be sure to note the characteristics of typical, functional neurons. This information will help you identify diseased neurons in Exercise 10.6 (Disorders of the Nervous System). Compare your focused microscope slides to the photomicrograph provided below, and answer the review questions that accompany each slide.

Motor Neuron
(Derived from a Section of the Spinal Cord)

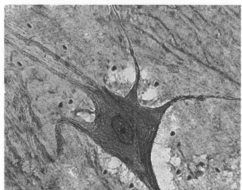

Label the dendrites on the neuron above. Do dendrites *release* or *receive* neurotransmitters? _____

Label the nucleus on the neuron above. Is the nucleus found within the dendrites, cell body, or axon? _____

Label the axon on the neuron above. How are messages conducted through the axon?_____

_____ are released from the terminal ends of axons. These chemicals diffuse across synapses to reach adjacent cells.

   Neuroglial cells are much smaller than neurons, but they greatly outnumber neurons in nervous tissue. Some of these support cells aid in immune defense and repair; others wrap themselves around axons, thereby creating a layer of electrical insulation. (The term *neuroglial* literally means *nerve glue*. In other words, neuroglial cells are the "glue" that hold the nervous system together.) Label a neuroglial cell on the image above.

Longitudinal Section of a Teased Peripheral Nerve

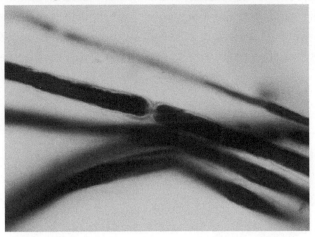

The axons shown above were originally bundled together in a nerve. These axons are covered by **myelin sheaths**: fatty layers of electrical insulation. Label the myelin sheath one of these axons. _____

_____

Nodes of Ranvier look like naked gaps along axons, since they are not covered by myelin. When a myelinated axon conducts a message, the current can jump from one node to another, thereby accelerating the speed of message conduction. Label a Node of Ranvier on the image above.

Some neurological disorders, such as multiple sclerosis, result from myelin destruction. How do you think **demyelination** (the loss of myelin) affects the speed, or velocity, of message conduction in affected neurons? Why?

_____

_____

STEP BY STEP 10.2   Nerve Conduction in the Human Body

During this exercise, you will use a voltmeter to detect nerve conduction in your own body. The basic parts of a digital voltmeter are labeled on Figure 10.3. Typically, this device is used to check the voltage of inanimate objects, such as batteries or electrical outlets. We will use a voltmeter to detect nerve conduction in the median nerve, which is located close to the surface of the skin. By placing the voltmeter's probes on top of the skin, you will be able to monitor nerve conduction in a noninvasive manner.

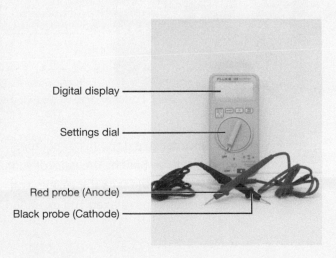

Digital display

Settings dial

Red probe (Anode)

Black probe (Cathode)

FIGURE 10.3   **Parts of a Handheld Digital Voltmeter**

Figure 10.4 shows the location of the median nerve in relation to the fingers, the palm of the hand, and the inner wrist. The median nerve runs through the carpal tunnel, which is a narrow passageway between the bones of the wrist. Carpal tunnel syndrome results from compression of this very nerve.

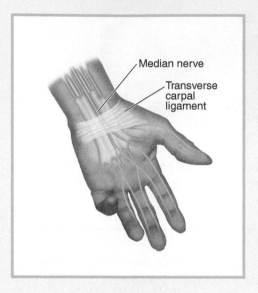

FIGURE 10.4 **Location of the Median Nerve**

Procedure

1. Turn on the voltmeter. If necessary, adjust the voltmeter's settings so it will take readings in volts direct current (VDC).

   Note: Be sure to take readings in millivolts (mV) during this experiment.

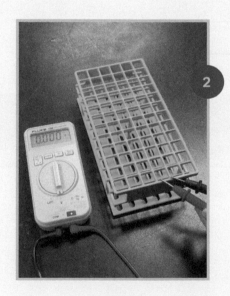

2. Place the probes on an object that is *not* electrically active, such as a plastic test tube rack. Record the voltage detected in Table 10.1.

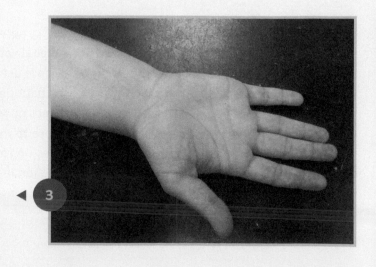

3. Ask your partner to extend his or her forearm out on the lab bench. The palm of your partner's hand should face the ceiling, and the fingers should be relaxed and extended.

4. Optional: Spread a small quantity of conductive gel on your partner's inner wrist.

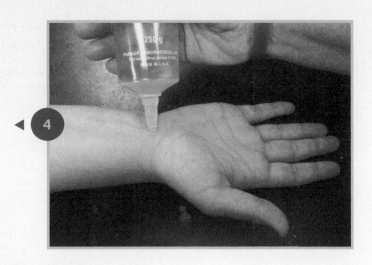

5. Place the probes on top of your partner's median nerve, as shown in the diagram below. The black probe and red probe should be approximately one inch apart from each other. Adjust the location of each probe, as needed, until a circuit is established. (Once a circuit is established, the voltage stops hovering at zero millivolts.) Record the voltage detected in Table 10.1.

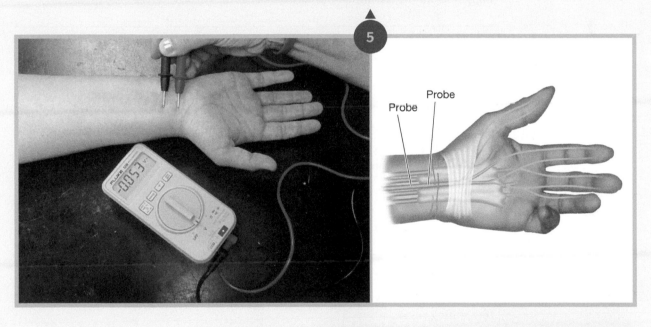

Probe

Probe

6. While the circuit is in place, ask your partner to flex his or her fingers and create a tight fist. Record the voltage detected during this step in Table 10.1.

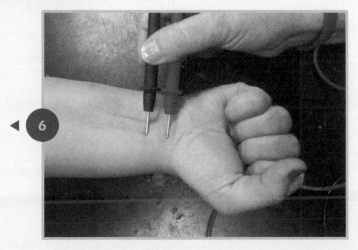

7. Exchange roles with your partner and repeat this experiment. Record your personal data in Table 10.1.

TABLE 10.1 **Results from Nerve Conduction Experiment**

Location of Probes	Voltage Detected in Millivolts (mV)
Inanimate Object—**Control**	
Median Nerve with Relaxed Hand—**Self**	
Median Nerve as Hand Clenches into Fist—**Self**	
|Change in Voltage|*—**Self**	
Median Nerve with Relaxed Hand—**Partner**	
Median Nerve as Hand Clenches into Fist—**Partner**	
|Change in Voltage|*—**Partner**	

*Record Absolute Values (Do Not Include Positive or Negative Signs)

Review Questions for Step by Step 10.2

Did you detect any voltage in the inanimate object you tested? If so, what does this suggest about the reliability of your results?

_____

_____

During this experiment, we monitored nerve conduction indirectly (outside of neurons), not directly (within neurons). When an axon becomes electrically active, its voltage spikes roughly 100 millivolts (mV) over the course of 2 milliseconds (ms). When your partner clenched his or her fist, did you detect any signs of nerve conduction in the median nerve? If so, what does this tell you about the function of the median nerve?

_____

_____

Did you, or any other students, have trouble detecting voltage near the median nerve? Based on your class results, did certain variables affect the outcome of this experiment? If so, which ones?
   * **Rolling of Median Nerve in Wrist**
   * **Gender**
   * **Body Frame**
   * **Other Anatomical Differences**
   * **Other Variables**

_____

_____

What are some potential sources of error in this experiment? What improvements could be made to this experiment?

_____

_____

GROSS ANATOMY OF THE NERVOUS SYSTEM

Given the remarkable complexity of the brain and spinal cord—not to mention the billions of neurons housed within them—it is surprising to learn how little each organ weighs. By adulthood, the human brain weighs approximately 3 pounds, while the spinal cord weighs in at less than one-tenth of a pound. Fortunately, these delicate organs are surrounded by a number of protective barriers: the **meninges** (three distinct membranes), **cerebrospinal fluid** (a shock-absorbing fluid that is produced within the brain), and **bones** (the cranial bones and vertebrae, respectively). The internal anatomy of the brain and spinal cord is illustrated in Figure 10.5. Key functions of the cerebrum, cerebellum, diencephalon, and brain stem are also summarized in Table 10.2.

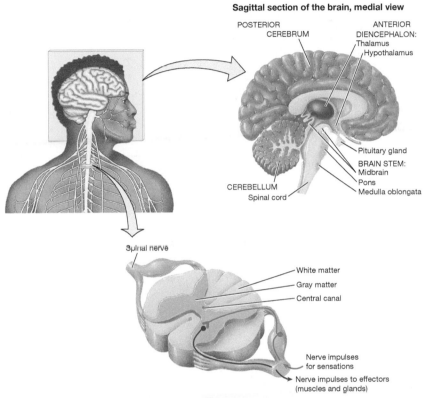

Sagittal section of the brain, medial view

POSTERIOR
CEREBRUM

ANTERIOR
DIENCEPHALON:
Thalamus
Hypothalamus

Pituitary gland

BRAIN STEM:
Midbrain
Pons
Medulla oblongata

CEREBELLUM
Spinal cord

Spinal nerve

White matter
Gray matter
Central canal

Nerve impulses
for sensations

Nerve impulses to effectors
(muscles and glands)

Transverse section of thoracic spinal cord

FIGURE 10.5 **Gross Anatomy of the Nervous System**

TABLE 10.2   **Principal Regions of the Human Brain**

Region	Components	Functions Regulated by the Region (Abbreviated List)
Cerebrum	Left and Right Hemispheres Lobes: Frontal, Parietal, Temporal, and Occipital*	Voluntary Movement Speech Sensations Thinking
Cerebellum	Left and Right Hemispheres	Balance Posture
Diencephalon	Thalamus Hypothalamus	Relay Messages to and from the Cerebrum Body Temperature Hunger and Thirst Smooth Muscle Contraction
Brain Stem	Midbrain Pons Medulla Oblongata	Breathing Rate Heart Rate Blood Pressure

* Brain lobes share their names with overlying cranial bones.

When you look at a dissected brain or spinal cord, two distinct substances are seen: white matter and gray matter. In the brain, gray matter is superficial to white matter, but the opposite is true in the spinal cord. This phenomenon is illustrated in Figure 10.6. **White matter** is composed primarily of myelinated axons. Since myelin is rich in fatty tissue, it imparts a white color to this substance. In contrast, **gray matter** consists primarily of cell bodies and shorter, unmyelinated axons. In the spinal cord, gray matter contains neurons that run horizontally to and from the spinal nerves. However, the white matter contains long axons that run vertically to and from the brain.

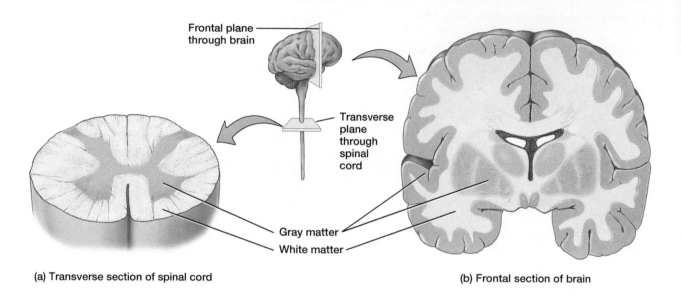

Frontal plane through brain

Transverse plane through spinal cord

Gray matter
White matter

(a) Transverse section of spinal cord

(b) Frontal section of brain

FIGURE 10.6 **Arrangement of White Matter and Gray Matter in the CNS**

Without cranial nerves, the brain would be isolated from a number of organs in the head, neck, thoracic cavity, and abdominal cavity. Most cranial nerves innervate the head and neck, where they play a role in sensation, muscle contraction, or a combination of the two. The longest cranial nerve—the vagus nerve—also innervates organs in the thorax and abdomen, including the heart, lungs, and stomach. (The name *vagus* was given to this nerve since it wanders through the body like a vagabond.) As shown in Figure 10.7, 12 pairs of cranial nerves are connected to the brain; these nerves must pass through holes (foramina) in the skull to reach peripheral tissues.

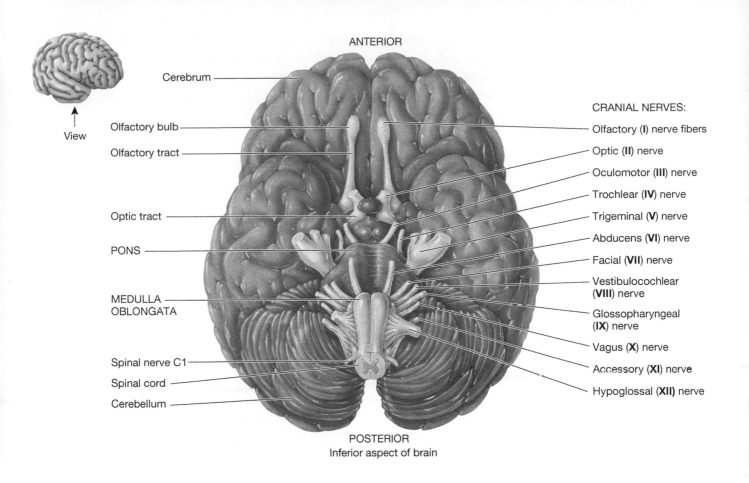

ANTERIOR

Cerebrum

Olfactory bulb

Olfactory tract

Optic tract

PONS

MEDULLA OBLONGATA

Spinal nerve C1

Spinal cord

Cerebellum

View

CRANIAL NERVES:

Olfactory (**I**) nerve fibers

Optic (**II**) nerve

Oculomotor (**III**) nerve

Trochlear (**IV**) nerve

Trigeminal (**V**) nerve

Abducens (**VI**) nerve

Facial (**VII**) nerve

Vestibulocochlear (**VIII**) nerve

Glossopharyngeal (**IX**) nerve

Vagus (**X**) nerve

Accessory (**XI**) nerve

Hypoglossal (**XII**) nerve

POSTERIOR
Inferior aspect of brain

Nerve	Type	Functions
I	Sensory	Smell
II	Sensory	Vision
III	Mixed	Sensory: proprioception Motor: movement of eyelid and eyeball; accommodation of lens
IV	Motor	Movement of eyeball
V	Mixed	Sensory: touch, pain, temperature, proprioception Motor: chewing
VI	Mixed	Sensory: proprioception Motor: movement of eyeball
VII	Mixed	Sensory: taste, proprioception Motor: facial expressions, secretion of tears and saliva
VIII	Mixed	Sensory: equilibrium and hearing Motor: sensitivity of receptors in ear
IX	Mixed	Sensory: taste, touch, pain on tongue, O_2, CO_2 and blood pressure levels Motor: swallow, speech
X	Mixed	Sensory: taste and pharynx sensations Motor: swallow, cough, speech, GI movements
XI	Mixed	Sensory: proprioception Motor: swallow, head and shoulder movements
XII	Mixed	Sensory: proprioception Motor: tongue movement

FIGURE 10.7 **Cranial Nerves**

Spinal nerves connect peripheral tissues directly to the spinal cord and indirectly to the brain itself. As shown in Figure 10.8, the name and number of each spinal nerve refers to its site of attachment along the spinal cord:

- Eight pairs of **cervical nerves** (C1-C8) connect to the spinal cord within the neck region.

- Twelve pairs of **thoracic nerves** (T1-T12) connect to the spinal cord within the chest region.

- Five pairs of **lumbar nerves** (L1-L5) connect to the spinal cord between the chest and the hips.

- Five pairs of **sacral nerves** (S1-S5) connect to the spinal cord within the hip region.

- One small pair of **coccygeal nerves** connect to the bottom of the spinal cord, which is located near the tailbone (coccyx).

Spinal nerves also share their names with surrounding vertebrae; this makes sense, since the terms *cervical*, *thoracic*, *lumbar*, *sacral*, and *coccygeal* refer to anatomical locations.

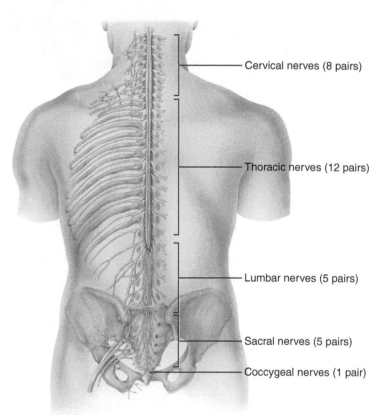

Cervical nerves (8 pairs)

Thoracic nerves (12 pairs)

Lumbar nerves (5 pairs)

Sacral nerves (5 pairs)

Coccygeal nerves (1 pair)

FIGURE 10.8   **Spinal Nerves**

EXERCISE 10.3   Gross Anatomy of the Nervous System

During this exercise, you will identify key structures in the CNS and PNS on anatomical models of the nervous system. Alternatively, your instructor may ask you to identify these structures on a virtual cadaver dissection, a dissected sheep brain, or a dissected human cadaver. Label the structures listed below on Figures 10.9–10.11, respectively, and answer the review questions provided at the end of this exercise.

THE BRAIN AND SELECTED CRANIAL NERVES

1. Cerebrum
   a. Left Cerebral Hemisphere
   b. Right Cerebral Hemisphere
2. Cerebellum
3. Diencephalon
   a. Thalamus
   b. Hypothalamus

4. Brain Stem
   a. Midbrain
   b. Pons
   c. Medulla Oblongata
5. Selected Cranial Nerves
   a. Olfactory (I) Nerves
   b. Optic (II) Nerves

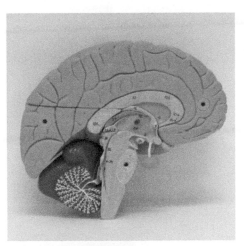

Sagittal section

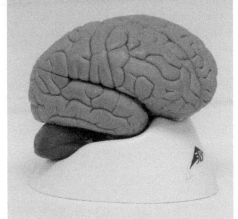

Lateral view

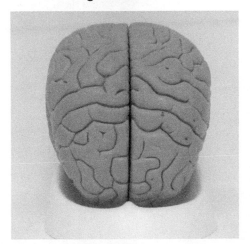

Superior view

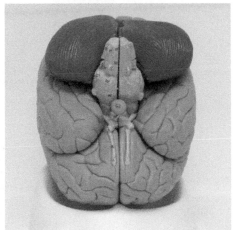

Inferior view

FIGURE 10.9 Identifying Regions of the Human Brain and Selected Cranial Nerves

THE SPINAL CORD (TRANSVERSE SECTION)

1. White Matter
2. Gray Matter
3. Central Canal
4. Spinal Nerves

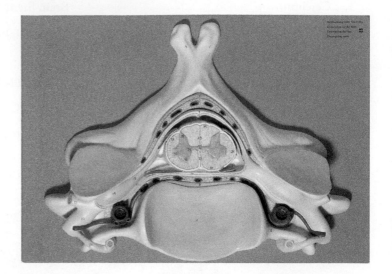

FIGURE 10.10 **Identifying Structures in the Spinal Cord**

SPINAL NERVES

1. Cervical Nerves
2. Thoracic Nerves
3. Lumbar Nerves
4. Sacral Nerves

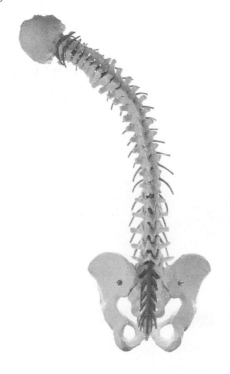

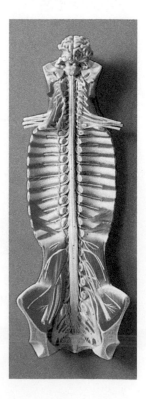

FIGURE 10.11 **Identifying Spinal Nerves**

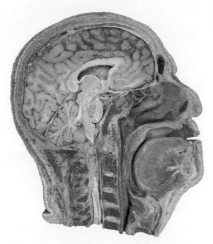

Sagittal Section of Head, Medial View

Label the cerebrum, cerebellum, and brain stem on the image above.

Where is the brain stem located in relation to the cerebrum? For instance, is the brain stem superior (above) or inferior to (below) the cerebrum?

_____

Label the spinal cord and the cranium on the image above.

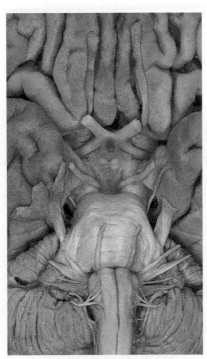

Inferior View of the Brain

Label the olfactory (I) nerves on the image above. Which sensory organs are innervated by them? _____

Label the optic (II) nerves on the image above. Which sensory organs are innervated by them? _____

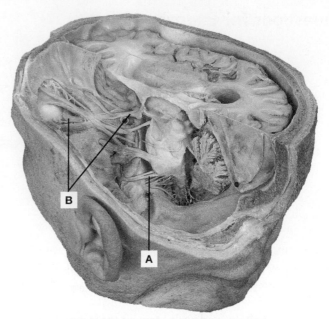

Cranial Nerve Dissection, Posterolateral View

Label the cerebellum on the image above.

Label the white matter and the gray matter within the cerebrum.

Within the cerebrum, is white matter superficial or deep to gray matter? _____

The vestibulocochlear (VIII) nerve is labeled A on the picture above. Which sensory organ does it innervate? _____

Which cranial nerve is labeled B on the image above? _____

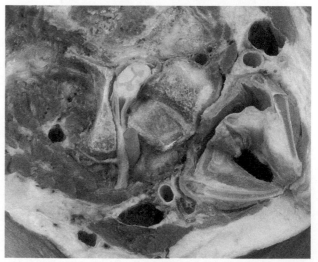

Cervical Spinal Cord and Nerve Dissection, Superolateral View

Label the vertebra bone on the image above.

Label the white matter and gray matter in the spinal cord.

Within the spinal cord, is white matter superficial or deep to gray matter? _____

_____

Label a spinal nerve on the image above. Do spinal nerves belong to the CNS or PNS?

_____

PSYCHOACTIVE DRUGS

Psychoactive drugs—including caffeine, nicotine, and alcohol—can alter mood, behavior, perception, cognition, sensation, and in some cases, even consciousness. The chemicals in **psychoactive drugs** alter communication patterns between neurons at synapses. This is typically accomplished in one of the following ways:

- Altering the concentration of a particular neurotransmitter

- Altering the number of receptors present for a particular neurotransmitter

- Altering the activity of receptors that bind a specific neurotransmitter

Table 10.3 summarizes the mechanisms of action used by nicotine, caffeine, and other psychoactive drugs. Certain psychoactive drugs are prescribed by physicians to manage pain or alleviate the effects of emotional disorders. When prescribing psychoactive drugs, physicians normally weigh the potential benefits of the drug against the potential risks for addiction, abuse, or toxic side effects.

TABLE 10.3   Mechanisms of Action for Selected Psychoactive Drugs

Psychoactive Drug	Primary Mechanisms of Action
Nicotine	Activates nicotinic cholinergic receptors Increases dopamine synthesis and release
Caffeine	Adenosine receptor antagonist
Most Antidepressants (SSRIs)	Inhibits the reuptake of serotonin
Amphetamines	Increases the release of dopamine Inhibits the reuptake of dopamine
Cocaine	Blocks the uptake of dopamine, thereby prolonging its effects
Alcohol	Increases the inhibitory effects of GABA Decreases the excitatory effects of glutamate
Morphine, Codeine, Heroin	Activates opioid receptors in the mesolimbic dopamine pathway
MDMA (Ecstasy)	Increases serotonin release Blocks reuptake of serotonin
Cannabinoids (example: THC)	Activate cannabinoid receptors Increase dopamine activity in the mesolimbic pathway
Hallucinogens	Different substances in this class affect different receptors: serotonin receptors, glutamate receptors, acetylcholine receptors, etc.

Source: _Neuroscience of Psychoactive Substance Use and Dependence_ (WHO 2004)
This report can be accessed online at http://www.who.int/substance_abuse/publications/en/Neuroscience_E.pdf

During this exercise, you will test the effects of caffeine consumption on three dependent variables: heart rate, blood pressure, and reaction time. **Reaction time** refers to the time that elapses between the onset of stimulation and a person's *response* to this stimulus. Before you begin this experiment, answer the questions below as a class.

Terminology Used in This Exercise

Independent Variable—the variable that is manipulated by researchers during a particular experiment. In this exercise, the independent variable is caffeine consumption (test group), or lack thereof (control group).

Dependent Variables—the variables that are measured by researchers to learn about the effects of the independent variable. In this exercise, the dependent variables are heart rate, blood pressure, and reaction time.

Blind Study—subjects in this study are not told whether they belong to the test group or the control group. The goal is to minimize the possibility of bias skewing the results.

Double-Blind Study—neither subjects *nor* researchers are told which subjects belong to the test group and control group. In theory, this should eliminate the possibility of bias skewing the results.

Create a hypothesis to test regarding the effects of caffeine on heart rate, blood pressure, and reaction time. Alternatively, you may formulate three separate hypotheses: one for each dependent variable.

_____

_____

Which beverage will be consumed by the test group? How much of this beverage will be consumed by each subject? (Refer to Table 10.4.)

_____

_____

TABLE 10.4 Caffeine Content of Various Beverages

Beverage	Quantity	Milligrams of Caffeine
Brewed Coffee	8 oz (240 mL)	95–200
Instant Coffee	8 oz (240 mL)	27–173
Black Tea	8 oz (240 mL)	40–120
Cola	12 oz (360 mL)	35–38
Diet Cola	12 oz (360 mL)	35–47
Other:		

Which beverage will be consumed by the control group? How much of this beverage will be consumed by each subject?

_____

_____

Are there any variables you wish to control among the subjects of the study? (Examples: gender, age, consumption of caffeine before class, volume of beverage consumed)

_____

_____

Will this experiment be performed as either a blind or double-blind study? If so, how will this be accomplished?

_____

_____

How many subjects will consume beverages in the control group and the test group? How many researchers will be assigned to each subject?

_____

_____

 SAFETY NOTE

If you experience adverse effects from caffeine consumption, please volunteer to assist with data collection.

Procedure

Collect the Following Data Prior to Beverage Consumption

1. Researcher: Measure the subject's initial pulse rate and record this information in Table 10.5. (Refer to Lab 6, as needed, to review how pulse rate readings are taken.)

2. Researcher: Measure the subject's initial blood pressure and record this information in Table 10.5. (Refer to Lab 6, as needed, to review how blood pressure readings are taken.)

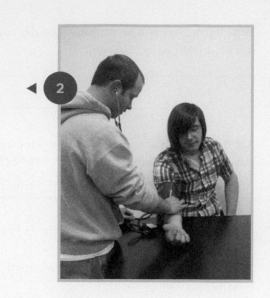

3. Using the procedure described below, determine the subject's initial reaction time.

   a. Subject: Sit down in a chair and extend the arm of your dominant hand. Spread your thumb and index finger approximately two inches apart.

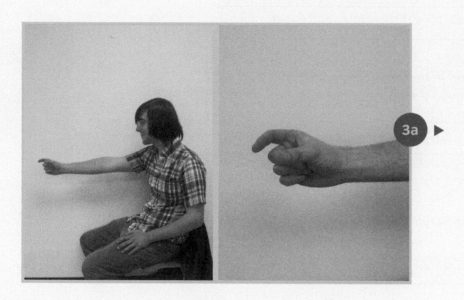

   b. Researcher: Hold a reaction time ruler (or a meter stick) directly above the subject's open fingers. The zero centimeter mark should be level with the subject's fingers.

c. Researcher: Release the meter stick at a time unknown to your partner.

d. Subject: Catch the meter stick as soon as possible with your thumb and index finger.

e. Researcher: Record the reaction time (or the centimeter mark covered by the subject's fingers) in Table 10.5. If you are using a meter stick, convert this distance into reaction time with the following equation:

$$t = \sqrt{2d/g}$$

t = reaction time (units = seconds)

d = distance ruler fell in centimeters

g = 980 cm/sec^2 (gravitational acceleration)

f. Repeat the reaction time test two additional times and record the results in Table 10.5.

g. Average the subject's initial reaction time and record the average in Table 10.5.

Beverage Consumption

 SAFETY NOTE

Since drinking and eating are not permitted in most science laboratories, this portion of the experiment may be performed outside of the laboratory space.

Subject: Drink the appropriate quantity of your assigned beverage. Note the time at which you finish drinking this beverage.

Fifteen Minutes after Beverage Consumption

1. Researcher: Measure the subject's final pulse rate and record this information in Table 10.5.

2. Researcher: Measure the subject's final blood pressure and record this information in Table 10.5.

3. Perform the reaction time test three times and average the subject's final reaction time. Record the results in Table 10.5.

4. Using the following equation, calculate the percent change in the subject's pulse rate, blood pressure, and reaction time. Record the results in Table 10.5.

$$\% \text{ change} = \left(\frac{\text{Initial Reading} - \text{Final Reading}}{\text{Initial Reading}} \right) \times 100\%$$

If a double-blind study was performed, ask your instructor which students belonged to the test group and the control group.

TABLE 10.5   Data Collected during Caffeine Experiment

Dependent Variable	Results	Averaged Results for Control Group	Averaged Results for Test Group
Initial Pulse Rate (bpm)			
Final Pulse Rate (bpm)			
Change in Pulse Rate (%)			
Initial Blood Pressure (mmHg)			
Final Blood Pressure (mmHg)			
Change in Blood Pressure (%)			
Initial Reaction Time—Trial 1 (sec)			
Initial Reaction Time—Trial 2 (sec)			
Initial Reaction Time—Trial 3 (sec)			
Initial Reaction Time—Average (sec)			
Final Reaction Time—Trial 1 (sec)			
Final Reaction Time—Trial 2 (sec)			
Final Reaction Time—Trial 3 (sec)			
Final Reaction Time—Average (sec)			
Change in Reaction Time (%)			

Review Questions for Step by Step 10.4

Based on the class results, does caffeine consumption have a noticeable effect on pulse rate, blood pressure, and/or reaction time? Explain your answer.

_____

_____

_____

_____

Was your initial hypothesis supported or rejected by the results of this experiment? If necessary, revise your initial hypothesis in the space provided below.

_____

_____

How do your results compare with those obtained by other students in the test (or control) group?

_____

_____

What are potential sources of error in this experiment? What improvements (if any) could be made to this experiment?

_____

_____

FUNCTIONAL CLASSES OF NEURONS

Neurons are classified into three distinct categories based on their functions within the nervous system: sensory neurons, interneurons, and motor neurons. Figure 10.12 illustrates the relationship between neurons in these three functional categories. **Sensory neurons** collect sensory information about the environment and send this information to the CNS. As indicated by their name, **interneurons** are located between (*inter-*) sensory neurons and motor neurons. Since interneurons are located exclusively within the CNS, they are responsible for interpreting sensory information and determining whether a response is necessary. If the CNS does implement a response, then **motor neurons** relay this information to the appropriate muscle(s) or gland(s) in the body. The desired response will likely involve one of the following actions: contracting a muscle, relaxing a muscle, releasing a substance from a gland, or *inhibiting* the release of a substance from a gland.

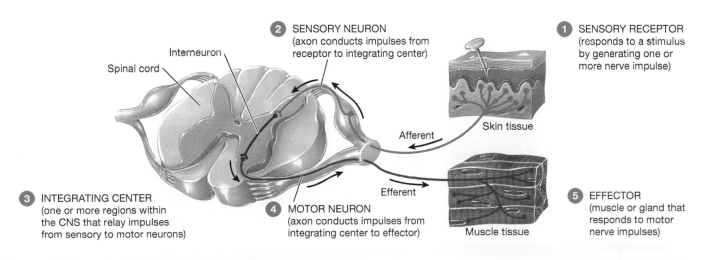

FIGURE 10.12 **Simple Reflex Arc**

Most of the actions we perform in everyday life require some degree of brain involvement. However, simple reflex arcs—such as the one displayed in Figure 10.12—are managed solely by interneurons in the spinal cord. This is why people pull away from hot objects (or painful objects) before they even realize what just happened! Amazingly, some reflex arcs don't even require an interneuron; the patellar (knee-jerk) reflex is one example. Simple reflex arcs aid in survival, since reaction time is much quicker without any input from the brain. During physical examinations, health professionals routinely check reflex responses in patients. An abnormal reflex response could indicate nerve damage, muscle damage, or perhaps some sort of injury to the brain or spinal cord.

Note: Since this exercise is not being conducted by a licensed health professional, the results are not intended to diagnose health disorders.

Patellar Reflex Test

1. Subject: Sit down on a surface that allows your legs to hang freely.

2. Researcher: Palpate the subject's left knee to find the location of the patellar tendon. As shown in Figure 10.13, the patellar tendon is located directly below the patella (knee cap).

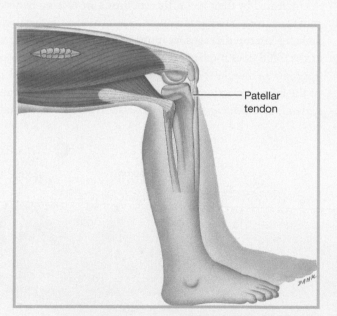

Patellar tendon

FIGURE 10.13   **Location of Patellar Tendon**

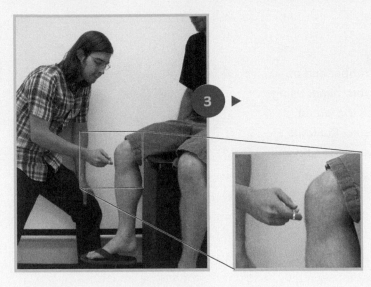

3. Researcher: Hold the reflex hammer horizontally in your dominant hand; the broad end of the hammer should face your partner's patellar tendon.

4. Researcher: Strike the subject's patellar tendon and observe the reflex action produced. Record the results in Table 10.6.

5. Switch roles with your partner and repeat this test.

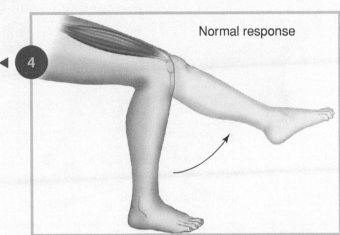

Normal response

Plantar Reflex Test

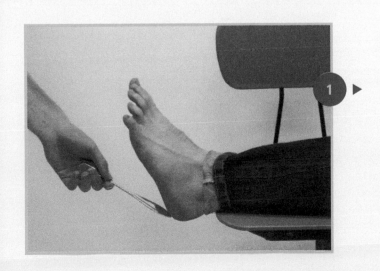

1. Subject: Remove your shoe and sock. Extend your leg and rest it on the seat of a chair.

2. Researcher: Hold the rubber end of the reflex hammer in your hand. Using the handle, firmly trace the lateral edge of the subject's sole. Continue tracing along the curve of the sole and toward the big toe.

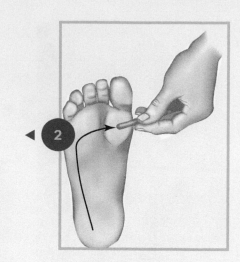

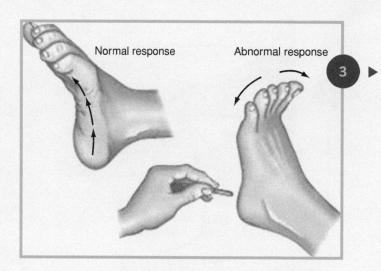

Normal response   Abnormal response

3. Observe the subject's reflex response and record the results in Table 10.6.

4. Switch roles with your partner and repeat this test.

TABLE 10.6 **Results of Reflex Tests**

Reflex Test	Action Produced by Reflex Test
Patellar Reflex—**Self**	
Plantar Reflex—**Self**	
Patellar Reflex—**Partner**	
Plantar Reflex—**Partner**	

During this exercise, you will use your knowledge of the nervous system to diagnose three patients. Monty, Alice, and Harriett are seeking medical attention for different neurological disorders. Use the symptoms and test results provided to make a logical diagnosis for each patient: Alzheimer's disease, Bell's palsy, or hydrocephalus.

Terminology Used in This Exercise

Magnetic Resonance Imaging (MRI)—medical imaging technique that uses magnets and radio waves to create detailed images of internal organs

Computerized Axial Tomography (CAT Scan or CT Scan)—X-rays are taken from many different angles to create thin, detailed images of internal body structures

Monty

When Monty rolled out of bed yesterday, he felt an odd, tingling sensation on the left side of his face. Wondering what was wrong, he rubbed his face as he stumbled toward the bathroom. Upon seeing his reflection in the mirror, Monty's grogginess was replaced by sheer panic; the left side of his face was drooping uncontrollably, as if it was paralyzed. As shown in Figure 10.14, Monty's smile was asymmetrical. In addition, he was unable to close his left eye.

FIGURE 10.14   Monty's Facial Paralysis

Monty's primary care physician referred him to a neurologist, who ordered an MRI of his brain. Monty's MRI results are shown in Figure 10.15. Swelling was detected in cranial nerve VII, which controls the movement of facial muscles. Fortunately, Monty found out that most patients with this disorder see signs of recovery in a few weeks.

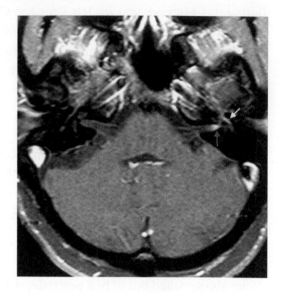

FIGURE 10.15 Monty's MRI

Which neurological disorder is suggested by Monty's symptoms and test results?

_____

In most cases, the right side of the brain controls the left side of the body, and vice versa. This is because most **tracts**—bundles of myelinated axons within the CNS—cross over and switch sides at specific locations, such as the brain stem or the spinal cord. Based on Monty's symptoms, is his swollen cranial nerve connected to the left or right side of his brain? _____

What treatment options (if any) exist for Monty's condition?

_____

_____

Alice

When Alice was born, everyone in the room—especially her first-time parents—expressed concern about the abnormally large size of her head (Figure 10.16). In addition, when Alice finally opened her eyes, she stared downward constantly. The attending physician ordered a broad spectrum of tests, including a CT scan of Alice's head. As shown in Figure 10.17, the **ventricles** (cavities) in Alice's brain were severely enlarged; according to the physician, too much cerebrospinal fluid had accumulated in her ventricles. Alice's parents were told that the problem needed to be resolved immediately, since the high intracranial pressure was compressing Alice's delicate brain tissues.

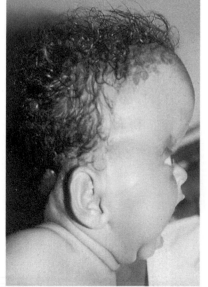

FIGURE 10.16 Alice's Head

CT Scan of Alice's Brain

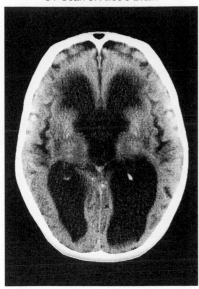

CT Scan of Normal Brain

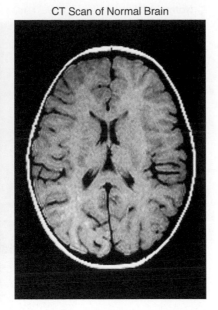

FIGURE 10.17 **Comparison of Normal CT Scan and Alice's CT Scan**

Which neurological disorder is suggested by Alice's symptoms and test results?

_____

What are some possible reasons why cerebrospinal fluid is accumulating within Alice's ventricles?

_____

_____

What treatment options exist (if any) for Alice's condition?

_____

_____

Harriett

For years, Harriett convinced her children that the forgetfulness she displayed was just part of the normal aging process. As time continued to pass, though, Harriett's kids grew more and more concerned about their mother's progressive memory loss. Harriett started misplacing objects on a regular basis, and the names of household items often escaped her. Eventually, she even started forgetting the names of close friends and family.

Harriett's kids—now adults with children of their own—made an appointment with a respected neurologist in the area. Following a series of cognitive tests, the neurologist ordered an MRI of Harriett's brain; the results of Harriett's MRI are shown in Figure 10.18. When compared to a normal MRI, it was clear that Harriett's brain had decreased in size at certain spots. According to the neurologist, the missing neurons had developed structural abnormalities, such as tangles; once these damaged neurons lost contact with neighboring cells, they perished. Plaques of hard, insoluble material had also accumulated in Harriett's brain tissue. Figure 10.19 compares the structure of healthy neurons with the structure of Harriett's damaged neurons.

Normal MRI   Harriett's MRI

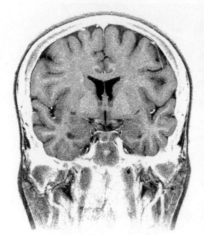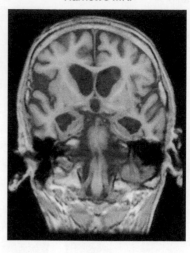

FIGURE 10.18 Comparison of
Normal MRI and Harriett's MRI

Normal Histology of the Cerebral Cortex   Histology of Harriett's Cerebral Cortex

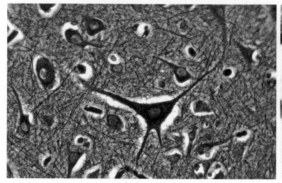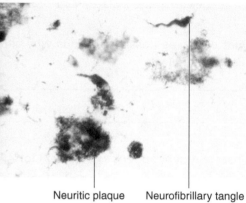

Neuritic plaque   Neurofibrillary tangle

FIGURE 10.19 Comparison of
Healthy Neurons and Damaged
Neurons

Which neurological disorder is suggested by Harriett's symptoms and test results?

_____

What treatment options (if any) currently exist for this condition?

_____

_____

A CLOSER LOOK AT NEUROLOGICAL DISORDERS

The National Institute of Neurological Disorders and Stroke (NINDS) is a division of the National Institutes of Health. The mission of NINDS is to "reduce the burden of neurological disease—a burden borne by every age group, by every segment of society, by people all over the world." If you would like to learn more about a specific neurological disorder, or the research programs sponsored by NINDS, you can visit their website at http://www.ninds.nih.gov.

REVIEW QUESTIONS FOR LAB 10

1. Which part of the neuron is labeled A below? What is the function of this long, thin extension?

   _____

   _____

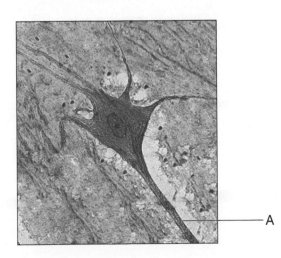

2. _____ relay(s) messages from one neuron to another.
   a. Hormones
   b. Electrical currents
   c. Glucose
   d. Neurotransmitters

3. What effects do stimulants (example: caffeine) typically have on reaction time, blood pressure, and heart rate?

   _____

   _____

4. Which regions of the brain are labeled A and B below? Name one function performed by each of these regions.

   _____

   _____

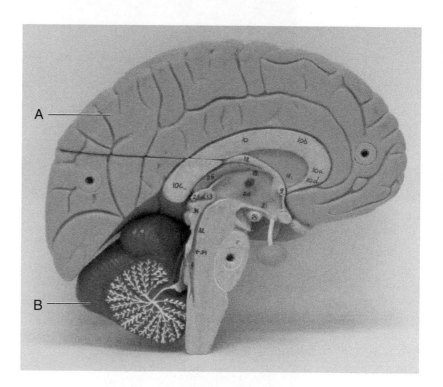

5. Where do cervical nerves connect to the spinal cord?
   a. In the neck area
   b. In the chest area
   c. Between the hips
   d. In the groin area

6. Label the white matter and gray matter on the spinal cord below.
   a. Are myelinated axons found in white matter or gray matter? _____
   b. In the spinal cord, do myelinated axons run vertically (to and from the brain) or horizontally (to and from spinal nerves)? _____

Cervical Spinal Cord Cross Section,
Superior View

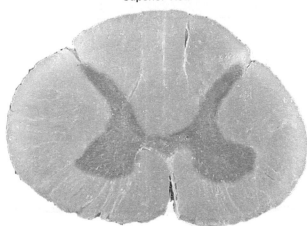

7. What role does the brain play in simple reflex arcs? How do reflexes aid in survival?

   _____

   _____

   _____

8. Label the sensory neuron, interneuron, and motor neuron on the diagram below.

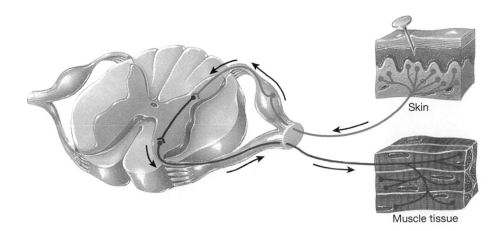

Skin

Muscle tissue

9. Name one mechanism that is used by psychoactive drugs to alter the activity of neurons.

_____

_____

10. Why do health professionals check reflex responses during physical exams?

_____

_____

LAB 11:
The Special Senses

By the End of This Lab, You Should Be Able to Answer the Following Questions:

- Where are sensory receptors located in the eyes, ears, nose, tongue, and skin?

- Which senses are dependent on chemoreceptors? Mechanoreceptors? Photoreceptors?

- What role does the sense of smell play in taste perception?

- What is the two-point discrimination test? Why does skin vary in its sensitivity to touch?

- Which parts of the ear can be examined with an otoscope? Why do health professionals routinely perform otoscopy on patients?

- Are visual clues required for accurate sound localization?

- Why does each retina possess a blind spot?

- Which parts of the eye can be examined with an ophthalmoscope?

- What factors may contribute to the deficiency or absence of a particular sense?

INTRODUCTION

The special senses can bring many joys to our everyday lives, such as tasting exotic cuisines, watching the sunrise, listening to live music, or even getting a massage. Although we tend to associate the senses with pleasure, the fact is they also play a critical role in our survival. How would we be aware of hazards that lurk around us without the eyes, ears, nose, tongue, and skin? Unfortunately, it is easy to take our senses for granted, especially when they are functioning properly. A deficiency or complete lack of a sense can bring serious challenges to daily activities, such as crossing the street, ordering a meal, or finding an entry to a building.

Sensory perception relies on an intimate association between the sensory organs and the nervous system. Some **sensory receptors** are composed entirely of nerve endings—a.k.a. the dendrites of sensory neurons—while others are encapsulated by additional support cells. A variety of sensory receptors are scattered throughout the body, and each class responds to a specific stimulus: light, injury, sound waves, chemicals, temperature changes, or some other event that physically distorts a receptor's shape. Once a group of sensory receptors is appropriately stimulated, messages fire through the adjoining nerve(s) to reach the spinal cord and/or the brain for interpretation.

Throughout the course of this lab, you will explore the structure and function of the major sensory organs in the body. By the end of this period, you should be able to answer the following questions for each sense and its respective sensory organ(s):

- Are mechanoreceptors, photoreceptors, thermoreceptors, chemoreceptors, or nociceptors involved in this sensation?

- How do environmental changes physically activate the sensory receptors in this organ?

- Where exactly are sensory receptors located in this organ?

- Which nerve(s) is (are) responsible for transmitting information on this sense to the brain?

- How can a deficiency or complete absence of this sense develop?

TASTE AND SMELL

The flavors you associate with different foods—such as cheesecake, oranges, and chicken noodle soup—are detected by different combinations of taste and smell receptors in the tongue and nose, respectively. By alerting the brain of hazardous chemicals, such as spoiled foods and toxic fumes, these sensory receptors also help protect the body from danger. The term **gustation** refers to the sense of taste, while **olfaction** refers to the sense of smell. These terms may sound odd at first, but they are actually derived from the Latin root words *gustare* (to taste) and *olfacere* (to smell). The tongue and nose contain chemoreceptors that are involved in gustation and olfaction, respectively. **Chemoreceptors** are activated by chemicals (*chemo-*) dissolved in nearby fluids, such as saliva or mucus.

As depicted in Figure 11.1, the tongue is covered in small, raised bumps called **papillae**. (The term *papilla* translates to mean a nipple-like structure.) Many of these papillae contain taste buds that are involved in gustation. Amazingly, there are only five taste categories known to date: sweet, sour, bitter, salty, and **umami**, which is a meaty or savory taste sensation. As soon as taste buds are stimulated by the presence of a chemical, such as sucrose, this information travels through adjoining cranial nerves to reach various parts of the brain, such as the gustatory area. The facial nerve (VII), glossopharyngeal nerve (IX), and the vagus nerve (X) are all involved in transmitting taste information to the brain.

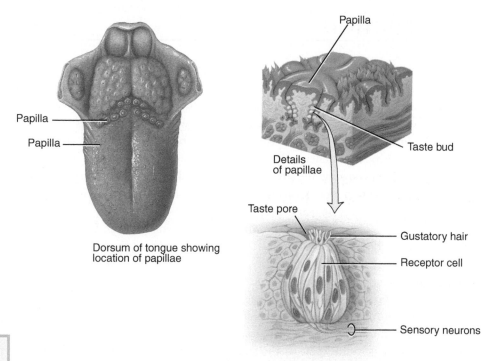

Dorsum of tongue showing location of papillae

Details of papillae

Structure of a taste bud

FIGURE 11.1 **Taste Bud Anatomy**

If there are only five different taste categories, then how on earth do we recognize hundreds of foods based on their unique flavor profiles? It turns out that smell, or olfaction, plays a major role in our perception of taste. Most of us have experienced this phenomenon while fighting a head cold, since congestion tends to diminish the sense of smell. As shown in Figure 11.2, olfactory receptors are located in the olfactory epithelium, which lines the roof of the nasal cavity. When odor molecules bind to these chemoreceptors, messages travel through the olfactory nerves (I) to reach the olfactory cortex of the brain. Amazingly, there are hundreds of different types of olfactory receptors; as a result, we can identify a wide variety of odors.

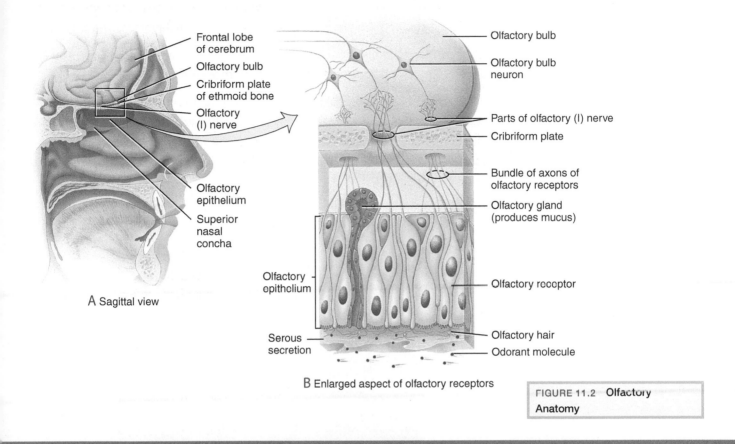

A Sagittal view

B Enlarged aspect of olfactory receptors

FIGURE 11.2 Olfactory Anatomy

EXERCISE 11.1   Testing the Relationship between Taste and Smell

During this exercise, you will work in pairs to examine taste perception without the sense of smell. Before you perform this experiment, create a hypothesis to test on the ability—or inability—of the brain to accurately identify flavors without olfactory information.

Hypothesis:

_____

_____

 SAFETY NOTE
Since drinking and eating are not permitted in most science laboratories, this experiment may be performed outside of the laboratory space.

If you have food allergies or are unable to consume sugar, please notify your instructor before participating in this exercise.

Procedure

1. Subject: Collect a blindfold from the materials bench.

2. Researcher: Collect three, individually wrapped candies to test on your partner (a.k.a. the subject). Do not let the subject know which flavors you have chosen.

3. Subject: Place the blindfold over your eyes while you are seated. Once you are blindfolded, pinch your nostrils together with your fingers.

4. Researcher: Hand the subject one wrapped candy. If the candy has a hard texture, ask the subject to suck on this candy—*without swallowing it*—and identify its flavor. If the candy has a soft texture, ask the subject to chew the candy—*without swallowing it*—and identify its flavor. Record the results in Table 11.1.

5. Researcher: Ask the subject to open his or her nostrils, continue sucking or chewing on the candy, and determine its flavor once more. Record the results in Table 11.1.

6. Repeat steps 4–5 with the two remaining candy flavors.

7. Switch roles with your partner and repeat this experiment.

TABLE 11.1 **Results of Flavor Experiment**

Flavor Tested	Flavor Detected without Smell	Flavor Detected with Smell
1.		
2.		
3.		

Review Questions for Exercise 11.1

Based on your results, is smell required for an accurate perception of taste? How do your results compare to those obtained by other students?

_____

_____

Was your initial hypothesis supported or rejected by the results you collected? If necessary, revise your initial hypothesis in the space provided below.

_____

_____

What steps were taken to minimize human bias during this experiment?

_____

_____

What are potential sources of error in this experiment? What improvements (if any) could be made to this experiment?

_____

_____

SENSORY RECEPTORS IN THE SKIN

Most organs multitask to keep the body functioning properly, and skin—the largest organ—is no exception. In addition to its role as a sensory organ, skin also prevents dehydration and protects underlying tissues from environmental hazards. As shown in Figure 11.3, there are two main tissue types within skin. The **epidermis** lies on top (*epi-*) of the dermis, and it contains stratified squamous epithelial cells. Many of these cells are actually dead, including the ones at the surface of your skin. Compared to the dermis, the epidermis contains a low number of sensory receptors. No blood vessels are found throughout the epidermis, either. Have you ever gotten a paper cut that didn't start to bleed? This means the paper cut penetrated the epidermis, but the dermis was still intact!

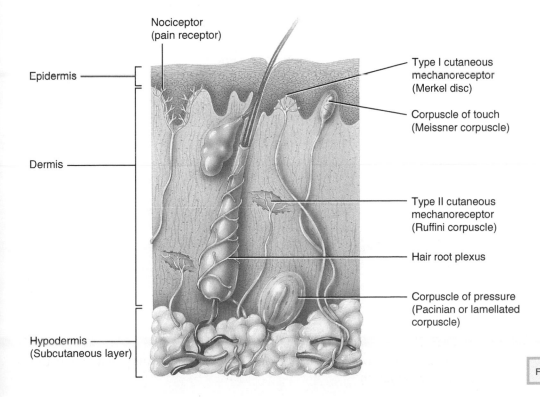

FIGURE 11.3 **Skin Anatomy**

The **dermis** is found below the epidermis, and it is composed of loose fibrous connective tissue. Blood vessels are present in the dermis, along with the overwhelming majority of sensory receptors. Let's say you cut your skin while shaving, and as a result, you feel pain *and* the cut starts to bleed. Based on these symptoms, you know the cut penetrated both the epidermis and the dermis. The **hypodermis**, or **subcutaneous layer**, is located underneath (*hypo-* or *sub-*) the skin. Adipose cells (fat cells) are found within the hypodermis, as well as a few sensory receptors.

The sensory receptors in skin constantly collect information about temperature changes, pressure, touch, and pain:

- **Nociceptors** are activated by harmful conditions, such as extreme heat or severe pressure. When nociceptors are stimulated, pain is felt within associated parts of the body. (The root word *nocere* means *to hurt* in Latin.)

- **Thermoreceptors** are activated by temperature changes.

- Touch and pressure receptors are considered **mechanoreceptors** since they are activated (physically distorted) by mechanical pressure.

When the sensory receptors in skin are stimulated, information travels through adjoining nerves to reach the somatosensory cortex of the brain. (In Greek, the word *soma* means *body*.) Depending on the nature of the message, it may be passed to additional regions of the brain, as well, such as the motor cortex.

EXERCISE 11.2   The Two-Point Discrimination Test

Did you know that certain patches of skin are more sensitive to touch than others? As a result, your ability to discriminate touch varies significantly from head to toe. Variations in touch sensitivity arise due to one or more of the following factors:

- Variations in the density of touch receptors.

- Variations in receptive field size. A **receptive field** is the total area innervated by a sensory neuron. Interestingly, a sensory neuron can't discriminate between two events—such as two separate points of touch—that occur within its receptive field at the same time.

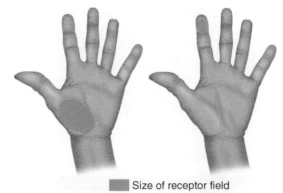

■ Size of receptor field

- **Neuropathy**, which is a disease (*-pathy*) or abnormality that affects the nervous system.

During this exercise, you will conduct a two-point touch discrimination test on the fingertip, forearm, calf, and back of the neck. Before you perform this experiment, create a hypothesis to test on the relative sensitivity of these four body parts to touch.

Hypothesis:

_____

_____

Procedure

1. Collect a blindfold and a pair of calipers from the materials bench. A pair of scissors and a metric ruler can be used in place of calipers.

2. Subject: Cover your eyes with the blindfold.

3. Researcher: Separate the tips of the calipers by 50mm.

4. Researcher: Gently press the tips of the calipers on the subject's fingertip. Both tips should touch the subject's skin at the same time.

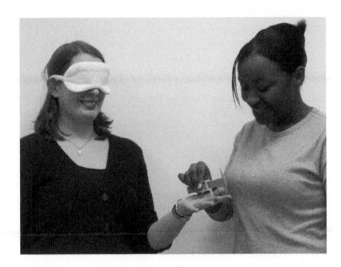

5. Researcher: Ask the subject how many points are felt on the fingertip. If the answer is two, narrow the distance between the tips by 5mm.

   Hint: Keep the subject on his or her toes as you cycle between steps 4 and 5! Every now and then, touch the subject's skin with only one caliper tip.

6. Repeat steps 4–5 until the subject feels one point of touch, even though two caliper tips are touching the skin. Now, widen the tips in 1 mm increments until your partner feels two separate points of touch again. Record this distance in Table 11.2.

7. Repeat the touch discrimination test on the forearm, calf, and back of the neck. Record the distances required for two-point touch discrimination in Table 11.2.

8. Switch roles with your partner and repeat this procedure.

TABLE 11.2 Results of Two-Point Touch Discrimination Test

Location on Body	Distance Required for Two-Point Touch Discrimination (mm)
Fingertip	
Forearm	
Back of the Neck	
Calf	

Review Questions for Exercise 11.2

In regard to two-point touch discrimination, which part of the body showed the greatest sensitivity? Which part of the body showed the least sensitivity? Were consistent results obtained throughout the class?

_____

_____

_____

Was your initial hypothesis supported or rejected by the results of this experiment? If necessary, revise your initial hypothesis in the space provided below.

_____

_____

What steps were taken to minimize human bias during this experiment?

_____

_____

What are some potential sources of error in this experiment? What improvements (if any) could be made to this experiment?

_____

_____

A CLOSER LOOK AT TACTILE DYSFUNCTION

Some people experience a heightened sensitivity to touch, which may result in the avoidance of clothes, objects, or physical interactions that others perceive as pleasant. This phenomenon is commonly called **tactile defensiveness**. Tactile defensiveness may be experienced as a feature of Tourette syndrome, attention deficit hyperactivity syndrome (ADHD), or autism spectrum disorders.

On the other hand, some people experience a diminished sensitivity to touch, or the complete absence of tactile sensation in certain regions of the body. Numbness, tingling, or pain may be felt in the affected area; in some cases, temperature and pain cannot be perceived in affected regions, either. This phenomenon is commonly associated with **peripheral neuropathies** that impair or destroy sensory neurons. Some peripheral neuropathies are inherited, while others are acquired due to trauma, autoimmune diseases, alcoholism, or complications from diabetes mellitus. Neurologists use blood tests, skin biopsies, nerve biopsies, and sensory tests (example: the two-point discrimination test) to diagnose peripheral neuropathy. If you would like to learn more about peripheral neuropathy, you can access the National Institute of Neurological Disorders and Stroke at **http://www.ninds.nih.gov/**.

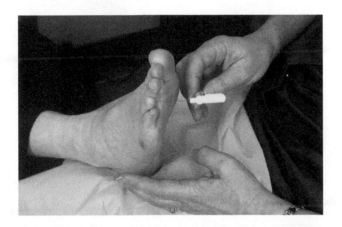

HEARING

Have you ever been startled by a loud noise, such as a wailing siren, exploding fireworks, or a blood-curdling scream? Did you jump, breathe rapidly, or feel your heart pounding in your chest? If you have experienced such an event, you know from personal experience that the ears do more than just bring pleasure to life. In fact, the opposite is true when you hear startling sounds, because the body anticipates danger and prepares itself accordingly.

When you look at your ears in the mirror, you can only see the **pinnae** in their entirety. (The parts of the outer ear, middle ear, and inner ear are illustrated in Figure 11.4.) Each pinna is considered part of the outer ear, along with a tube-like passageway called the **external auditory canal**. The outer ears collect sound waves and carry them toward the **tympanic membranes**, which are commonly called the eardrums. The tympanic membranes vibrate when they are hit by sound waves, and this generates movement in three adjoining bones: the malleus (hammer), the incus (anvil), and the stapes (stirrup). These **auditory ossicles** are actually the smallest bones in the human body. Sound waves are thereby converted into mechanical motion within the middle ear.

Beyond the stapes lies a membrane-covered opening called the **oval window**. The oval window vibrates when it is struck by the stapes, and as a result, fluid waves are produced within the **cochlea**. (The word *cochlea* means *snail shell* in Latin.) The sensory receptors within the cochlea are mechanoreceptors, since they are physically bent by fluid waves. The sensory receptors for hearing are called hair cells, and they are located in a portion of the cochlea called the organ of Corti. Once these hair cells are stimulated, sensory information travels through the vestibulocochlear nerve (VIII) to reach the auditory association area of the brain. The inner ear also contains sensory receptors for balance, which are located within the **vestibule** and the **semicircular canals**. Information about movement and spatial orientation is collected by these hair cells and sent to the brain through the vestibulocochlear (VIII) nerve.

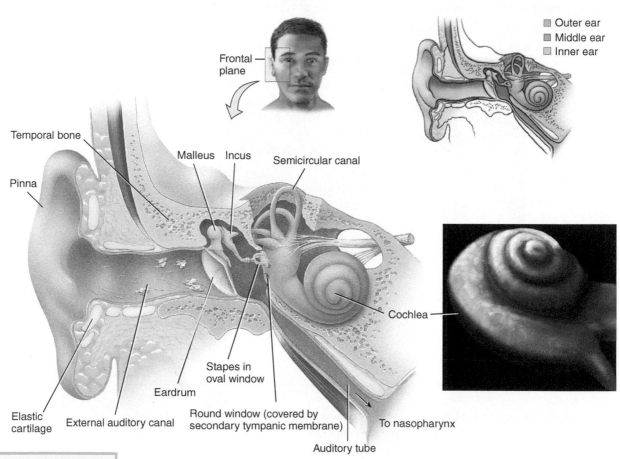

Frontal plane

Outer ear
Middle ear
Inner ear

Temporal bone

Pinna

Malleus   Incus

Semicircular canal

Cochlea

Stapes in oval window

Eardrum

Elastic cartilage

External auditory canal

Round window (covered by secondary tympanic membrane)

To nasopharynx

Auditory tube

FIGURE 11.4 **Ear Anatomy**

EXERCISE 11.3   Ear Anatomy

During this exercise, you will identify key structures on an anatomical model of the human ear. Alternately, your instructor may ask you to identify these structures on a dissected ear or a virtual cadaver dissection. Label the structures listed below on Figure 11.5, and answer the review questions that are provided at the end of this exercise.

1. Pinna (Auricle)
2. External Auditory Canal (External Acoustic Meatus)
3. Tympanic Membrane
4. Malleus
5. Incus
6. Stapes
7. Oval Window
8. Vestibule
9. Semicircular Canals
10. Cochlea
11. Vestibulocochlear Nerve
12. Auditory Tube (Eustachian Tube or Pharyngotympanic Tube)

FIGURE 11.5   **Identifying Parts of the Ear**

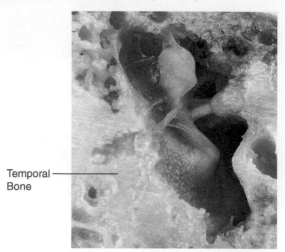
Dissection of Tympanic Cavity, Superior View

Temporal Bone

Label the tympanic membrane and the malleus on the image above. Are these structures located in the outer, middle, or inner ear? _____

When the malleus starts to vibrate, what happens to the incus and the stapes? _____

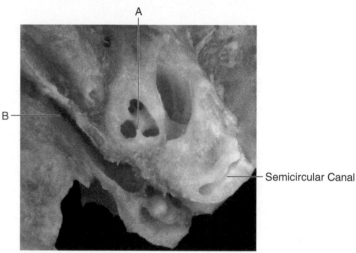
Dissection of middle and inner ear, superior view

A

B

Semicircular Canal

Which structure is labeled A on the image above? This structure contains receptors for hearing. _____

Which structure is labeled B on the image above? This tube connects the middle ear to the throat _____.

A semicircular canal is labeled on the image above. The sensory receptors in this structure are involved in _____.

Are mechanoreceptors, photoreceptors, or chemoreceptors involved in hearing and balance? Why are they classified this way?

_____

_____

Health professionals routinely check the health of the ears with a device called an **otoscope**. (The prefix *oto-* originates from the Greek term for *ear*.) Otoscopes illuminate and magnify structures in the outer and middle ear, such as the tympanic membrane, auditory ossicles, and external auditory canal. Health professionals look for signs of infection, inflammation, or injury while performing otoscopy. Figure 11.6 shows the appearance of a normal tympanic membrane, as well as symptoms for the following disorders:

- A perforated (pierced) tympanic membrane
- Acute otitis media, which is inflammation (*-itis*) of the middle (*media*) ear. This usually results from a middle ear infection.
- Collection of fluids (effusion) in the middle ear

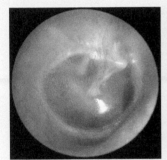 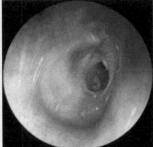 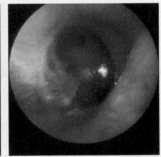 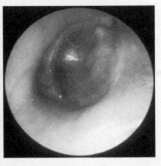

Normal   Perforated Tympanic Membrane   Acute Otitis Media   Effusion

FIGURE 11.6   **Otoscopy Results**

During this exercise, you will work in pairs to examine the ears with an otoscope. The parts of a typical otoscope are labeled on Figure 11.7; be sure to familiarize yourself with this piece of equipment before performing otoscopy. Alternatively, your instructor may elect to perform video otoscopy or virtual otoscopy. You can access a virtual otoscopy tool called *Diagnostics 101*, which is provided for students by Welch Allyn at **http://www.welchallyn.com/wafor/ students**.

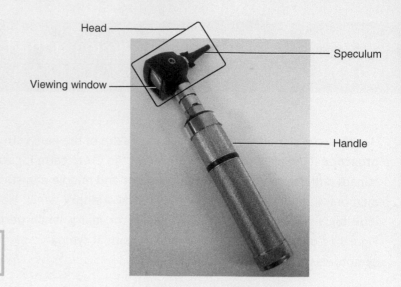

Head

Speculum

Viewing window

Handle

FIGURE 11.7 **Parts of a Typical Otoscope**

●●● **SAFETY NOTE**

Before you perform otoscopy, be sure to familiarize yourself with the parts and proper usage of an otoscope.

Since this exercise is not being performed by a licensed health professional, the results are not intended to diagnose health disorders.

Procedure

1. Ask your partner to sit down and tip his or her head toward one shoulder. You will evaluate the ear that is tilted up toward the ceiling.

2. Observe your partner's outer ear with the naked eye and record your observations in Table 11.3.

   If any abnormalities are observed with the naked eye, such as discharge or inflammation, do not perform otoscopy on your partner.

3. Place a clean pair of latex or nitrile gloves on your hands.

1

4. Attach a new, disposable speculum to the head of the otoscope.

5. Turn on the otoscope's light source.

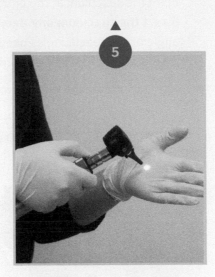

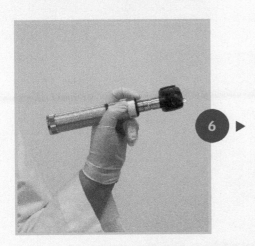

6. Hold the otoscope in your dominant hand. Securely grip the otoscope near the top of the handle with your thumb, index finger, and middle finger. This grip is similar to the one that secures a pencil in your hand for writing.

7. Gently pull your partner's pinna upward and backward to straighten the external auditory canal.

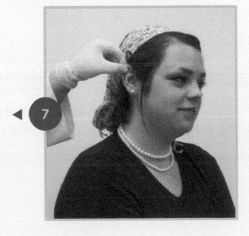

To avoid the risk of ear injury, the examiner must securely stabilize the otoscope during steps 8–9. The subject must also sit still during steps 8–9.

8. To stabilize the otoscope, rest your ring finger and pinky finger on your partner's cheek. As you watch carefully through the viewing window, place the tip of the speculum at the entryway to the external auditory canal. **Do not insert the speculum any deeper into your partner's ear canal.** Adjust the angle of the speculum, as needed, to view the tympanic membrane. **Stop immediately if otoscopy is painful to your partner.**

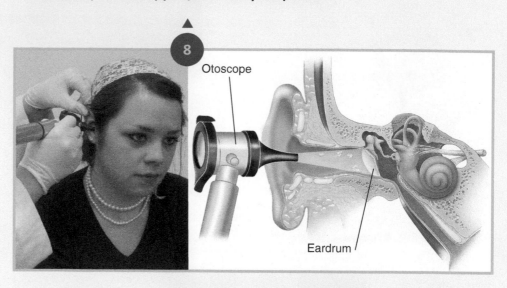

Otoscope

Eardrum

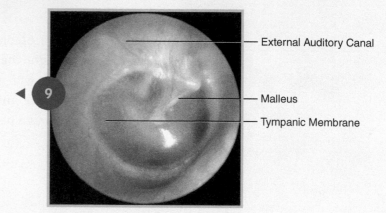

External Auditory Canal

Malleus

Tympanic Membrane

9. Inspect the color and opacity of the tympanic membrane. Can you see the ossicles in the middle ear? Normally, the malleus is resting on the tympanic membrane.

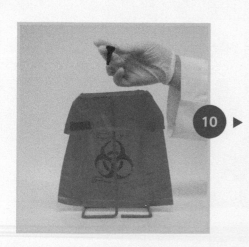

10. Carefully remove the speculum from your partner's ear, and place the used speculum in a biohazard bag. Record your observations in Table 11.3.

11. Switch roles with your partner and repeat this procedure.

TABLE 11.3   Observations from Ear Examination

Ear Structure	Viewing Method	Observations
Pinna	Naked Eye	Color: Presence/Absence of Inflammation (circle one) Additional Observations:
External Auditory Canal	Naked Eye	Presence/Absence of Hairs (circle one) Presence/Absence of Discharge (circle one) Additional Observations:
External Auditory Canal	Otoscope	Presence/Absence of Hairs (circle one) Presence/Absence of Ear Wax (circle one) Additional Observations:
Tympanic Membrane	Otoscope	Color: Opacity: Location of Malleus: Additional Observations:

Review Questions for Step by Step 11.4

Label the external auditory canal, the tympanic membrane, and the malleus on the image below.

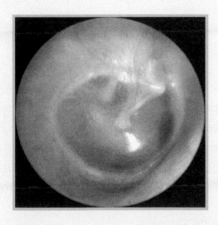

A healthy tympanic membrane is usually translucent and pearly white in color. Why would a health professional be concerned if a patient's tympanic membrane was red, yellow, or opaque?

_____

_____

Do you think a patient's hearing would be affected by middle ear effusion? Why or why not?

_____

_____

A person is said to have **binaural hearing** if both (*bi-*) ears *(-aural)* are able to detect sounds. When both ears are functioning properly, sounds can be distinguished based on their relative directions from the ears, as well as the relative distances they have traveled. This phenomenon is called **sound localization**. As shown in Figure 11.8, sound waves often reach the left and right ears at different time points. Sound intensity also plays a role sound localization, since it relates to the distance that separates the ears from the source of sound. When a person suffers from hearing loss in one ear, the ability to localize sound is either diminished or absent, depending on the degree of hearing loss.

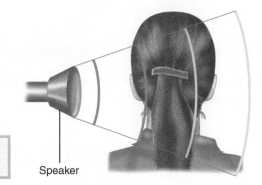

FIGURE 11.8 **Sound Localization**

Speaker

During this exercise, you will perform a sound localization test. Although this experiment is written as a group exercise, it can also be performed in pairs, assuming quiet space is available for each student group. Before you perform this exercise, create a hypothesis to test on the ability—or inability—of blindfolded subjects to accurately localize sounds that originate from the following sites:

- Above and below the ears
- In front of the ears and behind the ears
- To the left and right of the ears

Hypothesis:

_____

_____

Procedure

1. Pick one **subject** for the sound localization test. The subject sits in the center of the classroom with a blindfold over his or her eyes. When the subject hears a sound, he or she will point in the perceived direction of sound origin.

2. Pick six **researchers** for the sound localization test. Each researcher needs to collect a tuning fork and a rubber mallet. When prompted to do so, each researcher will strike the tuning fork with the mallet. (If you are working in pairs, the researcher will simply walk from one location to another.)

  - Researcher 1 stands approximately 10 feet in front of the subject.
  - Researcher 2 stands approximately 10 feet to the right of the subject.
  - Researcher 3 stands approximately 10 feet behind the subject.
  - Researcher 4 stands approximately 10 feet to the left of the subject.
  - Researcher 5 holds a tuning fork above the subject's head.
  - Researcher 6 holds a tuning fork under the subject's chair.

3. Pick one student to serve as the **leader** for this test. The leader will prompt sound production by randomly pointing to different researchers.

4. The remaining students will serve as **data recorders**. Data recorders are responsible for recording the location of sound origin *and* the direction pointed toward the subject.

5. **Leader**: Start the sound localization test by pointing to a researcher.

6. **Data Recorders**: Record the location of sound origination (front, back, right, left, above, or below) and the direction pointed toward by the subject in Table 11.4.

7. **Leader:** Continue pointing to researchers until all directions have been tested.

8. Repeat the sound localization procedure with two additional subjects.

TABLE 11.4 Results from Sound Localization Test

Name of Subject	Location of Sound	Direction Subject Pointed
Subject 1:		
Subject 2:		
Subject 3:		

Review Questions for Exercise 11.5

Based on the results from this experiment, does sound localization require the sense of vision? Explain your answer.

_____

_____

Was your initial hypothesis supported or rejected by the results you collected? If necessary, revise your initial hypothesis in the space provided below.

_____

_____

If an ear plug was placed in one of the subject's ears, do you think the results would be altered? Why or why not?

_____

_____

What are some potential sources of error in this experiment? What improvements (if any) could be made to this experiment?

_____

_____

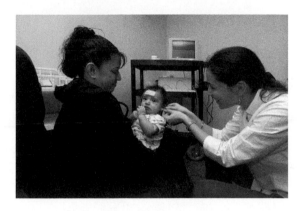
VISION

What aspects of life are impacted by vision loss? The answer is simply too vast to summarize on a single page. In fact, the words *on* this page have no meaning unless the eyes can decipher shapes, letters, and contrasting hues. Reading a lab manual is not essential for survival, but avoiding bodily injury definitely is. The eyes allow humans to assess the surrounding environment, anticipate danger, and get out of harm's way, whenever possible. Have you ever experienced a day without vision? If not, imagine crossing the street, approaching a staircase, or finding a bus stop without the sense of sight.

Similar to film within a camera, **photoreceptors** are activated by light. If the eyes are functioning properly, they will (1) adjust the amount of entering light and (2) focus light onto the retina, which is where photoreceptors are located. Figure 11.9 depicts the anatomy of the eye, and Table 11.5 summarizes the role of each structure in vision. While looking at your eyes in a mirror, you can identify the following structures: the **sclera** (white part of the eye), the **iris** (colored portion of the eye), and the **pupil** (black dot in the center of the eye). Light rays travel through the pupil, which is basically an opening in the iris. The size of the pupil is adjusted, as necessary, to regulate the amount of light that gains admission. This is accomplished through contraction or relaxation of muscles within the iris. Light rays are also **refracted** (bent) by the curved cornea and lens as they enter the eye. The proper degree of refraction is crucial, since it allows light rays to converge on the retina.

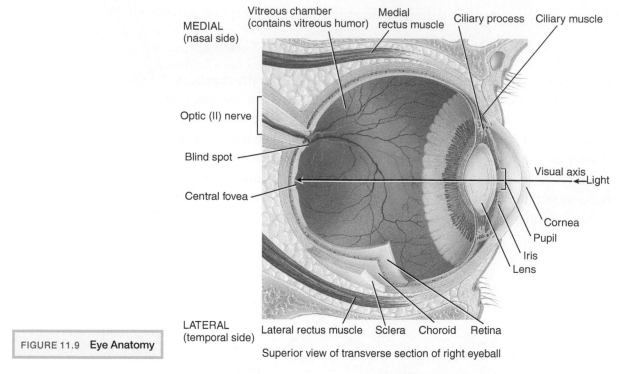

MEDIAL (nasal side)

Vitreous chamber (contains vitreous humor) Medial rectus muscle Ciliary process Ciliary muscle

Optic (II) nerve

Blind spot

Central fovea

Visual axis
Light

Cornea
Pupil
Iris
Lens

LATERAL (temporal side) Lateral rectus muscle Sclera Choroid Retina

FIGURE 11.9 **Eye Anatomy**

Superior view of transverse section of right eyeball

TABLE 11.5 **Eye Structures and Their Functions**

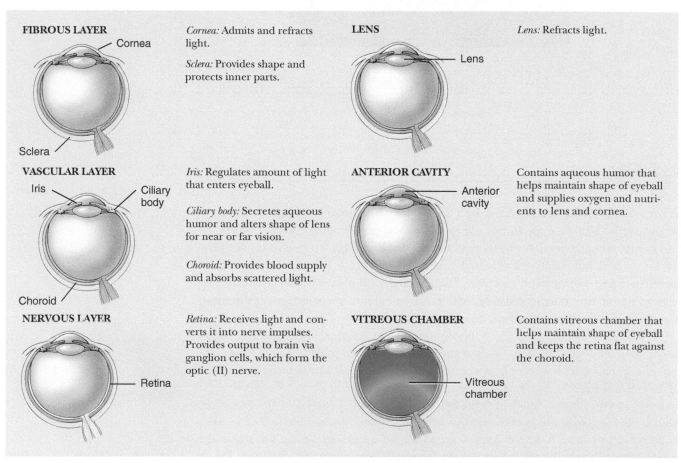

FIBROUS LAYER

Cornea

Sclera

Cornea: Admits and refracts light.

Sclera: Provides shape and protects inner parts.

LENS

Lens

Lens: Refracts light.

VASCULAR LAYER

Iris

Ciliary body

Choroid

Iris: Regulates amount of light that enters eyeball.

Ciliary body: Secretes aqueous humor and alters shape of lens for near or far vision.

Choroid: Provides blood supply and absorbs scattered light.

ANTERIOR CAVITY

Anterior cavity

Contains aqueous humor that helps maintain shape of eyeball and supplies oxygen and nutrients to lens and cornea.

NERVOUS LAYER

Retina

Retina: Receives light and converts it into nerve impulses. Provides output to brain via ganglion cells, which form the optic (II) nerve.

VITREOUS CHAMBER

Vitreous chamber

Contains vitreous chamber that helps maintain shape of eyeball and keeps the retina flat against the choroid.

Two special types of photoreceptors are found within the retina: rods and cones. Rods are responsible for night vision, since they are sensitive to low light intensities. However, rods cannot discriminate between different colors (wavelengths) of visible light. Cones are less sensitive to light, but they *can* distinguish colors. A blind spot is also present in each retina at the site where the optic nerve enters; this spot is commonly called the **optic disc**. The optic nerve sends information from photoreceptors to the primary visual cortex and visual association area of the brain. It is interesting to note that images are flipped and inverted due to light refraction; this phenomenon is illustrated in Figure 11.10. As a result, the brain must flip *and* invert visual information to accurately interpret it.

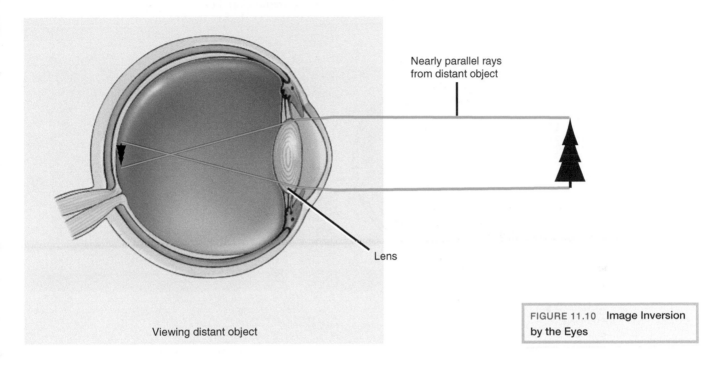

Nearly parallel rays
from distant object

Lens

Viewing distant object

FIGURE 11.10 **Image Inversion by the Eyes**

EXERCISE 11.6 Eye Anatomy

During this exercise, you will identify key structures on an anatomical model of the human eye. Alternatively, your instructor may ask you to identify these structures on a dissected cow eye or a virtual cadaver dissection. Label the structures listed below on Figure 11.11, and answer the review questions that are provided at the end of this exercise.

If you would like to watch a cow eye dissection, you can access the Exploratorium® museum online at **http://www.exploratorium.edu/learning_studio/cow_eye/**.

1. Cornea	7. Choroid
2. Sclera	8. Retina
3. Pupil	9. Optic disc
4. Iris	10. Anterior cavity
5. Ciliary muscles	11. Vitreous chamber
6. Lens	

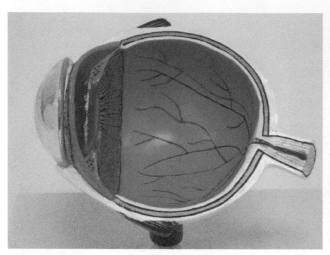
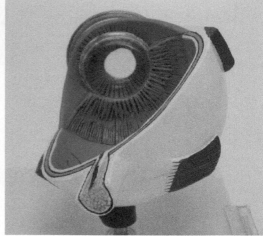

FIGURE 11.11 **Identifying Structures in the Eye**

Review Questions for Exercise 11.6

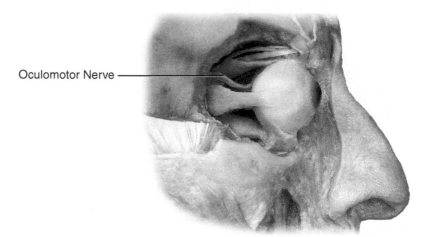

Oculomotor Nerve

Lateral View of the Eye

Label the sclera, iris, and pupil on the image above.

How is the size, or diameter, of the pupil adjusted? How does the size of the pupil change in response to varying light intensities?

_____

_____

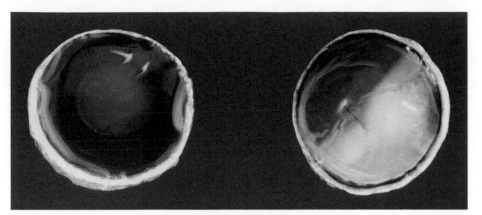

Dissected Cow Eye

Label the lens, the retina, and the vitreous humor (vitreous jelly) on the image above.

What is the function of the lens? _____

Where are photoreceptors located within the eye? _____

During eye exams, an **ophthalmoscope** is routinely used to examine the interior of the eyes. (The prefix *ophthalmo-* originates from the Greek term for *eye*.) Traditionally, an ophthalmoscope contains a light source, a tilted mirror, and glass lenses for magnification. While looking through the viewing hole, the examiner sees light rays reflected from the patient's inner eye. This allows the retina to be viewed by the examiner, as well as the choroid, macula, and other internal structures. While performing ophthalmoscopy, health professionals look for signs of injury or disease. Figure 11.12 shows the appearance of a normal retina, as well as signs of retinal detachment and macular degeneration.

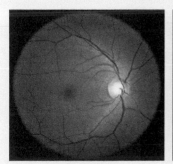 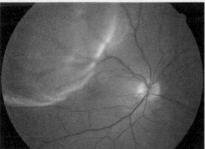 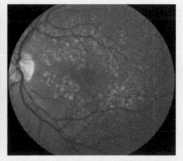

Normal Retina   Retinal Detachment   Age-Related Macular Degeneration

FIGURE 11.12 **Viewing the Retina with Ophthalmoscopy**

During this exercise, you will work in pairs and examine the retina with an ophthalmoscope. The parts of a typical ophthalmoscope are labeled on Figure 11.13; be sure to familiarize yourself with this piece of equipment before performing ophthalmoscopy. Alternatively, your instructor may elect to perform video ophthalmoscopy or virtual ophthalmoscopy during class. You can access a virtual ophthalmoscopy tool called *Diagnostics 101*, which is provided for students by Welch Allyn, at **http://www.welchallyn.com/wafor/students**.

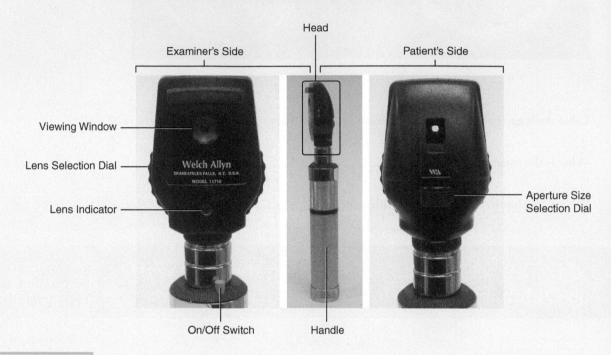

Head

Examiner's Side

Patient's Side

Viewing Window

Lens Selection Dial

Welch Allyn
SKANEATELES FALLS, N.Y. U.S.A.
MODEL 11710

Lens Indicator

Aperture Size
Selection Dial

On/Off Switch

Handle

FIGURE 11.13 **Parts of a Typical Ophthalmoscope**

●●● **SAFETY NOTE**

Before performing ophthalmoscopy, be sure to familiarize yourself with the parts and proper usage of an ophthalmoscope.

Since this exercise is not being performed by a licensed health professional, the results are not intended to diagnose health disorders.

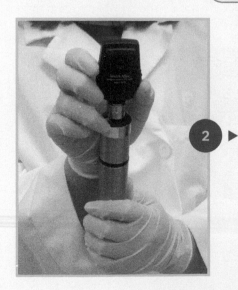

Procedure

1. Darken the room prior to performing ophthalmoscopy.

2. Turn on the ophthalmoscope and adjust the lens selection dial to zero diopters. Shine the light beam on your hand and adjust the light intensity, if necessary.

3. Hold the ophthalmoscope approximately one foot away from your partner's eye; the light beam should shine directly into your partner's pupil. Note the color of your partner's pupil in Table 11.6.

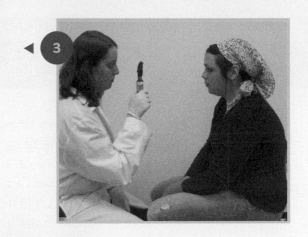

4. Move the ophthalmoscope toward your partner's eye until it is approximately one inch away from the eye. **Adjust the intensity of the light beam if it is uncomfortable for your partner.**

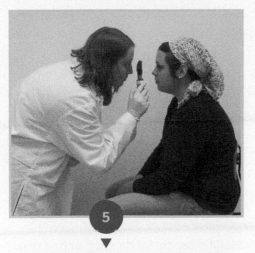

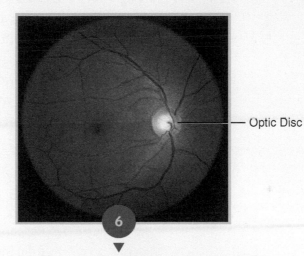

Optic Disc

5. Observe the blood vessels inside of the retina. Turn the lens selection dial, as needed, to bring the image into sharper focus.

6. Follow the blood vessels toward their point of convergence: the optic disc. Note the color and general appearance of the optic disc in Table 11.6.

7. Switch roles with your partner and repeat this procedure.

TABLE 11.6   Observations from Ophthalmoscopy

Eye Structure	Observation Notes (Color, General Appearance)	Sketch of this Eye Structure
Pupil		
Retina and Optic Disc		

When working properly, the eyes allow us to focus on objects that vary in distance from the body: street signs, words in novels, the moon, and so on. Sometimes, these objects are moving toward or away from your body as you view them. Perhaps your body is moving at the same time! To bring images into sharp focus, the lenses in the eyes must be flexible enough to change shape rapidly. The term **accommodation** refers to the ability of the lenses to flatten or bulge, as necessary, and bring light rays into sharp focus on the retina. Visual problems, such as presbyopia, may result if the lenses are unable to accommodate properly. Since the lenses get thicker and stiffer with age, their flexibility tends to diminish with age as well.

During this exercise, you will work in pairs and examine how well the eyes can focus on close objects. If you wear reading glasses, you can perform this procedure twice: once with your glasses and once without your glasses.

Procedure

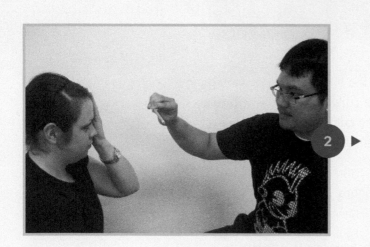

1. Researcher: Collect a pencil and a metric ruler from the materials bench. Sit down in a chair that is facing the subject's chair.

2. Researcher: Hold the pencil approximately one foot away from the subject's eyes. (Make sure that the words on the side of the pencil face the subject.) Adjust the height of the pencil, as necessary, so that it is parallel with the subject's eyes.

3. Subject: Cover one eye and focus on the pencil with your uncovered eye. Tell your partner when the words on the pencil start looking blurry.

4. Researcher: Slowly move the pencil toward the subject's eyes. Stop moving the pencil as soon the words appear blurry to the subject.

5. Researcher: Pick up the metric ruler with your free hand. Using the ruler, measure the distance between the pencil and the subject's eyes. Record this distance in Table 11.7.

6. Repeat the near point accommodation test on the subject's other eye.

7. Switch roles with your partner and repeat this procedure. Record the results in Table 11.7.

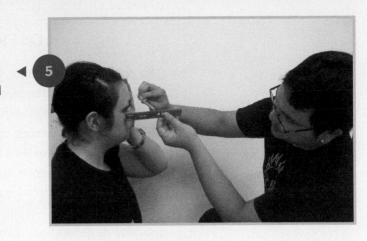

TABLE 11.7   Results from Near Point Accommodation Test

Eye Tested	Distance (mm)
My Right Eye without Reading Glasses	
My Left Eye without Reading Glasses	
My Right Eye with Reading Glasses (if applicable)	
My Left Eye with Reading Glasses (if applicable)	
Partner's Right Eye without Reading Glasses	
Partner's Left Eye without Reading Glasses	
Partner's Right Eye with Reading Glasses (if applicable)	
Partner's Left Eye with Reading Glasses (if applicable)	

Review Questions for Step by Step 11.8

How do your distances for near-point accommodation compare with those of other students in the class? Based on the class results, does age seem to play a role in near-point accommodation? Explain your answer.

_____

_____

Imagine you perform the near-point accommodation test on a 90-year old man who is not wearing reading glasses. Do you think his test results would be longer or shorter than yours? Explain your answer.

_____

_____

Imagine you perform the near-point accommodation test on the same 90-year-old man; this time, he is wearing reading glasses. How would reading glasses affect his test results?

_____

_____

A CLOSER LOOK AT EYE DISORDERS

The National Eye Institute (NEI) was established in 1968 as part of the National Institutes of Health. The mission of NEI is to "conduct and support research, training, health information dissemination, and other programs with respect to blinding eye diseases, visual disorders, mechanisms of visual function, preservation of sight, and the special health problems and requirements of the blind." If you would like to learn more about the diagnosis, treatment, and prevention of a specific eye disease — or a specific visual disorder—you can access the NEI website at **http://www.nei.nih.gov**.

You can also view simulations of the visual impairments caused by glaucoma, macular degeneration, diabetic retinopathy, cataracts, and other eye diseases at **http://www.nei.nih.gov**.

REVIEW QUESTIONS FOR LAB 11

1. Match each sensory receptor on the left side with the appropriate sense(s) on the right side.

   ___ Photoreceptor   a. Pain

   ___ Nociceptors   b. Vision

   ___ Chemoreceptor   c. Touch, Hearing, and Balance

   ___ Mechanoreceptor   d. Taste and Smell

2. Label the tympanic membrane and the malleus on the image below. What happens to the tympanic membrane when sound waves enter the ear?

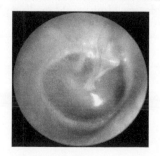

3. Which structures in the ears contain sensory receptors for hearing? How is sound information transmitted to the brain for interpretation?

_____

_____

4. Label the epidermis and dermis on the skin model below. Are blood vessels—and the majority of sensory receptors—located in the dermis or the epidermis? _____

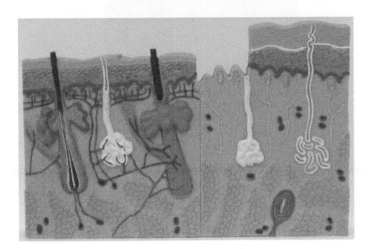

5. Why does skin vary in its sensitivity to touch?

_____

_____

6. Use the accompanying diagram to answer questions 6a–c.
   a. Which structures (A–H) are called the auditory ossicles? Are they located in the outer, middle, or inner ear?_____
   b. Where (A–H) are receptors located for balance? _____
   c. What is the name of structure H? _____

□ A
□ B
□ C

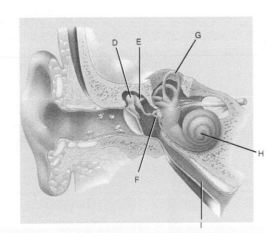

7. Photoreceptors are located within the _____ of the eye.
   a. choroid
   b. retina
   c. sclera
   d. lens

8. Label the optic disc on the image below. What structure is found in this part of the retina? _____

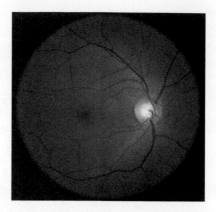

9. What is sound localization? Without visual clues, how can the ears localize sound waves?

   _____

   _____

   _____

10. Use the accompanying diagram to answer questions 10a and 10b.
    a. The iris is labeled _____, and the optic disc is labeled _____.
    b. What is the function of structure C? _____

    _____

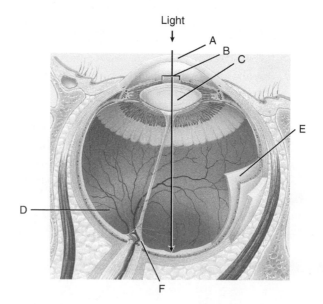

LAB 12:
The Skeletal System

By the End of This Lab, You Should Be Able to Answer the Following Questions:

- Where are compact bone, spongy bone, and articular cartilage found in the skeleton? How do these tissues differ in structure, function, and microscopic appearance?

- What role do protein fibers play in bone tissue? How is bone strength affected by a collagen deficiency?

- How does bone mineral density change during development and the aging process? How do these changes affect the resilience and strength of bone?

- Are sutures considered synovial (freely movable) joints, amphiarthrotic (slightly movable) joints, or synarthrotic (immovable) joints?

- How are cranial bones, limb bones, and vertebrae classified based on their shapes?

- Which bones belong to the pectoral girdles? Which bones belong to the pelvic girdle?

- How do simple, greenstick, and comminuted fractures differ in appearance on X-rays?

INTRODUCTION

Imagine you are a forensic anthropologist who specializes in identifying human remains. Today, you are visiting a sparsely populated town with a group of law enforcement officers; a routine construction project may have uncovered human remains in an otherwise barren field. When you arrive at the scene and inspect the remains, it is obvious that the soft body tissues decomposed months ago. You must rely on your training in **osteology**—the study of bones—to uncover clues in this investigation. Within a matter of minutes, you verify that the skeleton did indeed belong to a human. These bones undoubtedly originated from a male, given the proportions of the skull and pelvis. Additional measurements suggest this man was roughly 5'10" tall with Caucasian descent. A likely cause of death is jotted down by law enforcement officers when you discover evidence of blunt trauma to the back of the skull. By analyzing the sutures on the skull, along with the growth plates on several bones, you determine that this man was 35–45 years old at the time of death. Living cells once inhabited these bones, so DNA can also be harvested for laboratory analysis. Dental records will be used in conjunction with DNA evidence to confirm the identity of this man.

At first glance, the skeleton may look like one of the simplest organ systems in the body. We know that bones support the weight of the body so humans can stand erect and that delicate organs are protected by bony structures, such as the cranium. Bones are attached to one another by **ligaments**, and **articulations** (joints) provide flexibility for movement. What else is there to know, besides the name of every bone in the body? Appearances can be deceiving, and the skeletal system is no exception! For instance, hearing also relies on three tiny bones: the malleus, incus, and stapes. Figure 12.1 shows the size of these auditory **ossicles** (small bones) in relation to the size of a dime. When

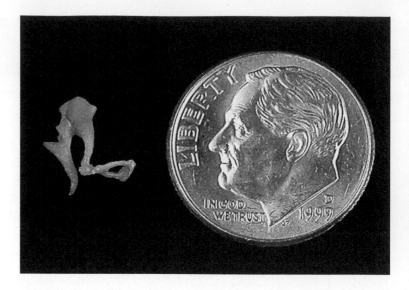

FIGURE 12.1   **The Auditory Ossicles**

the inner ear is stimulated by vibrating ossicles, sensory information races through an adjoining nerve to reach the brain for interpretation.

The bones in your body also store certain minerals, such as calcium and phosphorus. These minerals play a structural role in bones, but they also participate in other vital processes. Calcium ions, for instance, are involved in muscle contractions and nerve impulses. If your blood calcium levels dip too low, then bone rectifies the situation by releasing calcium into your bloodstream. Two special tissue types also live inside of bone cavities: yellow bone marrow and red bone marrow. Fat and fat-soluble vitamins are stored by yellow bone marrow, while red bone marrow constantly cranks out new blood cells. Last but not least, physical activities such as typing, jogging, and kissing would not be possible unless muscles interacted with bones. When a skeletal muscle contracts, it pulls on adjoining bones as it shortens in length. The result of this contraction depends on many factors, including the bone's range of motion at a particular joint.

TISSUE TYPES IN THE SKELETON

Bones are interlaced with cartilage throughout most of the human skeleton. Each individual bone is considered an organ, since it is constructed out of many different tissue types. As shown in Figure 12.2, the tough exterior of a bone contains **compact bone** tissue. When it comes to strength and rigidity, compact bone is only surpassed by one tissue type in the body: tooth enamel. Due to its honeycomb-like structure, **spongy bone** imparts a certain degree of lightness and flexibility to the skeleton. As shown in Figure 12.2, spongy bone fills the epiphyses (ends) of long bones; it is also sandwiched between compact bone in other parts of the skeleton, such as the flat bones in the skull. If bone marrow is present, it fills the hollow spaces in spongy bone and the medullary (central) cavities of long bones.

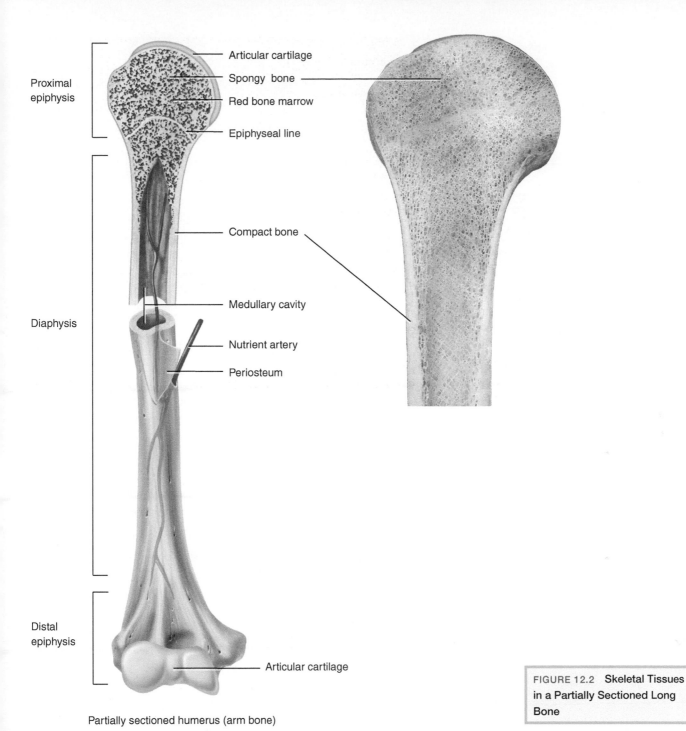

Proximal
epiphysis

Articular cartilage

Spongy bone

Red bone marrow

Epiphyseal line

Diaphysis

Compact bone

Medullary cavity

Nutrient artery

Periosteum

Distal
epiphysis

Articular cartilage

Partially sectioned humerus (arm bone)

FIGURE 12.2 **Skeletal Tissues in a Partially Sectioned Long Bone**

Compact bone and spongy bone are both examples of connective tissue, since they contain **osteocytes** (bone cells) lodged in a matrix of protein fibers and mineral salts. Amazingly, osteocytes produce the tough matrix material that ultimately surrounds them. Elastic cartilage, hyaline cartilage, and fibrocartilage are also examples of connective tissue. As with osteocytes, **chondrocytes** (cartilage cells) produce the matrix material that gives each type of cartilage its characteristic structure. Figure 12.3 shows the locations of hyaline cartilage and fibrocartilage throughout the skeleton.

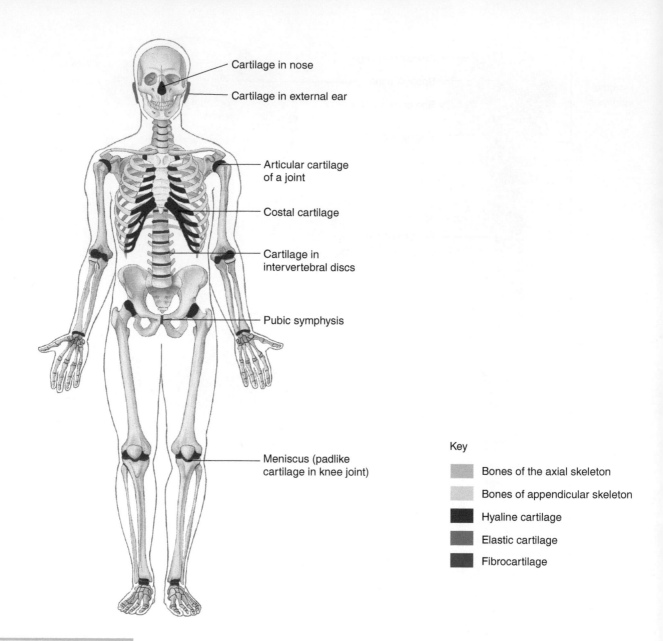

Cartilage in nose

Cartilage in external ear

Articular cartilage of a joint

Costal cartilage

Cartilage in intervertebral discs

Pubic symphysis

Meniscus (padlike cartilage in knee joint)

Key

Bones of the axial skeleton

Bones of appendicular skeleton

Hyaline cartilage

Elastic cartilage

Fibrocartilage

FIGURE 12.3 **Locations of Hyaline Cartilage and Fibrocartilage in the Skeleton**

EXERCISE 12.1   Histology of Skeletal Tissues

To understand how the skeleton functions, it is helpful to view its structure at the microscopic level. During this exercise, you will examine prepared microscope slides of compact bone, spongy bone, hyaline cartilage, and fibro—cartilage. Compare your focused microscope slides to the corresponding photomicrographs and answer the review questions that accompany each slide.

COMPACT BONE

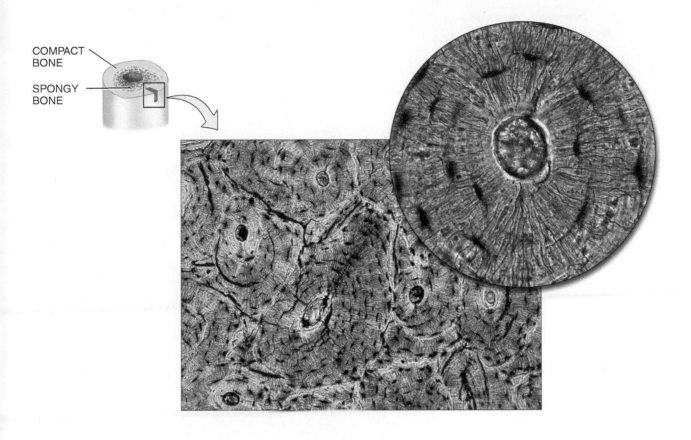

COMPACT BONE

SPONGY BONE

Compact bone is composed of long, cylindrical structures called osteons. The concentric rings (lamellae) in each osteon resemble the rings inside of a tree trunk. Label an osteon on the photomicrograph above.

Blood vessels and nerves run through the central canal of each osteon. Label a central canal on the photomicrograph above.

Calcium phosphate crystals and protein fibers are abundant in compact bone's rigid matrix. Ironically, this matrix is inhospitable to the osteocytes that produce it. As a result, bone cells are sequestered inside of tiny chambers called lacunae. (The word *lacuna* means *gap* or *pit* in Latin.) Each lacuna looks like a dark pit embedded in the matrix. Label a lacuna on the photomicrograph above.

Osteocytes are connected to the central canal—and each other—by tiny canals called canaliculi. Canaliculi resemble thin, dark lines that radiate away from the central canal. Label a canaliculus on the photomicrograph above. What materials must travel through these canals to keep bone cells alive?

_____

_____

SPONGY (CANCELLOUS) BONE TISSUE

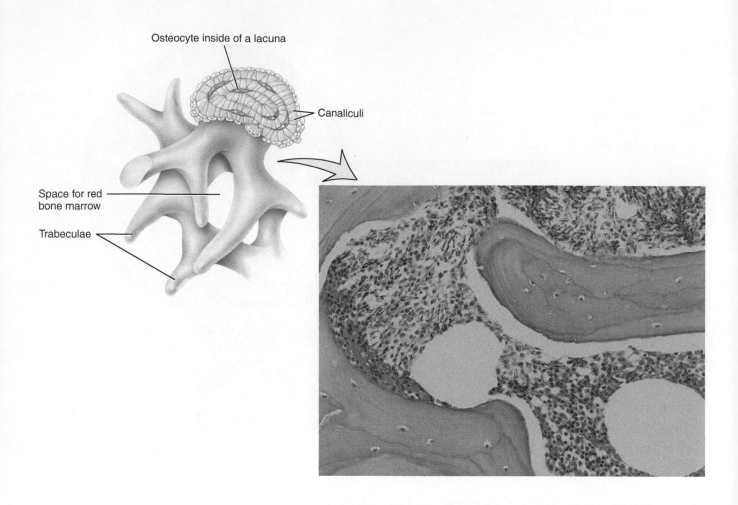

Osteocyte inside of a lacuna

Canaliculi

Space for red bone marrow

Trabeculae

Spongy bone contains small, open spaces and intertwining beams called trabeculae. (The word *trabecula* means *little beam* in Latin.) These beams are studded with osteocytes that live inside of lacunae. Label a trabecula, a lacuna, and an osteocyte on the photomicrograph above.

Bone marrow may fill the open gaps in spongy bone. Label the bone marrow on the photomicrograph above.

How does the presence of spongy bone affect the weight of the skeleton? Why is this important?

_____

_____

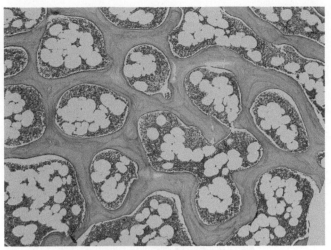

Normal Spongy Bone

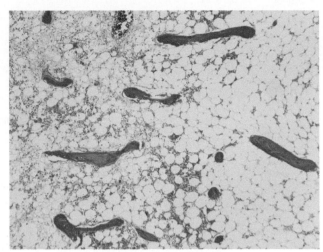

Osteoporotic Spongy Bone

Did you know that the earliest stages of osteoporosis affect spongy bone tissue? Compare the histology of normal and osteoporotic spongy bone. How is the structure of spongy bone altered by osteoporosis?

_____

_____

HYALINE CARTILAGE

Hyaline cartilage was named for its beautiful, glassy appearance under the microscope. (The term *hualos* actually means *glass* in Greek.) Label a lacuna on the image to the right, as well as the chondrocyte (cartilage cell) within it.

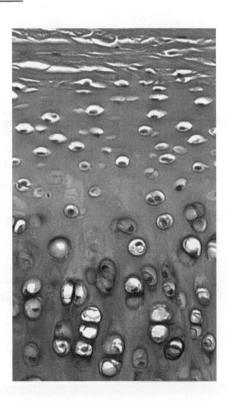

The matrix of hyaline cartilage contains protein fibers that are relatively large in size. As you scan the microscope slide, do you see any protein fibers in the matrix? If so, how do their sizes relate to that of a chondrocyte?

_____

_____

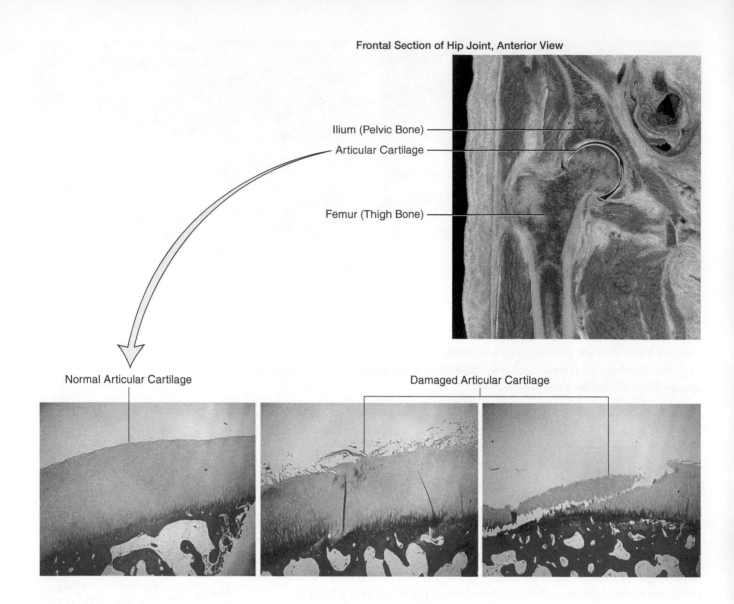

Frontal Section of Hip Joint, Anterior View

Ilium (Pelvic Bone)

Articular Cartilage

Femur (Thigh Bone)

Normal Articular Cartilage

Damaged Articular Cartilage

Hyaline cartilage is found on the articular (joint-forming) surfaces of many bones, including those in the hip joints. Osteoarthritis is caused by the deterioration of articular cartilage. Compare the histology of normal and damaged articular cartilage. How do these tissues differ in structure from one another? Why would the deterioration of articular cartilage lead to inflammation (-*itis*) and joint (*arthr-*) pain?

_____

_____

Imagine that the articular cartilage is completely destroyed within a hip joint. What will happen to the bones in this joint, since they are now grinding against each other?

_____

_____

Unlike most tissues, normal adult cartilage does not contain blood vessels or nerves. Since nerves are absent, cartilage can absorb shock and pressure without pain. Name one disadvantage that cartilage faces without its own supply of blood vessels and nerves.

_____

_____

FIBROCARTILAGE

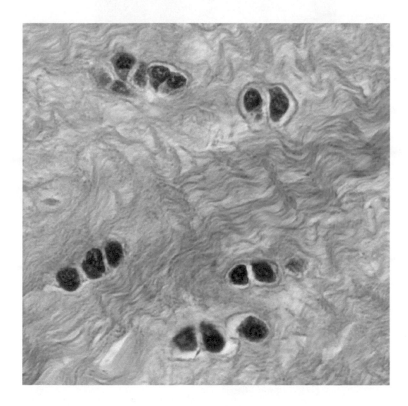

Pads of fibrocartilage are found within joints that withstand great pressure, such as the knee joints. As suggested by its name, fibrocartilage contains a high density of protein fibers (*fibro-*) in its matrix. Label several protein fibers on the photomicrograph above.

Label a lacuna on the photomicrograph above, as well as the chondrocyte within it.

Besides the knee joints, where is fibrocartilage found in the skeleton?_____

Knee Joint, Posterior View

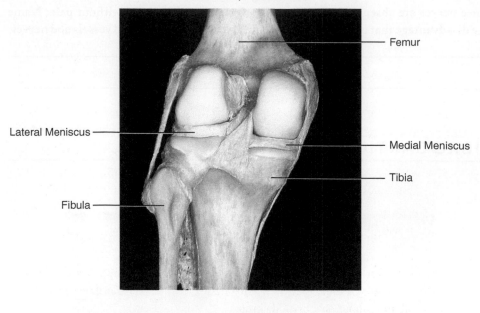

Femur

Lateral Meniscus

Medial Meniscus

Tibia

Fibula

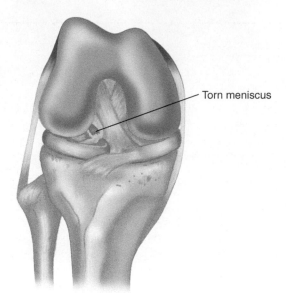

Torn meniscus

Within each knee joint, the lateral meniscus and medial meniscus are composed of fibrocartilage. Compare the microscopic structure of fibrocartilage to that of hyaline cartilage. Why is fibrocartilage well suited to absorb shock in weight-bearing joints, such as the knees?

_____

_____

Compare the location of menisci, bones, and ligaments in the knee joint shown above. Why do meniscal tears commonly lead to joint instability and a decreased range of motion?

_____

_____

We know that protein fibers, such as collagen, and calcium phosphate crystals are abundant in bones. What happens to bones when one of these substances is deficient or absent? Do you think one component—protein fibers or mineral salts—is more important for strength than the other? During this exercise, you will expose chicken bones to several chemical treatments. Some treatments (example: bleach) will remove protein fibers, while others (example: acids) will remove mineral salts. Distilled water will be used as a negative control for comparison.

Create a hypothesis to test regarding the effects of bleach on the appearance, texture, and strength of chicken bones. Bleach is expected to remove protein fibers from bone, but leave the calcium phosphate crystals intact.

_____

_____

Create a hypothesis to test regarding the effects of acids on the appearance, texture, and strength of chicken bones. Acids are expected to remove calcium phosphate crystals from bone, but leave the protein fibers intact.

_____

_____

Note: Your instructor may elect to complete this experiment in one period. If the chicken bones have already been treated with chemicals, skip to the _Procedure for Collecting Results_.

●●● SAFETY NOTE ON BLEACH

Bleach contains corrosive chemicals that are potentially toxic to living tissues. Use extreme care when handling bleach to avoid irritation of the eyes, skin, and respiratory tract. Safety goggles, gloves, and proper ventilation should be used when handling open containers of bleach. If bleach comes in direct contact with the skin or eyes, flush the area with water for 15 minutes. Notify your lab instructor of the situation immediately so proper medical attention can be provided.

Procedure for Chemically Treating Bones

1. Collect a glass jar from the materials bench.

2. Label the jar with the liquid your group is testing: vinegar, bleach, lemon juice, or distilled water.

3. Put on safety goggles and a pair of disposable gloves.

4. Fill your jar approximately 3/4 full with the appropriate liquid.

5. Screw the lid back onto the jar.

6. Collect a chicken leg, a small pair of surgical scissors, and a dissection tray from the materials bench.

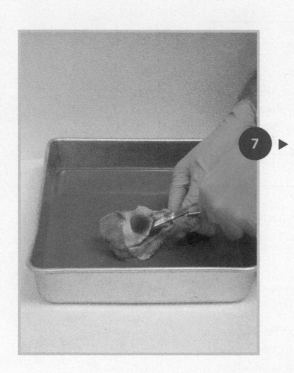

7. Use the scissors and your gloved hands to remove the meat from the chicken bone. **Use care to avoid cutting your hands during this step; the chicken legs may be slippery.**

8. Rinse the bone off with water and blot it dry with a paper towel.

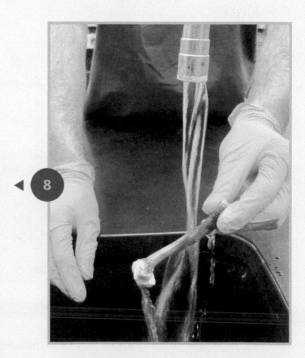

9. Use a scale to determine the initial weight of this bone. Record this information in Table 12.1.

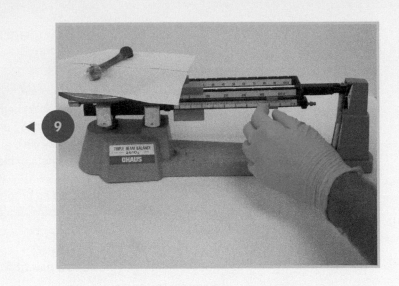

10. Assess the color, texture, and strength of the bone as you manipulate it with your gloved hands. Record your initial observations in Table 12.1.

11. Place the bone into the liquid-filled jar and screw on the lid securely.

12. Set your glass jar in the area designated by your instructor. Each bone will be chemically treated for one week.

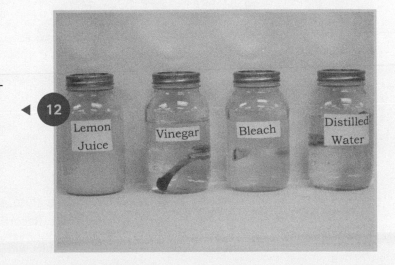

Procedure for Collecting Results

> ●●● **SAFETY NOTE ON BLEACH**
>
> Bleach contains corrosive chemicals that are potentially toxic to living tissues. Use extreme care when handling bleach to avoid irritation of the eyes, skin, and respiratory tract. Safety goggles, gloves, and proper ventilation should be used when handling open containers of bleach. If bleach comes in direct contact with the skin or eyes, flush the area with water for 15 minutes. Notify your lab instructor of the situation immediately so proper medical attention can be provided.

1. Put on safety goggles and a pair of disposable gloves.

2. Using tongs, carefully remove your treated bone from its jar.

3. Rinse the bone off with water and blot it dry with a paper towel.

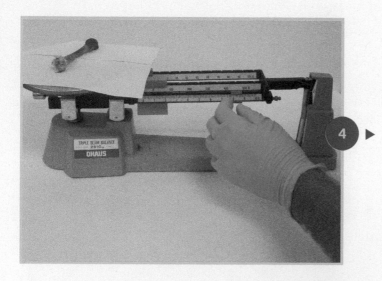

4. Use a scale to determine the final weight of the bone. Record this information in Table 12.1.

5. Assess the color, texture, and strength of the bone as you manipulate it with your gloved hands. Record your final observations in Table 12.1.

6. Compare your results with those obtained by other students in the class. Record the pooled class results in Table 12.1.

TABLE 12.1   Bone Strength before and after Chemical Treatment

Treatment	Initial Observations	Final Observations	Initial Weight of Bone (grams)	Final Weight of Bone (grams)
Distilled Water	Color: Texture: Strength:	Color: Texture: Strength:		
Vinegar	Color: Texture: Strength:	Color: Texture: Strength:		
Lemon Juice	Color: Texture: Strength:	Color: Texture: Strength:		
Bleach	Color: Texture: Strength:	Color: Texture: Strength:		

Review Questions for Step by Step 12.2

Bleach degrades organic molecules, such as protein fibers, in the extracellular matrix of bone. Based on your results, what role do protein fibers play in bone structure?

_____

_____

Inorganic salts, such as calcium phosphate, are dissolved away by certain acids. Based on your results, what role do calcium phosphate crystals play in bone structure?

_____

_____

Was your initial hypothesis supported or rejected by the results of this experiment? If necessary, revise your initial hypothesis in the space provided below.

_____

_____

Did the weight of your bone significantly change following its chemical treatment? How does this compare to the weight change observed by other treatment groups?

_____

_____

What are potential sources of error in this experiment? What improvements (if any) could be made to this experiment?

_____

_____

A CLOSER LOOK AT BONE COMPOSITION

As you witnessed during the previous experiment, minerals such as calcium and phosphorus play crucial roles in bone strength. Due to the prevalence of osteoporosis in our society, many of us already know that bone density is an important parameter of bone health. When bone mineral density (BMD) drops too low, as in osteoporosis, a person becomes more susceptible to developing bone fractures. Many people actively monitor their daily calcium intake for this very reason. But what about the density of protein fibers in our bones? Does this have an impact on bone health as well? Ask a person who is living with osteogenesis imperfecta: the answer is a resounding yes! Osteogenesis imperfecta (OI) is actually a group of genetic disorders that affects collagen production. OI either results in a collagen deficiency or the production of abnormal collagen fibers. Without the proper amount of functional collagen in their bones, people with OI are very susceptible to bone fractures. OI is commonly called "brittle bone disease" for this very reason. Since hearing also relies on bones—the auditory ossicles—hearing loss may also develop as a result of OI.

Just as a person's bone mineral density can change throughout life, so too can the density of collagen fibers in bone. We will return to this topic in Exercise 12.5 (Identifying Bone Fractures), since it relates to greenstick fractures.

GROSS ANATOMY OF THE HUMAN SKELETON

When the skeleton first develops in a human embryo, it is composed primarily of hyaline cartilage. Over time, however, **ossification** (bone formation) converts most of this hyaline cartilage into bone. Although infants are born with 300 to 350 bones, the adult skeleton only contains about 206 bones. How on earth could the number of bones in your skeleton change over time? Many bones actually fuse together during development, thereby decreasing the number that remains in your body. By adulthood, roughly half of your bones—more than 100 bones total—are found within your hands and feet. As a result, the majority of joints are also found within these body parts.

As shown in Figure 12.4, the human skeleton is composed of two divisions: the axial skeleton and the appendicular skeleton. Bones in the **axial skeleton** lie along the midline (axis) of the body; they construct the framework of the skull, ribcage, and vertebral column. The axial skeleton protects the organs that reside within it cavities: the brain, the heart, the lungs, and the spinal cord. The vertebral column also transfers pressure from the upper body to the pelvis, legs, and ultimately the feet. The bones of the **appendicular skeleton** reside in the upper limbs, the lower limbs, and **girdles** that connect these limbs to the axial skeleton. The **pectoral** (shoulder) girdles connect the arms to the axial skeleton, while the **pelvic** (hip) girdles connect the legs to the axial skeleton.

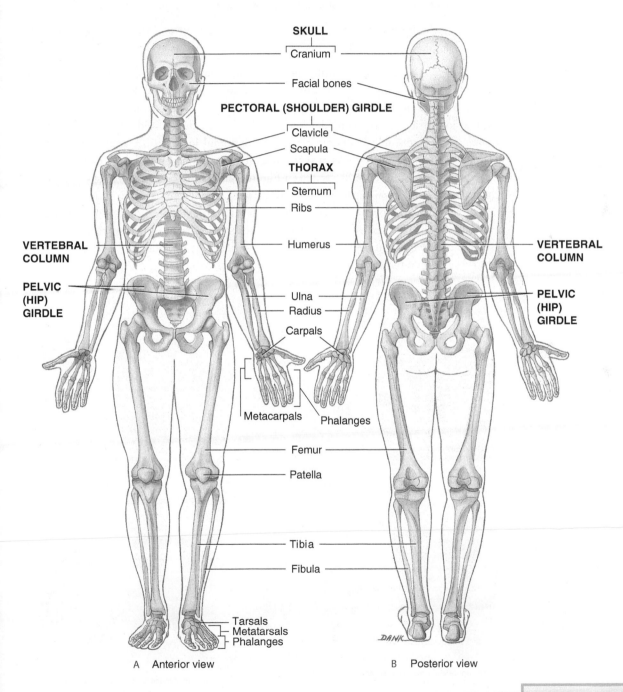

A   Anterior view   B   Posterior view

FIGURE 12.4   The Human Skeleton

The size and shape of each bone in the body—whether it is a skull bone, arm bone, or auditory ossicle—is directly related to its function. For example, the unique shape of the humerus (arm bone) allows it to articulate with bones in the shoulder and the forearm. The projections on the humerus also provide attachment sites for skeletal muscles. Each bone in the body can be placed into one of four categories based on shape:

- A **flat bone** is thin, flat, and usually curved. The bones that surround the top of the brain are examples of flat bones.

- A **long bone** is cylindrical in the middle. As the name suggests, its length also exceeds its width. The humerus is an example of a long bone.

- **Short bones**, such as wrist and ankle bones, are shaped liked cubes.

- **Irregular bones** fit into no other category due to their odd shapes; vertebrae are examples of irregular bones.

When a skeletal muscle contracts, it shortens in length and tugs on adjoining bones. The result of this contraction—turning your head, bending your knee, smiling, etc.—depends in part on the joint(s) involved in the process. If the bone belongs to a freely movable joint, such as the shoulder joint or knee joint, then a wide range of motion is possible. Freely movable joints are also called **synovial joints**, since they are lubricated with synovial fluid. Several types of synovial joints—including hinge joints, ball and socket joints, and pivot joints—are illustrated in Figure 12.5. If the bone of interest belongs to a semimovable joint, such as the joints between vertebrae, then a smaller range of motion is possible. Slightly movable joints are also called **amphiarthrotic** joints. In contrast, the sutures between facial bones are considered immovable joints, or **synarthrotic joints**. Since these bones cannot be moved by contracting muscles, they provide the leverage necessary to move soft facial tissues and create facial expressions.

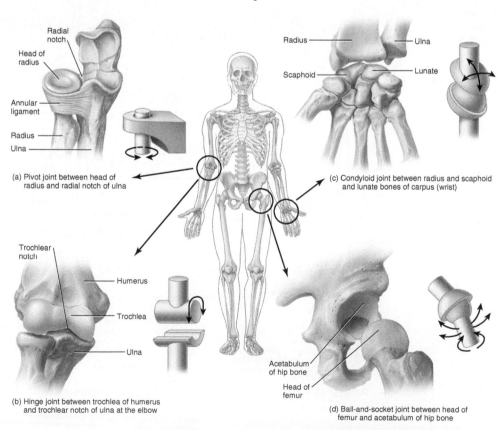

(a) Pivot joint between head of radius and radial notch of ulna

(b) Hinge joint between trochlea of humerus and trochlear notch of ulna at the elbow

(c) Condyloid joint between radius and scaphoid and lunate bones of carpus (wrist)

(d) Ball-and-socket joint between head of femur and acetabulum of hip bone

FIGURE 12.5 **Types of Synovial Joints**

EXERCISE 12.3   Gross Anatomy of the Axial Skeleton

During this exercise, you will identify selected bones in the skull, vertebral column, and ribcage. Depending on the nature of your course, these bones will either be identified on articulated skeletons, a virtual cadaver dissection, or in disarticulated (disjointed) form. Exercise 12.3 is split into the following subexercises:

- 12.3a Identifying Bones in the Skull
- 12.3b Identifying Vertebrae
- 12.3c Identifying Bones in the Ribcage

THE SKULL

By adulthood, the human skull typically contains a total of 22 bones. Some of these bones construct the framework of the **cranium**, which surrounds the brain, while others provide attachment sites for soft facial tissues. With one exception—the mandible bone—skull bones are connected to one another by sutures. As a result, the mandible is the only freely movable bone in the skull. Cavities also develop in the skull to accommodate sensory organs, such as the eyes and ears, not to mention air intake through the nasal cavities. In addition, the **foramina** (openings) in certain skull bones provide entryways for nerves, blood vessels, and the spinal cord. As shown in Figure 12.6, the names of certain cranial bones—frontal, temporal, parietal, and occipital—mirror the names of underlying brain lobes. Figure 12.6 also shows the hyoid bone, which is located under the tongue. The hyoid bone is the only bone in the human skeleton that doesn't articulate with any other bones.

FIGURE 12.6   **Skull Bones**

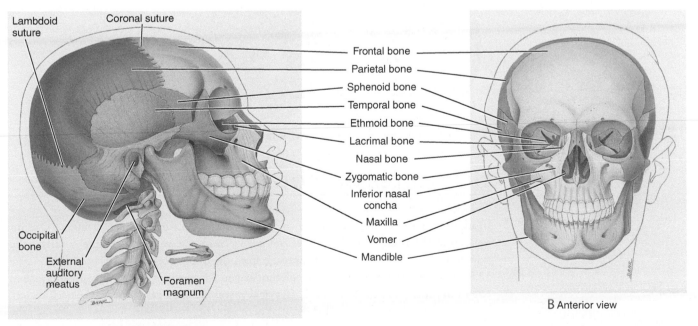

A Right lateral view

B Anterior view

Identify the bones listed below on a human skull, and label these bones on Figure 12.7.

1. Frontal bone
2. Parietal bones
3. Temporal bones
4. Occipital bone

5. Maxilla
6. Mandible
7. Zygomatic bones
8. Nasal bone

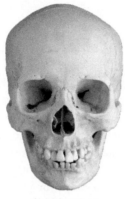
Anterior view

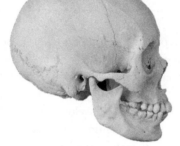
Right lateral view

Posterior view

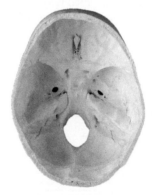
Inferior view

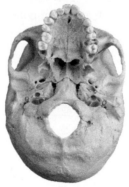
Superior view of floor of cranium

FIGURE 12.7 **Identifying Skull Bones**

THE VERTEBRAL COLUMN

In addition to protecting the delicate spinal cord, vertebrae also provide attachment sites for muscles in the neck, shoulders, back, and pelvis. Since the vertebral column connects the skull to the pelvis, it also transmits pressure from the upper body to the lower limbs. Figure 12.8 shows an articulated vertebral column; as you can see, vertebrae gradually increase in size from the neck to the lower back. **Cervical vertebrae** are located within the neck region. The first cervical vertebra is called the atlas because it holds the head, just like Atlas held the world in Greek mythology. (The term *cervical* refers to any narrow, neck-like structure in the body.) **Thoracic vertebrae** articulate with ribs that surround the thoracic cavity, and **lumbar vertebrae** are located in the lower back. The **sacrum** is composed of five vertebrae that fuse together during development, and it articulates with the coxal (hip) bones in the pelvis. Finally, the **coccyx** is composed of four to six coccygeal vertebrae that typically fuse together during development. The coccyx is commonly called the tailbone for good reason; it is a remnant of the postanal tail that each human possessed during embryonic development.

Figure 12.8 compares the structure of a typical cervical vertebra, thoracic vertebra, and lumbar vertebra. Although these vertebrae differ based on their sizes and shapes, they do share some anatomical features in common. The body of each vertebra bears the weight of the upper body, while the processes act as sights for muscle and ligament attachment. The vertebral foramen serves as a passageway for the spinal cord, while transverse foramina act as passageways for nerves and blood vessels. Fibrocartilage discs are wedged between adjacent vertebrae; these discs absorb shock and pressure while preventing vertebrae from grinding against each other.

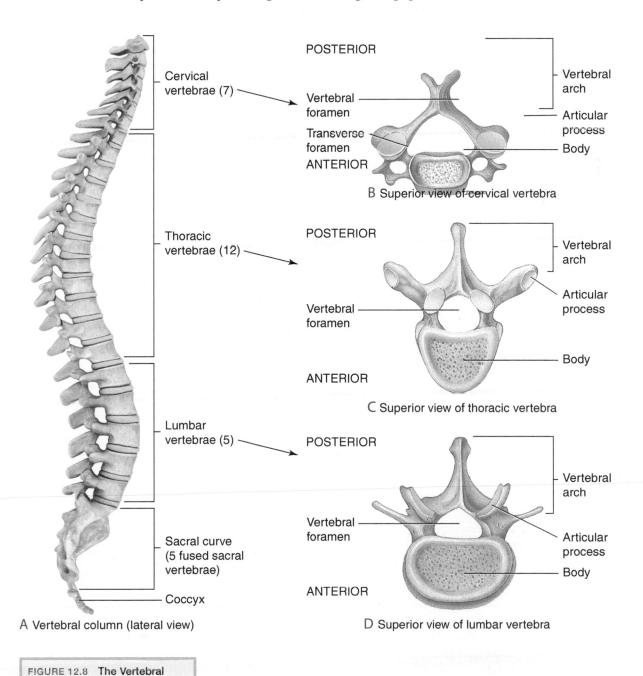

A Vertebral column (lateral view)

B Superior view of cervical vertebra

C Superior view of thoracic vertebra

D Superior view of lumbar vertebra

FIGURE 12.8   The Vertebral Column

Identify the bones (or bony structures) listed below on an articulated skeleton, and label them accordingly on Figure 12.9.

- Sacrum
- Coccyx
- Cervical Vertebrae
- Lumbar Vertebrae
- Thoracic Vertebrae

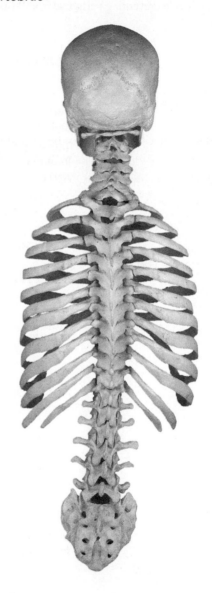

FIGURE 12.9 Identifying Vertebrae

THE RIBCAGE

In addition to protecting delicate internal organs, such as the heart and lungs, the ribcage also provides attachment sites for muscles in the neck, shoulders, back, abdomen, and **intercostal** spaces, which are located between (inter-) the ribs (-costal). As shown in Figure 12.10, the ribcage consists of thoracic vertebrae, the sternum (breastbone), costal cartilage, and

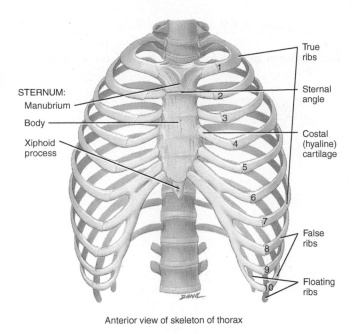

STERNUM:
Manubrium

Body

Xiphoid
process

True
ribs

Sternal
angle

Costal
(hyaline)
cartilage

False
ribs

Floating
ribs

Anterior view of skeleton of thorax

FIGURE 12.10   The Ribcage

the ribs themselves. **True ribs** are attached directly to the sternum by individual pieces of costal cartilage. A **false rib** may share costal cartilage with the rib above it, or it may act as a **floating rib** by not attaching to the sternum at all. People who are certified in CPR know they must avoid the sternum's xiphoid process while administering CPR. If the xiphoid process is broken off, it can damage underlying organs, such as the liver. The sternum is also of interest to surgeons, since it must be separated prior to an open chest surgery.

EXERCISE 12.3C   Identifying Bones in the Ribcage

Identify the bones (or structures) listed below on an articulated skeleton, and label them accordingly on Figure 12.11.

1. Ribs

   a. True Ribs

   b. False Ribs

   c. Floating Ribs

2. Costal Cartilage

3. Sternum

   a. Manubrium

   b. Body

   c. Xiphoid Process

4. Thoracic Vertebrae

FIGURE 12.11   Identifying Bones in the Ribcage

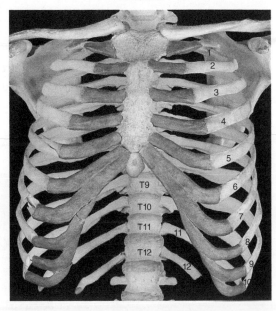

SUPERIOR

INFERIOR

Anterior view

The _____ bone (shown below) forms the bony framework of the forehead.

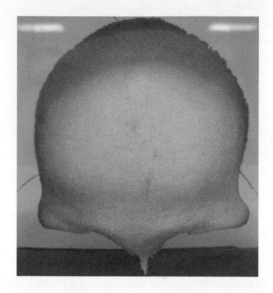

The _____ bone (shown below) is the only movable bone in the skull.

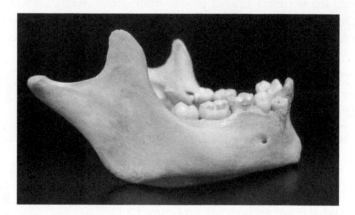

Label the vertebrae below as cervical, thoracic, or lumbar. What characteristics allow you to distinguish these vertebrae from one another?

_____

_____

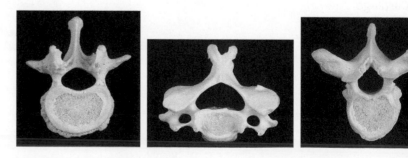

Are vertebrae considered flat, long, short, or irregular bones? _____

The _____ is labeled A on the sternum below.

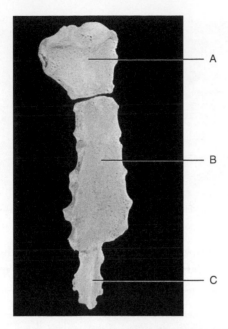

Compare the ribcage on a male and female skeleton. Does the total number of ribs differ between men and women? _____

A true rib is shown on the picture below. Which end (A or B) articulates with the sternum? How can you tell? _____

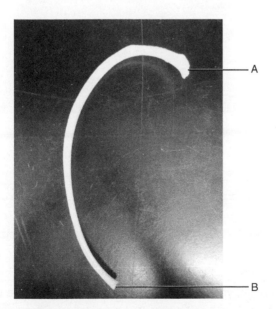

Are ribs considered flat, long, short, or irregular bones? Why? _____

The ear canals are located in the _____ bones, which are shown below.

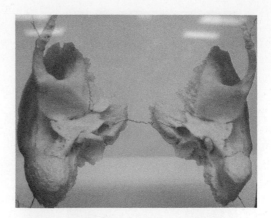

The _____ bone (shown below) is located in the floor of the cranium. What passes through the large foramen (aka the foramen magnum) in this bone?

_____

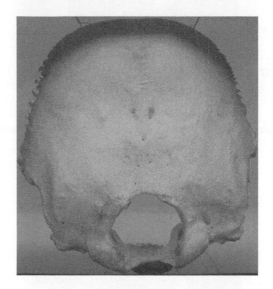

EXERCISE 12.4   Gross Anatomy of the Appendicular Skeleton

During this exercise, you will identify selected bones in the upper limbs, pectoral girdles, lower limbs, and pelvic girdle. Depending on the nature of your course, these bones will be either identified on articulated skeletons, a virtual cadaver dissection, or in disarticulated (disjointed) form. Exercise 12.4 is split into the following subexercises:

- 12.4a Identifying Bones in the Pectoral Girdles and Upper Limbs
- 12.4b Identifying Bones in the Pelvic Girdle and Lower Limbs

THE PECTORAL GIRDLES AND UPPER LIMBS

As shown in Figure 12.12, the pectoral (shoulder) girdles connect the arms to the axial skeleton. Amazingly, each pectoral girdle is only composed of two bones: the scapula (shoulder blade) and the clavicle (collar bone). The humerus is the only bone found within the anatomical arm; it articulates with the scapula at one end and with the forearm bones (radius and ulna) at the other end. The radius actually rotates in the forearm when you turn the palm of your hand back and forth. Regardless of palm orientation, though, the radius is always located on the same side as the thumb, and the ulna is always located on the same side as the pinky finger. Carpals, metacarpals, and phalanges are located in the wrist, hand, and fingers, respectively. Since a large number of bones are found in these parts of the body, a large number of joints are found here as well.

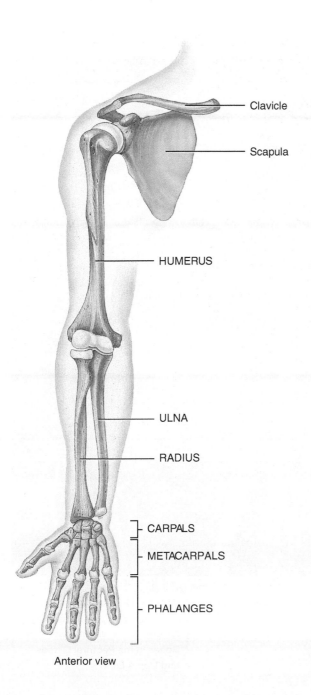

Clavicle

Scapula

HUMERUS

ULNA

RADIUS

CARPALS

METACARPALS

PHALANGES

Anterior view

FIGURE 12.12 **Bones in the Pectoral Girdle and Upper Limb**

Identify the bones listed below on an articulated skeleton, and label them accordingly on Figure 12.13.

1. Humerus

2. Carpals

3. Metacarpals

4. Phalanges

5. Scapula

6. Clavicle

7. Radius

8. Ulna

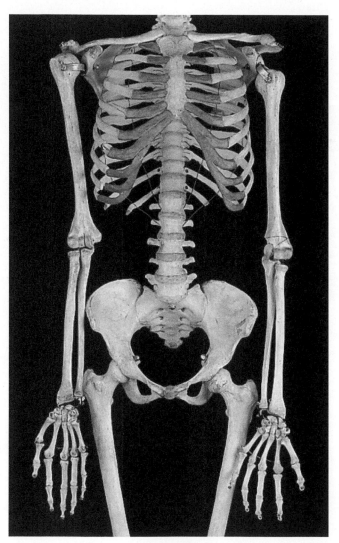

Anterior view

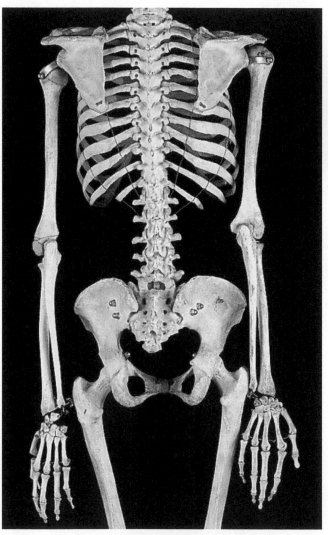

Posterior view

FIGURE 12.13 Identification of Bones in the Pectoral Girdles and Upper Limb

THE PELVIC GIRDLE AND LOWER LIMBS

As shown in Figure 12.14, the pelvic girdle connects each leg to the vertebral column. The bony pelvis is a basin-like structure that contains two coxal (hip) bones, the sacrum, and the coccyx. The coxal bones articulate with each other anteriorly (toward the front of the body) and with the sacrum posteriorly (toward the back of the body). The femur is the only bone found within each thigh, and it is by far the longest and heaviest bone in the body. Each patella (kneecap) is embedded inside of a tendon called the patellar tendon. Humans are not born with patellar bones, since they ossify later during development. Each lower leg contains two bones: the tibia (shin bone) and the fibula. The bumps you feel at the sides on your ankles are actually projections from these bones. Similar to the bones in the upper limbs, tarsals (versus carpals), metatarsals (versus metacarpals), and phalanges are found within the ankles, feet, and toes, respectively.

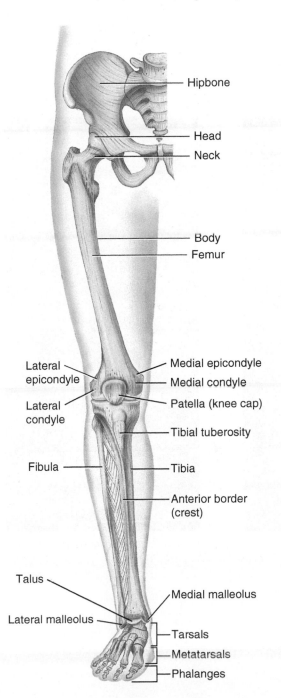

FIGURE 12.14 **Bones in the Pelvic Girdle and Lower Limb**

Identify the bones listed below on an articulated skeleton, and label them accordingly on Figure 12.15.

1. Femur

2. Bony Pelvis

    a. Coxal Bones

    b. Sacrum

    c. Coccyx (if present)

3. Patella

4. Tarsals

5. Metatarsals

6. Phalanges

7. Tibia

8. Fibula

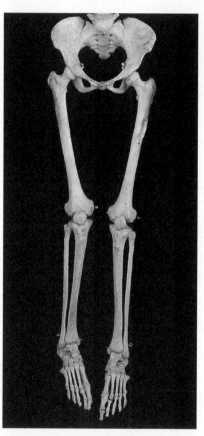

Anterior view

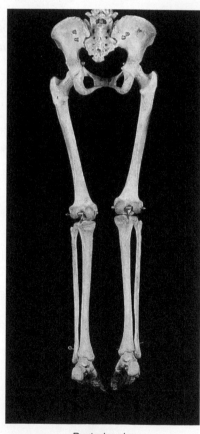

Posterior view

FIGURE 12.15 Identification of Bones in the Pelvic Girdle and Lower Limbs

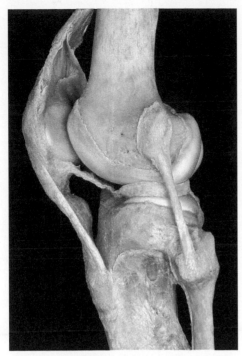

Knee Joint, Lateral View

Label the tibia, fibula, patella, and femur on the photograph above. Does this photograph show the right or left knee? How can you tell? (*Hint:* the terms *left* and *right* always refer to *patient's* side of the body.) _____

_____

Do all of your fingers contain the same number of phalanges? Explain. _____

_____

Label the glenoid cavity on the scapula below; the humerus articulates with this cavity. Did this scapula originate from the left or right side of the body? _____

Label the head of the humerus on the photograph below; this structure articulates with the glenoid cavity of the scapula.

Which bone is shown in the photograph below? One end articulates with the scapula, and the other end articulates with the sternal manubrium. _____

Label the radius and ulna below. Next, label the olecranon process on the ulna; this process articulates with a depression in the humerus called the olecranon fossa.

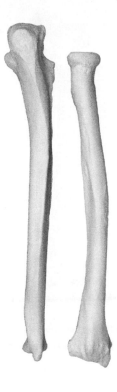

A male pelvis and a female pelvis are shown below. What differences do you notice between them? _____

_____

_____

Label an acetabulum (hip socket) on a pelvis below; the head of the femur articulates with this socket. Next, label the pubic symphysis on a pelvis below; this piece of fibrocartilage connects the coxal bones to each other.

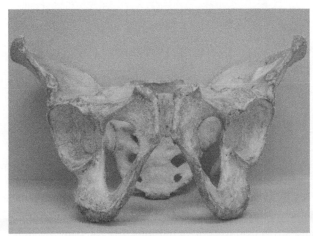

Male Pelvis

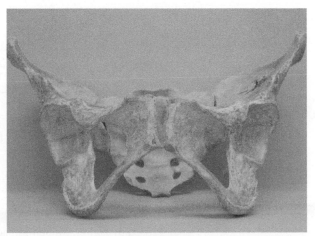

Female Pelvis

Which bone is shown in the photograph below? _____ **Label the head of this bone, which articulates with the acetabulum. Did this bone originated from the left or right side of the body?** _____

Label the tibia and fibula on the photograph below. Next, label the medial malleolus on the tibia; this projection creates a bulge in your inner ankle.

Are tarsals considered flat, short, long, or irregular bones? Why? _____

_____

BONE FRACTURES

Many people end up dealing with broken bones at some point in their lives; perhaps you already have personal experience on this topic. A bone is likely to break, or **fracture**, when it encounters more pressure than it can physically withstand. Bone fractures are commonly caused by falling, sudden impact, and stress from repetitive motions. Intense pain, swelling, bruising, and decreased mobility often accompany fresh bone fractures; in some cases, the site of injury will also look misshapen or expose a protruding bone. Table 12.2 summarizes the

TABLE 12.2   Classification of Bone Fractures

Type of Fracture	Description	Illustration
Open (Compound) Fracture	Fractured bone breaks skin and is exposed to the outside environment	Humerus / Radius / Ulna
Closed Fracture	Fractured bone does not break skin	Ulna / Radius / Wrist bones
Incomplete Fracture (Example: Greenstick Fracture)	Only one side of the bone is fractured, so the bone does not split into separate pieces. The nonfractured side of the bone typically bends, bulges, or buckles. Incomplete fractures are common in children because their bones are relatively flexible. This is due to the high density of collagen in immature bones.	Radius / Ulna / Wrist bones
Simple Fracture	Bone splits into two pieces	Humerus / Radius / Ulna
Comminuted (Multifragmentary) Fracture	Bone splits into three or more pieces	Humerus

characteristics of several common types of fractures: closed, open (compound), greenstick, simple, and comminuted (multifragmentary) fractures. X-rays are evaluated by radiologists to diagnose and assess the severity of bone fractures. In some cases, MRIs and CT scans are used in conjunction with X-rays. While assessing a bone fracture on an X-ray, radiologists address the following questions:

- What bone(s) is (are) fractured?

- Where specifically is the fracture line located on the bone? For instance, is the fracture located on the shaft, head, or neck of the bone?

- Did a fractured bone break the patient's skin? If so, the risk of infection must be addressed during treatment.

- Did a broken bone split into three or more pieces?

- What angle was produced by the fracture in relation to the bone? For instance, is the fracture parallel or perpendicular (transverse) to the axis of the bone?

In some cases, fractured bones can be **reduced** (realigned) without surgery. **Closed reduction** refers to bone realignment that is accomplished through external, manual manipulation. If the bone cannot be reduced in this manner, then surgery must be performed to realign the bone fragments. **Open reduction** refers to the surgical realignment of bone fragments; in some cases, the bone fragments are also reinforced with metal pins, screw, nails, or plates during surgery. Following realignment—if realignment was necessary—the fracture must be immobilized for several weeks or several months to promote healing and bone remodeling. A splint or a cast may be worn during this time to immobilize the joints surrounding the fracture. Once the risk of refracturing a bone has been reduced, normal motion and weight-bearing activities actually promote bone remodeling and help complete the healing process.

EXERCISE 12.5   Identifying Bone Fractures on X-Rays

During this exercise, you will evaluate X-rays of fractured bones from different regions of the body. Although several X-ray images are provided in this exercise, your instructor may provide additional radiographs to examine during lab. First, use your knowledge of the skeletal system to identify the bones on each X-ray. Next, use your knowledge of bone fractures to determine whether a simple, incomplete, or comminuted fracture is shown on the X-ray.

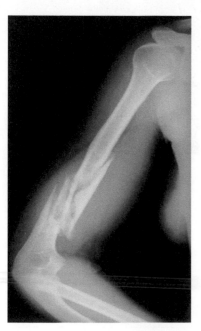

Label each bone on the X-ray to the left. Which bone is fractured? _____

Is this fracture incomplete, simple or comminuted? Explain your answer.

_____

_____

Do you think closed reduction or open reduction will be used to realign this bone? Explain your answer.

_____

_____

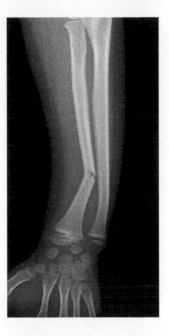

Label each bone on the X-ray above. Which bone is fractured?

_____

Is this fracture incomplete, simple, or comminuted? Explain your answer.

_____

Do you think this X-ray shows the bones of a child or the bones of an elderly individual?
Explain your answer._____

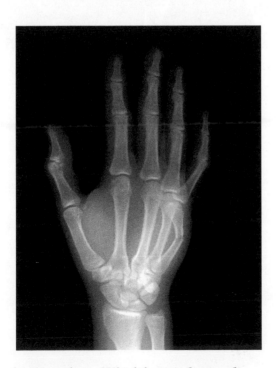

Label each bone on the X-ray above. Which bone is fractured?_____

Is this fracture incomplete, simple, or comminuted? Explain your answer.

_____

A CLOSER LOOK AT SKELETAL DISEASES

The National Institute of Arthritis and Musculoskeletal and Skin Diseases (NIAMS) is part of the National Institutes of Health. The mission of NIAMS is to "support research into the causes, treatment, and prevention of arthritis and musculoskeletal and skin diseases; the training of basic and clinical scientists to carry out this research; and the dissemination of information on research progress in these diseases." If you would like to learn more about the diagnosis, treatment, and prevention of a skeletal disease—such as osteoporosis, osteogenesis imperfecta, or rheumatoid arthritis—you can access NIAMS online at **http://www.niams.nih.gov**.

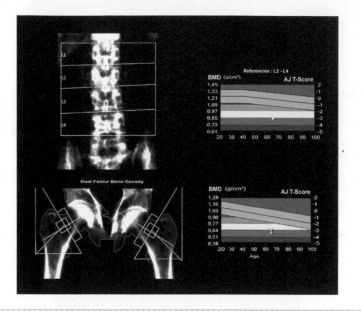

1. Label each tissue shown below as either compact bone, spongy bone, or hyaline cartilage.

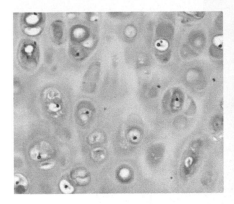 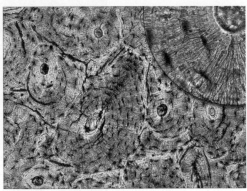 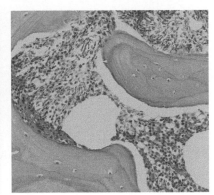

    a.   What similarities exist between these three tissue types?

       _____

       _____

    b.   Label one site where compact bone, spongy bone, and hyaline cartilage are found on the humerus below.

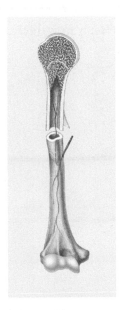

2. The _____ girdles connect the arms to the axial skeleton, while the _____ girdle connects the legs to the axial skeleton.
    a.   Which bones belong to the pelvic girdle? _____
    b.   Which bones belong to the pectoral girdles? _____

3. Is a simple, comminuted, or incomplete fracture shown on the X-ray below?
    _____

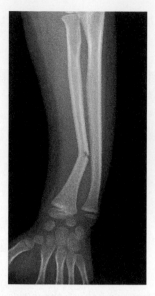

4. Match each joint on the left with the appropriate joint category on the right.
    ___ Shoulder Joint   a. Synarthrotic Joint
    ___ Cranial Suture   b. Amphiarthrotic Joint
    ___ Intervertebral Joint   c. Synovial Joint

5. Label the temporal bone, maxilla, and mandible on the photograph below.
    a. Which of these bones contains an ear canal? _____
    b. Which of these bones is freely movable? _____

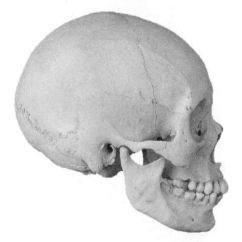

6. _____ vertebrae are found in the neck, while _____ vertebrae are found in the lower back.

7. How is bone mineral density (BMD) related to the development of osteoporosis?

    _____

    _____

8. Is a simple, comminuted, or incomplete fracture shown on the X-ray below?

   _____

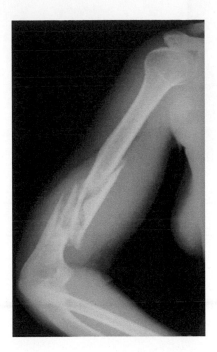

9. Label the bones shown below with their respective names and shape categories: flat, short, long, or irregular.

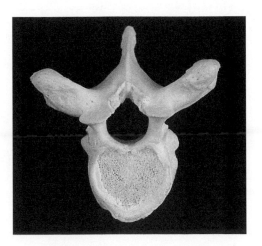

   _____

   _____

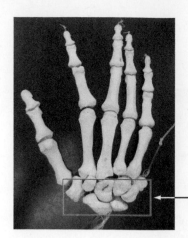

_____

_____

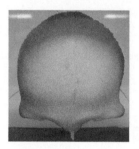

_____

_____

10. Label a tibia, humerus, clavicle, ulna, and cervical vertebra on the skeleton below.

a. Did this skeleton belong to a male or a female? How can you tell?

_____

b. Where are the carpals and tarsals located in this skeleton?

_____

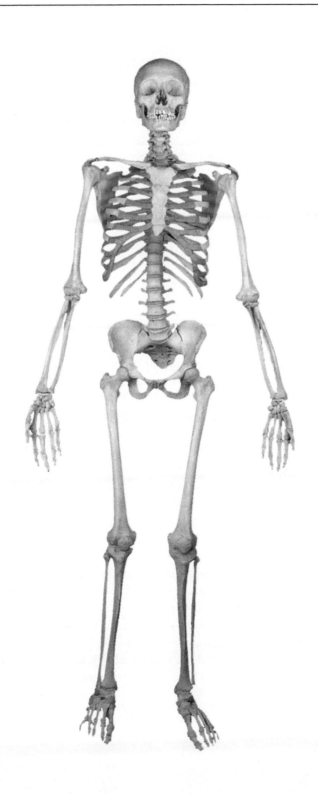

LAB 13:
The Muscular System

By the End of This Lab, You Should Be Able to Answer the Following Questions:

- What is the difference between a sarcomere, a muscle fiber, and a fascicle?

- When viewing a muscle biopsy, how can you tell whether the tissue looks healthy or necrotic? Which diseases are associated with skeletal muscle necrosis?

- Can glycerinated muscle contract outside of a living organism? If so, what chemicals are required for contraction?

- What is the difference between an agonist (prime mover), an antagonist, and a synergist?

- Which muscles contract when you smile? Blink? Raise your eyebrows?

- Where are the biceps brachii and biceps femoris located in the body? What attributes do these muscles have in common?

- What is the difference between a sprain and a strain? How can a muscle strain be diagnosed, treated, and prevented?

INTRODUCTION

Did you know that over six hundred muscles are attached to the bones in your body? Just the thought of memorizing every muscle's name is mind boggling to most people. (Luckily for medical students, most skeletal muscles come in pairs!) Surprisingly, this collection of muscles weighs a great deal more than the skeleton it covers and manipulates. Every movement you make during the day—the ones involved in typing, brushing your teeth, and perhaps even jogging—rely on the manipulation of bones by contracting skeletal muscles. When skeletal muscles shorten in length, they tug on adjoining bones and create the movements requested by the central nervous system. Body posture and facial expressions also rely on the triangle that unites the nervous system, muscular system, and skeletal system. Gestures like smiling, winking, and nodding—not to mention movement of the body itself—may be difficult or impossible to carry out when one component of this triangle is not functioning properly.

Although we tend to focus on movement when discussing skeletal muscles, their actions are certainly not confined to one specific task. Breathing would not be possible without the diaphragm muscle, which is attached to the bottom of the ribcage, as well as the intercostal muscles, which lie between (*inter-*) the ribs (*-costal*). As shown in Figure 13.1, the size of the thoracic cavity changes as these muscles contract and relax in a cyclical manner. Contracting muscles also emit heat energy, which gets picked up by the bloodstream and distributed throughout the body. Shivering actually exploits this process when the body is exposed to bitterly cold temperatures.

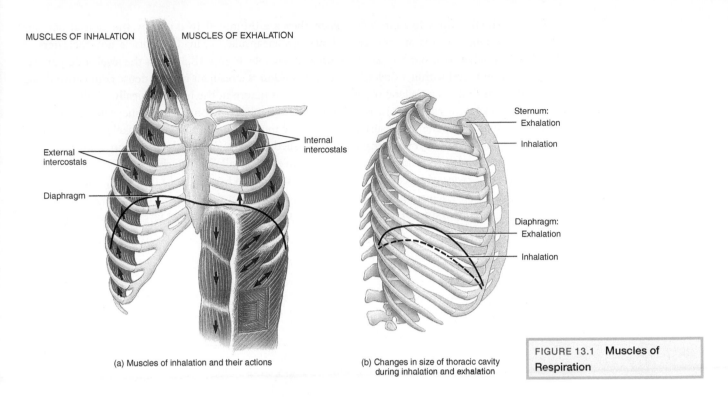

MUSCLES OF INHALATION   MUSCLES OF EXHALATION

External intercostals

Diaphragm

Internal intercostals

Sternum:
Exhalation
Inhalation

Diaphragm:
Exhalation
Inhalation

(a) Muscles of inhalation and their actions

(b) Changes in size of thoracic cavity during inhalation and exhalation

FIGURE 13.1 **Muscles of Respiration**

In addition to generating heat, certain muscles actually push blood through veins as they contract. This phenomenon is illustrated in Figure 13.2. When a person stands still for a long period of time, blood pools within the legs and cause a drop in blood pressure. If the brain receives inadequate blood flow due to hypotension (low blood pressure), then this person may start to feel dizzy and eventually wind up fainting.

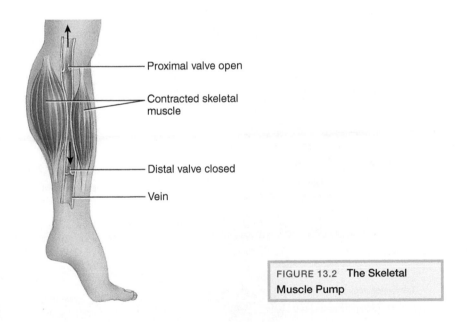

Proximal valve open

Contracted skeletal muscle

Distal valve closed

Vein

FIGURE 13.2 **The Skeletal Muscle Pump**

SKELETAL MUSCLE ORGANIZATION

The skeletal muscles in your body are considered organs, since they are constructed out of many different tissue types. Within each skeletal muscle, blood vessels supply **muscle fibers** (muscle cells) with the oxygen and nutrients required for contraction. Motor neurons stimulate muscle contraction when they are prompted to do so by the central nervous system. Sheaths of connective tissue also encapsulate individual muscle fibers, clusters of muscle fibers, and the skeletal muscle as a whole. Figure 13.3 shows the levels of organization found within a skeletal muscle. A **tendon** is a resilient rope of dense connective tissue that anchors a skeletal muscle to another structure in the body. Typically, tendons anchor skeletal muscles to bony projections. Another sheet of connective tissue called the **epimysium** covers the entire surface (*epi-*) of the muscle (*my-* or *myo-*). The skeletal muscle itself consists of long, cylindrical structures called fascicles. A **fascicle** is a group of muscle fibers surrounded (*peri-*) by a sheet of connective tissue called the **perimysium**. You can actually see fascicles while cutting into a cooked steak; each fascicle looks like a long, slender string of meat. Each muscle fiber is also surrounded by a thin layer of connective tissue called the **endomysium**. The prefix *endo-* tells us this is the innermost layer of connective tissue.

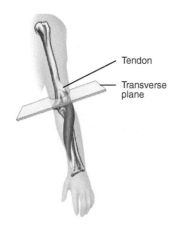

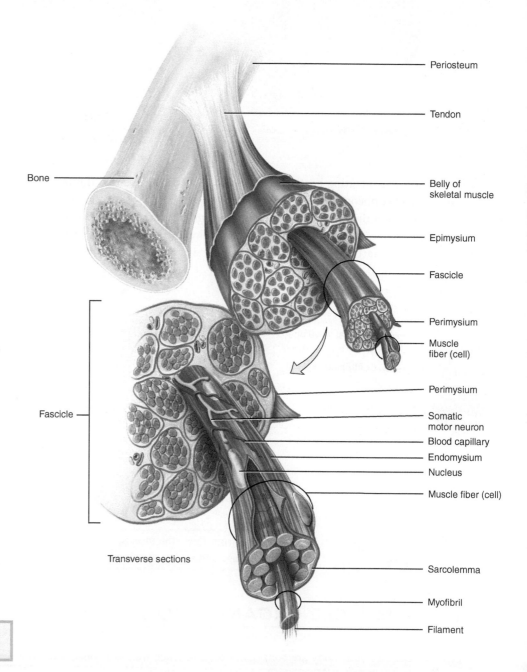

FIGURE 13.3 **Structure of a Typical Skeletal Muscle**

We know that muscle contraction is prompted by messages from the brain or the spinal cord. Nevertheless, this message must travel through a nerve to reach the appropriate muscle fibers. As shown in Figure 13.4, a **motor unit** consists of a motor neuron and every muscle fiber it innervates. Some motor units contain 10 muscle fibers, while others contain tens of thousands of muscle fibers. Regardless of size, each motor unit will contract in an all-or-none fashion. Therefore, the size of a motor unit affects its precision, as well as its force. Small motor units are involved in delicate movements, such as the turning of an eyeball. Large motor units produce the force required for walking, jumping, or lifting an 80-pound barbell. In reality, several factors affect the amount of force generated during a muscle contraction: the number of muscle fibers involved, the size of the muscle fibers involved, and how frequently the muscle fibers are stimulated to contract.

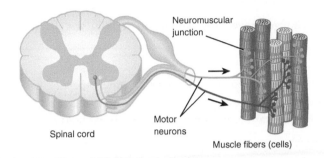

Neuromuscular junction

Spinal cord

Motor neurons

Muscle fibers (cells)

FIGURE 13.4 **Structure of a Motor Unit**

EXERCISE 13.1   Histology of Skeletal Muscles

To understand how an organ functions, it is helpful to view its structure at the microscopic level. During this exercise, you will examine prepared microscope slides of skeletal muscles. Each slide will show the tissue arrangement from a different perspective. As you view each microscope slide, be sure to note the characteristics of healthy, functional skeletal muscle fibers. This information will help you diagnose muscular disorders in Exercise 13.4 (Disorders of the Muscular System). Compare your focused microscope slides to the corresponding photomicrographs, and answer the review questions that accompany each slide.

Skeletal Muscle Fibers (Longitudinal Section)

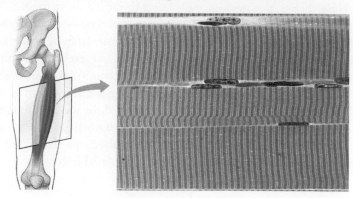

Label a skeletal muscle fiber on the image above. How would you describe the shape of these cells? _____

Cardiac muscle looks striated, or striped, when viewed under the microscope. These stripes reflect the arrangement of contractile proteins within each muscle fiber. Smooth muscle, on the other hand, does not contain any visible striations. Is skeletal muscle striated or nonstriated? _____

Label a nucleus on a muscle fiber shown above. Does each muscle fiber contain one nucleus or multiple nuclei? _____

Muscle Biopsy (Longitudinal Section)

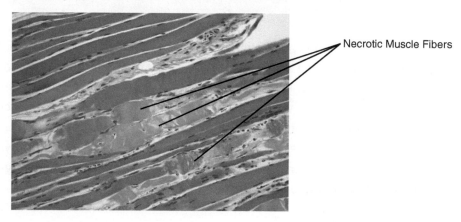

Necrotic Muscle Fibers

Necrotic (dying) muscle fibers are labeled on the photomicrograph above. Skeletal muscle necrosis may result from physical trauma, a muscle infection, or certain myopathies (muscle disorders) such as muscular dystrophy. How do healthy skeletal muscle fibers differ in appearance from necrotic ones? Be specific!

_____

_____

Unless necrosis occurs instantaneously, necrotic cells are still alive during certain stages of the process. Based on their structure, do you think necrotic muscle fibers can physically contract? Why or why not?

_____

_____

Skeletal Muscle (Cross Section)

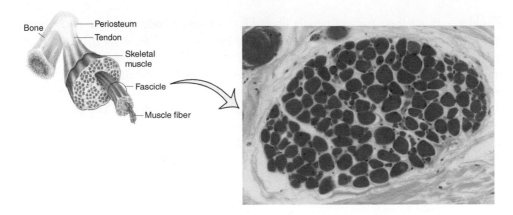

Label a muscle fiber and a fascicle on the image above.

Label the perimysium on the image above. Does this sheath surround a fascicle or a single muscle fiber? _____

Muscle Biopsy (Cross Section)

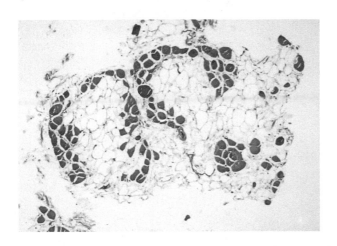

The muscle biopsy above was obtained from a patient with chronic muscle fatigue. How does this patient's skeletal muscles differ in structure from healthy skeletal muscle?

_____

_____

_____

Based on the biopsy results, why is this patient suffering from chronic muscle fatigue?

_____

_____

_____

SKELETAL MUSCLE CONTRACTION

Hundreds, if not thousands, of muscle fibers fill every single skeletal muscle in the body. We know that cells, such as muscle fibers, are the basic units of life. With that said, there are additional levels of organization *within* each muscle fiber. As shown in Figure 13.5, muscle fibers are filled with long, slender protein rods called **myofibrils**. (In anatomy and medicine, the prefix *myo-* is commonly used when referring to muscle.) The functional units of contraction—**sarcomeres**—span the length of each myofibril. (The prefix *sarco-* is derived from the Greek word *sarx*, which is used when referring to flesh.) Figure 13.5 shows how sarcomeres are arranged along a myofibril. When a muscle shortens in length at the macroscopic level, which is the level visible to the naked eye, this is because millions of sarcomeres have shortened in length at the microscopic level.

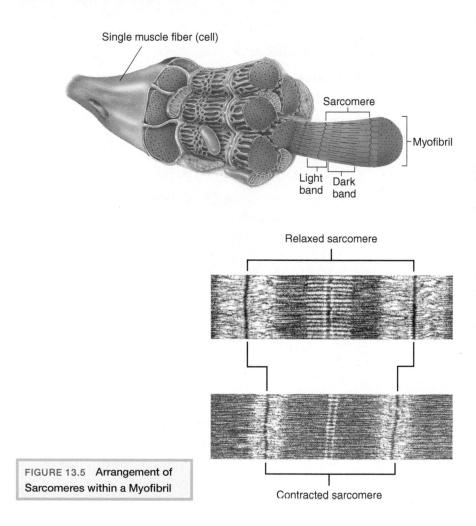

FIGURE 13.5 **Arrangement of Sarcomeres within a Myofibril**

To stimulate muscle contraction, motor neurons release a neurotransmitter called **acetylcholine**, which binds to acetylcholine receptors on neighboring muscle fibers. If an electrical current is subsequently produced by the muscle fiber, then stored calcium ions will flood into its cytoplasm. As a result, the protein filaments in sarcomeres can form cross-bridges and ultimately slide past one another. Figure 13.6 shows the steps involved in sarcomere contraction; this series of events is commonly called the **sliding filament model**. During this process, the chemical energy in ATP molecules gets converted into tension, force, and mechanical motion.

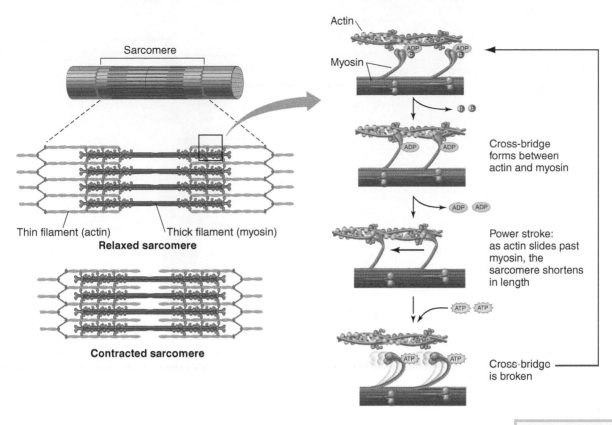

Cross-bridge forms between actin and myosin

Power stroke: as actin slides past myosin, the sarcomere shortens in length

Cross-bridge is broken

Sarcomere

Thin filament (actin)   Thick filament (myosin)

Relaxed sarcomere

Contracted sarcomere

Actin

Myosin

FIGURE 13.6   **The Sliding Filament Model**

STEP BY STEP 13.2   Observing Muscle Contraction

Visualizing
THE LAB

Oxygen gas, salt ions, and ATP molecules each play a unique role in muscle contraction. Do you think these substances can shorten muscle fibers that are no longer living? During this exercise, you will test the effects of ATP and salts on muscle fibers that originated from the psoas muscle. As shown in Figure 13.7, the psoas major connects the femur to vertebrae in the lower back. The psoas major is ideal for this experiment, given the length of its muscle fibers. ATP and salt ions were removed from this muscle by pretreating it with glycerol. Glycerol, or glycerin, is an alcohol that forms a viscous solution in water. Glycerination also increases the permeability of muscle fibers, so ATP and salt ions will easily enter them during the experiment.

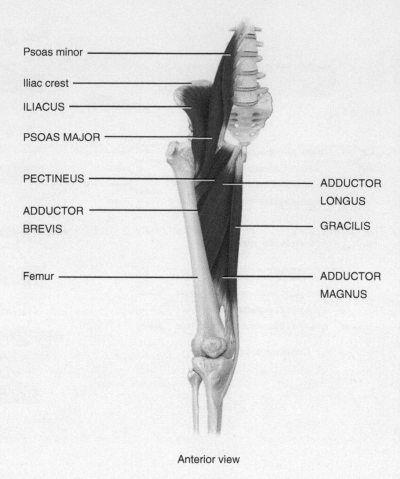

Psoas minor

Iliac crest

ILIACUS

PSOAS MAJOR

PECTINEUS

ADDUCTOR
BREVIS

Femur

ADDUCTOR
LONGUS

GRACILIS

ADDUCTOR
MAGNUS

Anterior view

FIGURE 13.7 **Location of the
Psoas Major Muscle**

Using your knowledge of muscle contraction, formulate a hypothesis to test during this experiment. What effect (if any) will ATP and salts—both individually and together—have on glycerinated muscle?

_____

_____

Procedure

1. Collect five microscope slides and a permanent marker from the materials bench.

2. Label your microscope slides as follows:

    - *Microscope*
    - *Control*
    - *Salts*
    - *ATP*
    - *Salts + ATP*

3. Using two dissecting needles — one in each hand — remove five fascicles from the glycerinated muscle.

Note: *A single fascicle resembles a long, slender string of meat. Be sure to tease the muscle thoroughly during this step: the thinner the better! If the bundles are too thick, then the experiment will not work properly.*

Glycerinated muscle   Teased fascicle   Dissecting needle   Teased fascicle

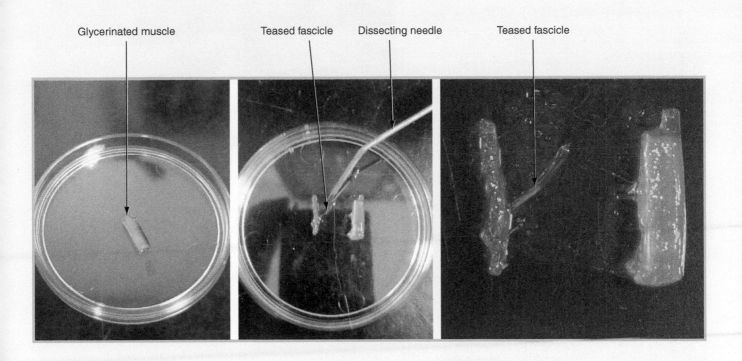

4. Using a dissecting needle, transfer one fascicle onto the *Microscope* slide.

5. Place this slide on a compound microscope and close the iris diaphragm.

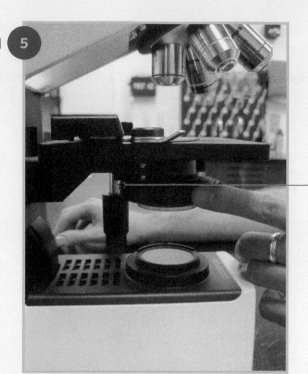

Iris diaphragm lever

6. Observe the fascicle's structure at different levels of magnification, and draw a sketch of the magnified fascicle below.

Appearance of Fascicle under _____ x Magnification

7. Using a dissecting needle, transfer one fascicle to each remaining slide:

- *Control*
- *Salts*
- *ATP*
- *Salts + ATP*

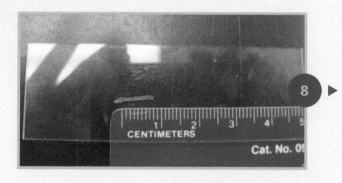

8. Measure the initial length of each fascicle, and record this information in Table 13.1.

9. Add the following solutions to the corresponding fascicles. Once each fascicle is immersed in fluid, observe it closely for one minute. Record any changes you witness in Table 13.1.

- *Control*: two drops of distilled water
- *ATP Only*: two drops of ATP solution
- *Salts Only*: two drops of salt solution
- *ATP + Salts*: one drop of ATP solution and one drop of salt solution

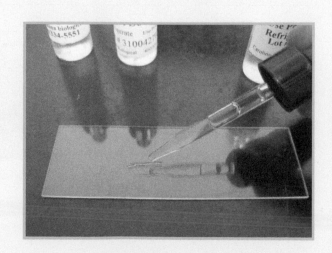

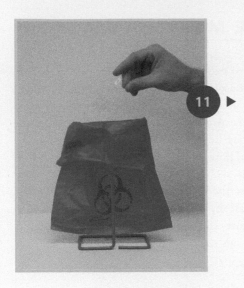

10. Measure the final length of each fascicle, and record the results in Table 13.1.

11. Place the used slides into a biohazard bag for disposal.

12. Using the following equations, calculate each fascicle's change in length. Record this information in Table 13.1.

Initial length (mm) − Contracted length (mm) = Change in Length (mm)

Change in Length (mm) × 100 = Percent Change in Length (%)

TABLE 13.1 Effects of ATP and Salt Ions on Fascicle Length

Slide	Initial Length (mm)	Final Length (mm)	Change in Length (mm)	Change in Length (%)	Visible Observations
Control					
ATP Only					
Salts Only					
ATP + Salts					

Review Questions for Step by Step 13.2

Can muscle fibers contract when they are no longer living? If so, what chemicals are required for muscle contraction?

_____

_____

Was your initial hypothesis supported or rejected by the results of this experiment? If necessary, revise your initial hypothesis in the space provided below.

_____

_____

How do your results compare to those obtained by other students? What does this suggest about the reliability of your results?

_____

_____

What are potential sources of error in this experiment? What improvements (if any) could be made to this experiment?

_____

_____

GROSS ANATOMY OF THE MUSCULAR SYSTEM

Skeletal muscles are actually layered on top of each other throughout the body. For this reason, it is impossible to show every muscle on a single photograph or illustration. The terms *superficial* and *deep* help us describe the location of one bodily structure—such as a tissue or an organ—in relation to others. Skeletal muscles are **superficial** to bones, for example, because they lie closer to the surface of the body than bones. Skeletal muscles are **deep** to skin, however, because they lie farther away from the surface of the body than skin. Since skeletal muscles are layered on top of each other, these terms can also be used to describe the location of one muscle in relation to another. To simplify muscle identification, we will focus primarily on superficial muscles in the upcoming exercise. Figure 13.8 shows selected muscles in the head and neck regions, while Figure 13.9 shows superficial muscles that are located throughout the entire body.

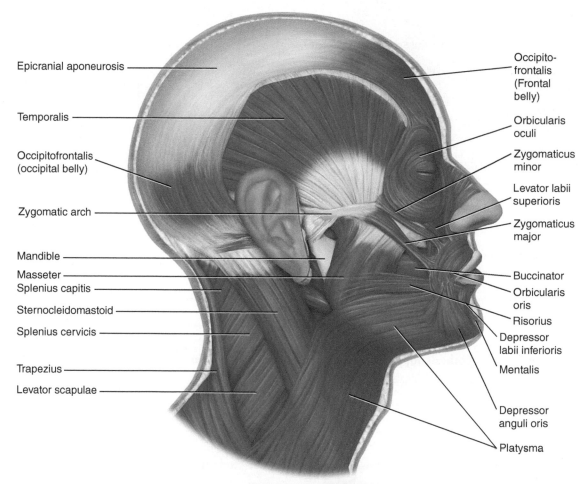

Right lateral superficial view

FIGURE 13.8 **Selected Muscles of the Head and Neck**

The names of some skeletal muscles might seem confusing at first, since they are derived from Latin root words. Once you learn how to translate these names, though, you discover that they refer to physical attributes—such as size, shape, and location—that help us distinguish muscles from one another. Table 13.2 describes the relationship between muscle names and the following attributes: location, shape, relative size, organization of muscle fibers, and/or points of attachment in the body. While characteristics such as size

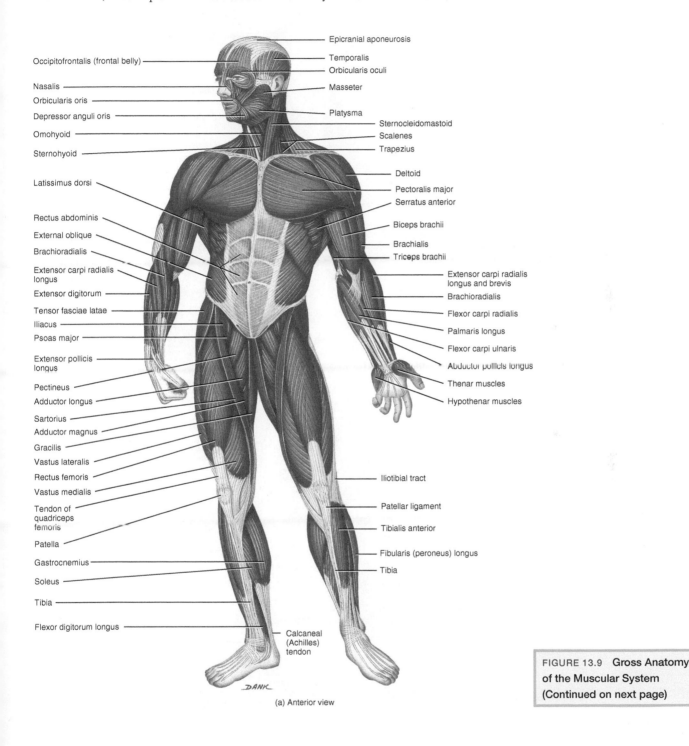

Epicranial aponeurosis
Occipitofrontalis (frontal belly)
Temporalis
Orbicularis oculi
Nasalis
Masseter
Orbicularis oris
Depressor anguli oris
Platysma
Omohyoid
Sternocleidomastoid
Scalenes
Sternohyoid
Trapezius
Latissimus dorsi
Deltoid
Pectoralis major
Serratus anterior
Rectus abdominis
Biceps brachii
External oblique
Brachialis
Brachioradialis
Triceps brachii
Extensor carpi radialis longus
Extensor carpi radialis longus and brevis
Extensor digitorum
Brachioradialis
Tensor fasciae latae
Flexor carpi radialis
Iliacus
Palmaris longus
Psoas major
Flexor carpi ulnaris
Extensor pollicis longus
Abductor pollicis longus
Pectineus
Thenar muscles
Adductor longus
Hypothenar muscles
Sartorius
Adductor magnus
Gracilis
Vastus lateralis
Rectus femoris
Iliotibial tract
Vastus medialis
Tendon of quadriceps femoris
Patellar ligament
Tibialis anterior
Patella
Fibularis (peroneus) longus
Gastrocnemius
Tibia
Soleus
Tibia
Flexor digitorum longus
Calcaneal (Achilles) tendon
DANK
(a) Anterior view

FIGURE 13.9 **Gross Anatomy of the Muscular System (Continued on next page)**

and shape are easy to understand, we need to discuss *points of attachment* in greater detail. Every skeletal muscle is attached to two or more structures in the body; these points of attachment are classified as either origins or insertions. The **origin** of a muscle remains stationary during contraction. The **insertion** is attached to a movable bone or some other type of movable structure. When a muscle contracts, its insertion ultimately moves closer to its site of origin; this phenomenon is illustrated in Figure 13.10.

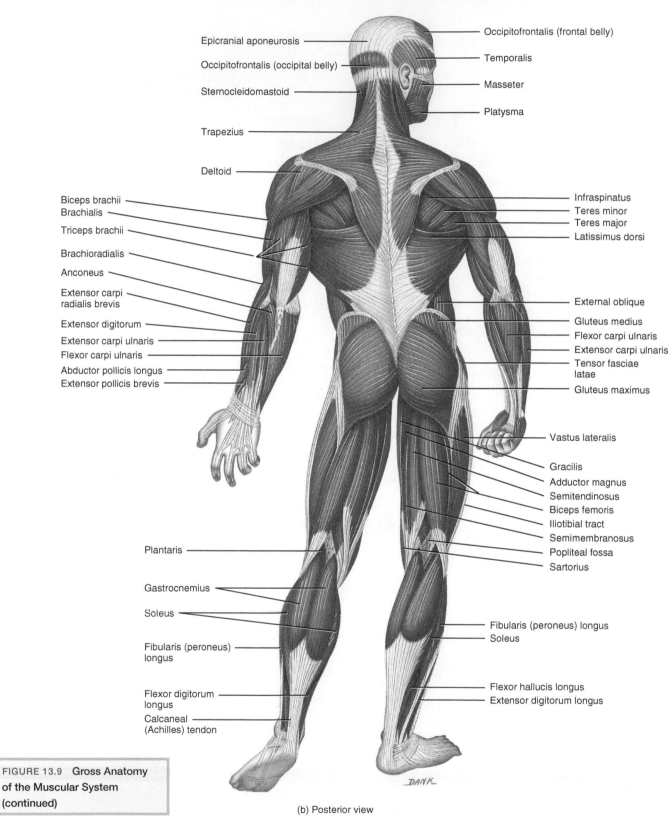

Epicranial aponeurosis

Occipitofrontalis (occipital belly)

Sternocleidomastoid

Trapezius

Deltoid

Biceps brachii
Brachialis
Triceps brachii
Brachioradialis
Anconeus
Extensor carpi radialis brevis
Extensor digitorum
Extensor carpi ulnaris
Flexor carpi ulnaris
Abductor pollicis longus
Extensor pollicis brevis

Plantaris

Gastrocnemius

Soleus

Fibularis (peroneus) longus

Flexor digitorum longus

Calcaneal (Achilles) tendon

Occipitofrontalis (frontal belly)

Temporalis

Masseter

Platysma

Infraspinatus
Teres minor
Teres major
Latissimus dorsi

External oblique

Gluteus medius
Flexor carpi ulnaris
Extensor carpi ulnaris
Tensor fasciae latae
Gluteus maximus

Vastus lateralis

Gracilis
Adductor magnus
Semitendinosus
Biceps femoris
Iliotibial tract
Semimembranosus
Popliteal fossa
Sartorius

Fibularis (peroneus) longus
Soleus

Flexor hallucis longus
Extensor digitorum longus

DANK

FIGURE 13.9 **Gross Anatomy of the Muscular System (continued)**

(b) Posterior view

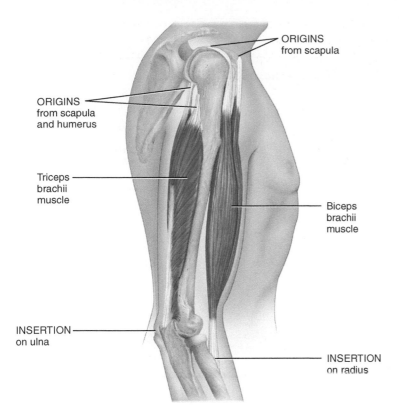

ORIGINS
from scapula

ORIGINS
from scapula
and humerus

Triceps
brachii
muscle

Biceps
brachii
muscle

INSERTION
on ulna

INSERTION
on radius

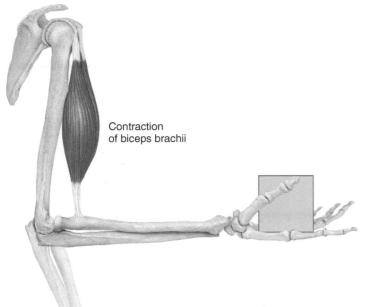

Contraction
of biceps brachii

FIGURE 13.10 **Origin(s) and Insertion(s) Skeletal Muscles**

The terms *prime mover*, *antagonist*, and *synergist* describe the role a muscle plays in a particular body movement. It is important to note that these are relative terms, just as the terms *superficial* and *deep* are. Therefore, a muscle can function as the prime mover in one scenario and as the antagonist in another scenario. A **prime mover** or an **agonist** is the main muscle involved in producing a movement; for instance, the biceps brachii muscle is the prime mover involved in forearm flexion. The **antagonist** for this movement is the triceps brachii muscle, since it reverses or opposes the action of the biceps brachii. In the case of forearm extension, however, the roles of these muscles are reversed. The triceps brachii is now the prime mover, and the biceps brachii is now the antagonist. **Synergists** either promote the action of the prime mover or stabilize a joint to prevent undesirable movements. Prime movers, agonists, and synergists all work together to prevent jerky body movements.

TABLE 13.2   Interpreting Muscle Names

Term	Definition	Example	Illustration
Relative Size			
Maximus	Largest	Gluteus maximus	
Medius	Middle	Gluteus medius	
Minimus	Smallest	Gluteus minimus	Posterior superficial view Posterior deep view
Longus	Long	Adductor longus	
Brevis	Short	Adductor brevis	Anterior deep view
Shape			
Deltoid	Triangle-shaped	Deltoid	Posterior view
Trapezius	Trapezoid-shaped	Trapezius	Posterior superficial view
Action			
Flexor	Decreases the angle at a joint (example: bends a limb)	Flexor carpi ulnaris	Wrist joint
Extensor	Increases the angle at a joint (example: straightens a limb)	Extensor carpi ulnaris	

TABLE 13.2   Interpreting Muscle Names (*Continued*)

Term	Definition	Example	Illustration
Adductor	Moves a structure toward the midline of the body (example: draws a limb toward the trunk)	Adductor longus	
Abductor	Moves a structure away from the midline of the body (example: draws a limb away from the trunk)	Abductor digiti minimi	Abduction Adduction Hip joint
Location			
Brachii	Arm	Biceps brachii	BICEPS BRACHII: LONG HEAD SHORT HEAD Anterior view
Femoris	Thigh	Biceps femoris	BICEPS FEMORIS: LONG HEAD SHORT HEAD Posterior superficial view
Pectoralis	Chest	Pectoralis major	Pectoralis major Anterior superficial view
Gluteus	Buttocks	Gluteus maximus	GLUTEUS MEDIUS GLUTEUS MAXIMUS Posterior superficial view

TABLE 13.2   Interpreting Muscle Names *(Continued)*

Term	Definition	Example	Illustration
	*Number of Origins*		
Biceps	Two points of origin	Biceps brachii	 Anterior view
Triceps	Three points of origin	Triceps brachii	 Posterior view
Quadriceps	Four points of origin	Quadriceps femoris	 Anterior views
	*Direction that Muscle Fibers Run*		
Rectus	Parallel to the midline of the body	Rectus abdominis	 Anterior deep view
Transverse	Perpendicular to the midline of the body	Transverse abdominis	
Oblique	Diagonal	External obliques	
Oribularis	Circular	Orbicularis oris	 (a) Anterior superficial view (b) Anterior deep view

During this exercise, you will identify skeletal muscles on anatomical models of the human body, a virtual dissection of the human body, or a dissected human cadaver. You will also work together as a class to identify the prime movers for certain facial expressions and body movements. Label the muscles of interest on Figures 13.11–13.14, respectively, and answer the review questions provided at the end of this exercise.

Facial Muscles

Label the following muscles on Figure 13.11:

- Front belly of occipitofrontalis
- Temporalis
- Orbicularis oculi
- Orbicularis oris
- Zygomaticus
- Masseter

Which muscles contract to produce the following actions?

Blinking: _____

Smiling: _____

Kissing and whistling: _____

Chewing: _____

Raising the eyebrows: _____

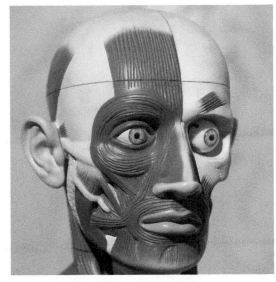

FIGURE 13.11   Identifying Facial Muscles

A CLOSER LOOK AT FACIAL MUSCLES AND BOTOX®

Botox® injections are well known for their ability to minimize facial wrinkles. Did you know that the active ingredient in Botox® is actually considered a neurotoxin? This ingredient is called botulinum toxin A, and it is produced by a bacterial species named *Clostridium botulinum*. When administered at low concentrations, this protein temporarily paralyzes facial muscles. Botox® works at the site of injection by blocking acetylcholine receptors on muscle fibers. Since the nervous system cannot tell these muscle fibers to contract, furrows (wrinkles) cannot be produced in the tissues that cover these muscles either.

Muscles of the Neck, Chest, Abdomen, and Back

Label the muscles listed below on Figure 13.12:

- Sternocleidomastoid
- Pectoralis major
- Rectus abdominis
- External obliques
- Internal obliques
- Transverse abdominis
- Trapezius
- Latissimus dorsi

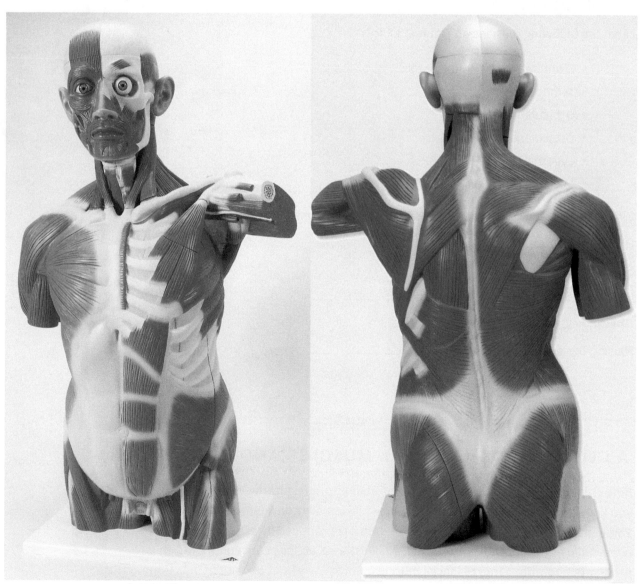

Anterior View   Poterior View

FIGURE 13.12 Identifying
Muscles of the Neck, Chest,
Abdomen, and Back

Which muscles contract to produce the following actions?

Flexing the arms: _____

Extending the arms: _____

Sit-ups or curls: _____

Rotating the trunk: _____

Extending the head: _____

Muscles of the Shoulder, Arm, and Forearm

Label the muscles listed below on Figure 13.13:

- Deltoid
- Biceps brachii

- Palmaris longus
- Triceps brachii

- Extensor digitorum

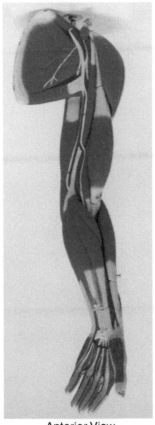
Anterior View

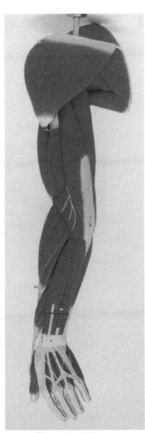
Posterior View

FIGURE 13.13 **Identifying Muscles of the Shoulder and Upper Limb**

Which muscles contract to produce the following actions?

Extending the forearm: _____

Flexing the forearm: _____

Flexing the wrist: _____

Extending the fingers: _____

Muscles of the Buttocks, Thigh, and Leg

Label the muscles listed below on Figure 13.14:

- Sartorius
- Quadriceps femoris group
  - Rectus femoris
  - Vastus intermedius (not shown, deep to rectus femoris)
  - Vastus lateralis
  - Vastus medialis

- Adductor longus
- Gracilis
- Tibialis anterior
- Gastrocnemius
- Extensor digitorum longus

- Gluteus maximus
- Hamstrings group
  - Biceps femoris
  - Semimembranosus
  - Semitendinosus

Anterior View   Posterior View

FIGURE 13.14 **Identifying Muscles of the Buttocks, Thigh, and Leg**

Which muscles contract to produce the following actions?

Flexing the thigh: _____

Extending the thigh: _____

Adducting the thigh: _____

Abducting the thigh: _____

Extending the toes: _____

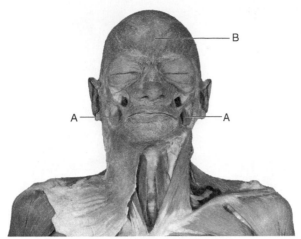

Facial Muscles, Anterior View

Label the orbicularis oculi muscles on the image above. What happens when these muscles contract? _____

Which muscles are labeled A above? The name of this muscle means *the chewer* in Greek. _____

The frontal belly of the occipitofrontalis is labeled B above. What happens to your face when this muscle contracts? _____

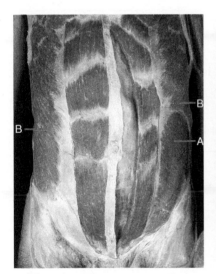

Abdominal Muscles, Anterior View

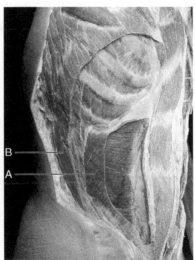

Abdominal Muscles, Lateral View

Label the rectus abdominis and the transverse abdominis on the images above.

Is the rectus abdominis superficial or deep to the transverse abdominis? _____

The internal and external obliques are labeled A and B, respectively. Name one movement that involves these muscles. _____

What does the name *rectus abdominis* tell you about this muscle?

_____

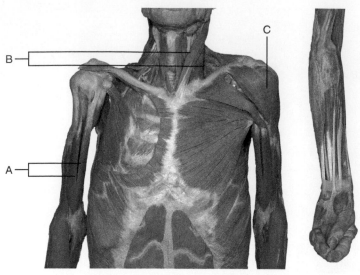

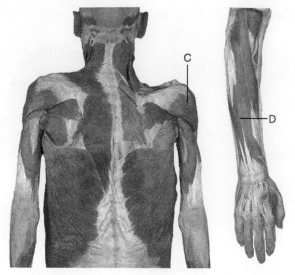

Muscles of the Torso and Upper Limbs, Anterior View   Muscles of the Back and Upper Limbs, Posterior View

Label the trapezius, pectoralis major, and latissimus dorsi muscles on the images above.

Which muscle is labeled A above? This muscle is the prime mover (agonist) involved in forearm flexion. _____

The sternocleidomastoid muscle is labeled B above. What does the name _sternocleido-mastoid_ tell you about this muscle? _____

_____

Which muscle is labeled C above? This muscle is named for its triangular shape.
_____

The extensor digitorum is labeled D above. What does this name tell you about the muscle? _____

Label the tendons that are attached to the extensor digitorum. Where are the insertions for this muscle located? _____

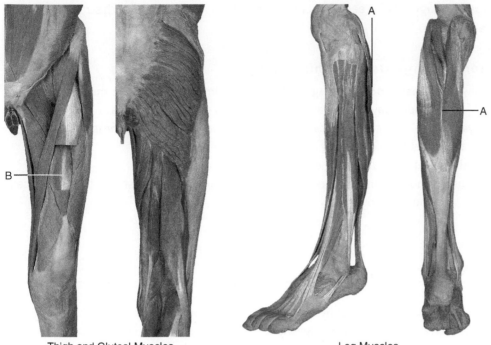

Thigh and Gluteal Muscles
Anterior and Posterior Views

Leg Muscles
Lateral and Posterior Views

Label the sartorius, biceps femoris, and tibialis anterior muscles on the images above. Which of these muscles belongs to the hamstring group? _____

What does the name *biceps femoris* **tell you about this muscle?** _____
_____

Which muscle is labeled A? The Achilles tendon connects this muscle to the calcaneus (heel bone). _____

The vastus intermedius is labeled B above. Where is this muscle located in relation to the vastus medialis and vastus lateralis? _____ **Is the vastus intermedius superficial or deep to the biceps femoris?** _____

During this exercise, you will use your knowledge of the muscular system to diagnose three patients. Paul, Jess, and Mary Jane are seeking medical attention for different muscular disorders. Use the symptoms and test results provided to make a logical diagnosis for each patient: myasthenia gravis, muscular dystrophy, a strain, or a sprain.

Terminology Used in this Exercise

Magnetic Resonance Imaging (MRI)—medical imaging technique that uses magnets and radio waves to create detailed images of internal organs

Myasthenia—muscle weakness (the term *asthenia* refers to *weakness*)

Sprain—stretching or tearing a ligament within a particular joint

Strain—stretching or tearing a particular muscle or tendon

Autoimmune—an immune response that is launched against healthy tissue cells in a person's own body

PAUL

Paul appeared perfectly healthy as an infant, but his motor skills developed much later than everyone expected. His parents were quite concerned about this developmental delay, but he did start sitting up and standing around the age of 2. By the time Paul entered preschool, he was falling on a regular basis and struggling to climb stairs. As shown in Figure 13.15, Paul's calf muscles were also very large in relation to the size of his body. Concerned about his health, Paul's pediatrician ordered a comprehensive panel of blood tests. DNA was also isolated from his white blood cells and subjected to genetic testing.

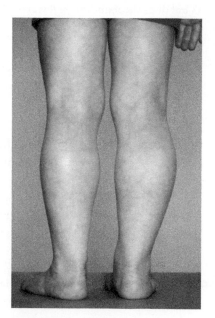

FIGURE 13.15 Paul's Calf Muscles

According to his test results, Paul had inherited a mutated gene from his mother. This gene codes for a structural protein that plays a crucial role in skeletal muscle fibers. To confirm Paul's diagnosis, a tissue biopsy was collected from one of Paul's muscles. As shown in Figure 13.16, muscle fibers were perishing inside of Paul's skeletal muscles; these muscle fibers were being replaced by adipose (fat) cells.

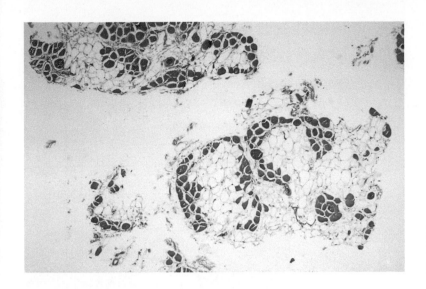

FIGURE 13.16 Paul's Muscle
Biopsy (Cross Section)

Which disorder is suggested by Paul's symptoms and test results? _____

Why do you think Paul's calf muscles are enlarged, even though his muscle fibers are perishing?

_____

_____

What treatment options (if any) exist for Paul's condition?

_____

_____

JESS

Jess was sprinting as hard as she could during the final track meet of the season. Before she reached the finish line, though, she fell to the ground with excruciating pain in her right hamstring. Although Jess's trainer immediately wrapped her leg in ice, an enormous bruise was forming on the back of her right thigh. X-rays were taken of her leg at the emergency room; luckily, these images showed no signs of broken bones or torn ligaments. However, an MRI clearly revealed the nature of her injury. As shown in Figure 13.17, one of the muscles in Jess's hamstring had torn during the race. To reduce swelling—as well as prevent additional injury—Jess was instructed to elevate her leg and continue icing it for the next several days. Unless her hamstring muscle shows signs of improvement at her follow-up exam, surgery may be required to mend this torn muscle.

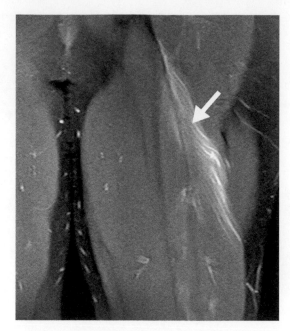

FIGURE 13.17 MRI of Jess's Right Thigh (Posterior View)

Which disorder is suggested by Jess's symptoms and test results? _____

Which muscle appears to be torn in Jess's MRI? _____

What strategies are commonly used to prevent this type of injury?

_____

_____

What is the difference between a strain and a sprain?

_____

_____

MARY JANE

During any sort of physical activity, Mary Jane develops such severe muscle fatigue that it ultimately becomes crippling in nature. Although her strength improves whenever she rests, her muscles rapidly weaken if she tries to stand up, walk around, or even play with a toy. As shown in Figure 13.18, Mary Jane's eye muscles even droop during times of fatigue. When she encounters extremely stressful situations, it also becomes difficult for her to speak, swallow, and breathe.

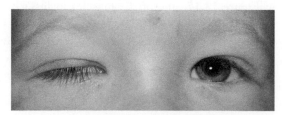

FIGURE 13.18 Mary Jane's Eyelids

Blood tests have indicated that Mary Jane's symptoms are caused by an autoimmune disease. Her immune system is destroying the acetylcholine receptors on her skeletal muscle fibers. Since her muscles struggle to receive messages from the nervous system, they have great difficulty contracting on demand.

Which disorder is suggested by Mary Jane's symptoms and test results?

_____

_____

What treatment options (if any) exist for Mary Jane's disorder?

_____

_____

Mary Jane's disorder is considered an autoimmune disease, since her immune system is actually responsible for her symptoms. Name another autoimmune disease that affects human health.

_____

_____

A CLOSER LOOK AT MUSCLE DISORDERS

The National Institutes of Health (NIH) is one of many operating divisions within the U.S. Department of Health and Human Services. NIH is the largest source of funding for medical research worldwide, and its mission is to "seek fundamental knowledge about the nature and behavior of living systems and the application of that knowledge to enhance health, lengthen life, and reduce the burdens of illness and disability."

If you would like to learn more about the diagnosis and treatment of a specific muscular disorder, you can access the NIH online at the following links:

Muscle Disorders: **http://health.nih.gov/topic/MuscleDisorders**

Movement Disorders: **http://www.health.nih.gov/topic/MovementDisorders**

Neuromuscular Disorders: **http://www.health.nih.gov/topic/NeuromuscularDisorders**

National Institute of Arthritis and Musculoskeletal and Skin Diseases: **http://www. niams.nih.gov.**

1. Which sarcomere below (A or B) is contracted? Which sarcomere is relaxed? How can you tell?

   _____

   _____

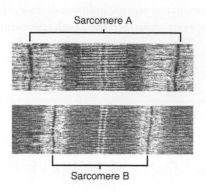

Sarcomere A

Sarcomere B

2. Glycerinated muscle fibers require _____ for contraction.
   a. ATP only
   b. Salt ions only
   c. Both a and b
   d. Glycerinated skeletal muscles cannot contract, since they are no longer living!

3. What is the difference between a prime mover, an antagonist, and a synergist?

   _____

   _____

4. Label the orbicularis oculus and orbicularis oris on the diagram below. Which muscles are involved in the following actions?
   a. Blinking _____
   b. Puckering the mouth _____
   c. Raising the eyebrows _____

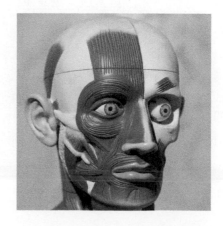

5. Label a skeletal muscle fiber and a fascicle on the photomicrographs below.

   a. What happens to the length of a muscle fiber when it contracts?

   _____

   b. What is the difference between a muscle fiber and a fascicle?

   _____

   _____

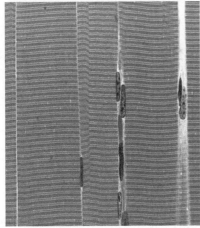

Skeletal Muscle (Longitudinal Section)

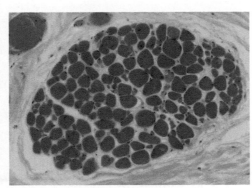

Skeletal Muscle (Cross Section)

6. Where are the biceps brachii and the biceps femoris located in the body? What attributes do these muscles have in common?

   _____

   _____

7. The _____ of a muscle remains stationary during contraction, while the _____ is attached to a movable bone or structure.

8. Label the sartorius, rectus femoris, and tibialis anterior on the diagram below.

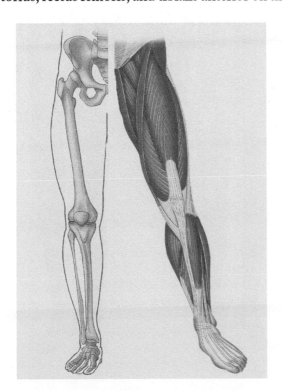

9. Label the rectus abdominis and transverse abdominis on the diagram below.
   a. Is the rectus abdominis superficial or deep to the transverse abdominis? What does this mean?_____
   _____

   b. What do the names *rectus abdominis* and *transverse abdominis* tell you about these muscles? _____
   _____

   c. Name one movement that involves these abdominal muscles.
   _____

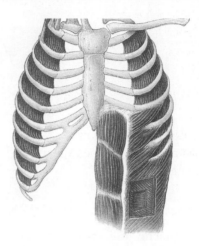

10. Label the trapezius, deltoid, gastrocnemius, and rectus femoris on the images below.

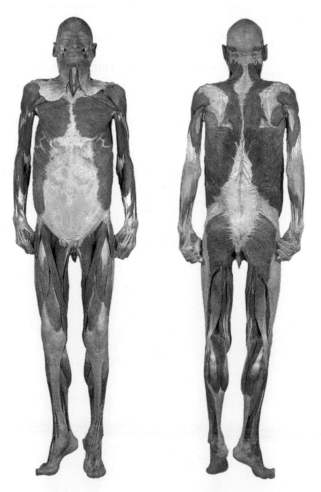

LAB 14:
Human Genetics

By the End of This Lab, You Should Be Able to Answer the Following Questions:

- What is the difference between a nucleotide, a gene, a chromosome, and a genome?

- What chemicals are used to extract DNA from cheek cells? What role does each chemical play in the extraction process?

- Can DNA be seen with the naked eye? If so, what does DNA look like?

- What is the difference between a dominant allele and a recessive allele? Are dominant traits always prevalent at the population level?

- For diseases that follow simple inheritance patterns, such as sickle cell anemia, how do genetic counselors calculate the risk of transmission from one generation to the next?

- Why are height, skin color, and eye color considered polygenic traits?

- Why is Type II diabetes considered a multifactorial disease?

- What is a karyotype? Why do geneticists examine karyotypes?

INTRODUCTION

Less than a hundred years ago, no one knew that DNA held the key to the mystery of heredity. In fact, many scientists suspected that proteins were the answer to this puzzle, for good reason. The DNA alphabet only contains four different letters (DNA bases), while the protein alphabet contains more than 20 different letters (amino acids). How could a molecule as simple as DNA code for something as complex as life? It wasn't until 1952 that the mystery started to unravel. During that year, two American scientists (Alfred Day Hershey and Martha Chase) demonstrated that DNA molecules—*not* proteins—pass heritable traits from one generation to the next. Since then, the field of genetics has exploded with countless discoveries and innovations. These include elucidating the structure of DNA (Watson and Crick in 1953), developing a novel technique for DNA sequencing (Sanger and Coulson 1975), DNA fingerprinting (Jeffreys in 1984), and sequencing the human genome (Collins and Venter in 2000), to name a few. Today, genetics plays a pivotal role in many aspects of society, such as agriculture, medicine, forensic science, paternity and maternity testing, genetic counseling, and even evolutionary research.

Before we delve into human genetics, we need to discuss how DNA is organized at the molecular level. As shown in Figure 14.1, DNA molecules are composed of **nucleotides**, which are the basic building blocks of all nucleic acids. (The abbreviation *DNA* stands for *deoxyribonucleic acid*, while *RNA* stands for *ribonucleic acid*.) Each DNA nucleotide contains two structural groups—a sugar called **deoxyribose** and a **phosphate** ion—that are used to link nucleotides together in chains. Although these groups carry no genetic information, they do form the structural backbone of DNA, which resembles a double helix or a spiral staircase. The **nitrogenous** (nitrogen-containing) **base** is the only part of a nucleotide that carries genetic information. Amazingly, there are only four different bases in the DNA alphabet: adenine (A), guanine (G), cytosine (C), and thymine (T).

With that said, though, a logical question arises. How could the blueprint for every protein in the body be written in the language of DNA? After all, there are more than 20 different amino acids, but only 4 different DNA bases. As puzzling as this may seem, there is actually a simple explanation. It takes not one, but three DNA bases (example: -CTT-) code for a single amino acid (example: lysine). Since insulin is built out of 51 amino acids, therefore, it takes three times this number of bases—153 in total—to code for the amino acids in insulin.

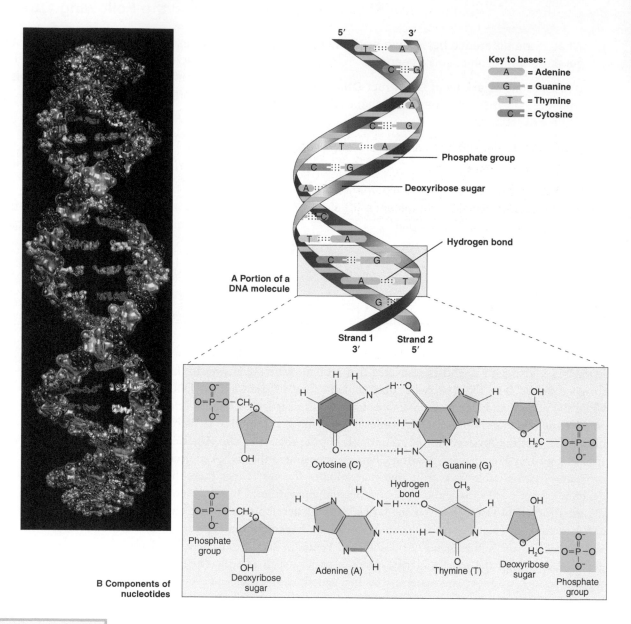

FIGURE 14.1 The Structure of DNA

In the English language, letters are strung together in different orders to create different words. The language of genetics is remarkably similar, although the letters (DNA bases) are chemical in nature. To record the genetic blueprint for a specific protein, such as insulin, the letters of the DNA alphabet must be strung together in a specific way. A region of DNA that codes for a particular protein or trait is called a **gene**. Many genes are studded along each DNA molecule, similar to the words that occupy the pages of a book. Each individual reference book, or DNA molecule, is called a **chromosome** when it is tightly coiled together. A full set of chromosomes—the 23 reference books you inherited from each parent—is called a **genome**. Figure 14.2 illustrates the relationship between a DNA base, a gene, a chromosome, and a genome.

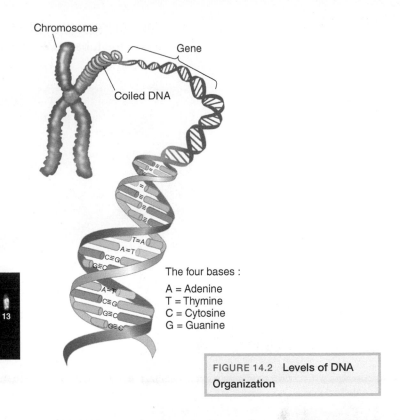

Chromosome

Coiled DNA

Gene

The four bases :
A = Adenine
T = Thymine
C = Cytosine
G = Guanine

Human Chromosome

| 1 | 2 | 3 | 4 | 5 | 6 | 7 | 8 | 9 | 10 | 11 | 12 | 13 |

| 14 | 15 | 16 | 17 | 18 | 19 | 20 | 21 | 22 |

Human Genome

FIGURE 14.2 **Levels of DNA Organization**

STEP BY STEP 14.1   Extracting DNA from Cheek Cells

Visualizing THE LAB

Did you know that household chemicals, such as salt, detergent, and alcohol, can be used to extract DNA from cells? By the end of this exercise, you will see DNA that originated from your own body.

Procedure

1. Pour 10 mL of saline solution (0.9% NaCl) into a disposable drinking cup.

2. Swish the saline solution in your mouth for approximately 30 seconds.

3. Spit the saline solution back into the cup.

4. Add 5 mL of detergent solution (3 parts water: 1 part liquid detergent) to the cup.

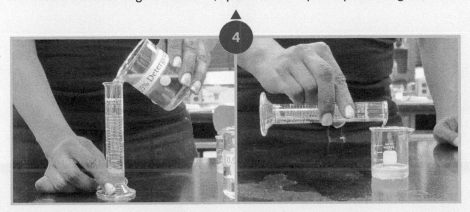

5. Pour the contents of the cup into a large test tube and cap tightly. Slowly invert the tube back and forth for two minutes.

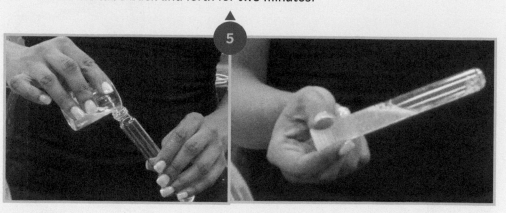

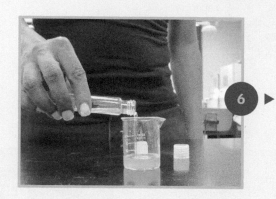

6. Pour the contents of the test tube back into the cup.

7. Slowly pour 5 mL of 95% ethanol down the side of the cup; the ethanol should form a layer on top of the aqueous solution. Note the thin white strands that are forming between the liquid layers; this is your DNA.

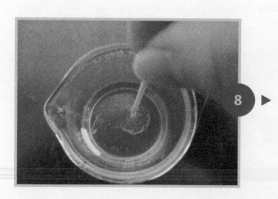

8. Insert a toothpick into the solution. Using a gentle swirling motion, slowly wind the DNA strands onto the toothpick.

Review Questions for Step by Step 14.1

Match each ingredient listed below with its corresponding function. Think about the outcome of each step in the procedure, as well as the chemical properties of each ingredient.

_____ Alcohol

_____ Detergent

_____ Saline

A) Suspends cheek cells in an aqueous solution

B) Dissolves lipids in the cell and nuclear membranes
   Denatures proteins

C) Precipitates DNA from the aqueous solution

Describe the color, texture, and quantity of DNA that was harvested from your cheek cells.

_____

_____

Cheek cells are microscopic, since they can't be seen with the naked eye. However, the DNA extracted from these cells *is* visible to the naked eye. How can this be explained?

_____

_____

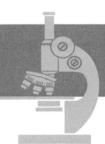

EXERCISE 14.2 Examining Chromosomes on a Microscope Slide

When a cell is not actively dividing, its DNA loosens up and fills the interior of the nucleus. When a cell is preparing to divide, though, each DNA molecule coils into a compact chromosome; the nuclear membrane also dissolves around the same time.

During this exercise, you will examine chromosomes on a prepared microscope slide. The cells on this slide originated from a human being, and the chromosomes were stained to increase their visibility. Compare your focused microscope slide to the photomicrograph shown below, and answer the review questions that accompany this exercise.

Stained Chromosomes in a White Blood Cell

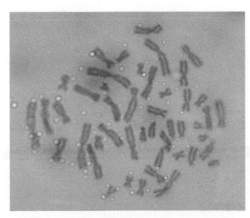

Label a chromosome on the photomicrograph. How would you describe the shape of a human chromosome? Do they all share the same basic shape, or do they vary in shape?

_____

_____

Do human chromosomes vary in size, or do they all share a similar size?

_____

Do you see a nucleus present in the cell you are viewing? What does the presence (or absence) of a nucleus tell you about this cell?

_____

_____

How many chromosomes are visible in the cell you are viewing? Is this the number you expect to find in a human cell?

_____

_____

THE LANGUAGE OF GENETICS

Like most disciplines, the field of genetics has a language all of its own. Unless you know the meanings of a few key terms, a lecture on genetics will sound like a foreign language! If a genetic term seems confusing at first, try to focus on its prefix, root word, and/or suffix for clues. For example, the term **genotype** refers to the genes (_geno-_) that a person inherited from his or her parents. Recall that humans inherit two complete sets of chromosomes: one from each biological parent. This means that humans also inherit two copies of most genes. (When it comes to genes on the sex chromosomes, though, this rule does not apply in males.) In contrast, the term **phenotype** refers to a person's physical traits. The prefix _pheno-_ actually means _showing_ or _displaying_. Notice how _physical_ and _phenotype_ sound similar to each other; this might help you remember the meaning of this term.

Figure 14.3 shows how a person's genotype and phenotype are related, assuming that a trait is determined solely by genetics. Although freckles, eye color, and other visible traits are part of a person's phenotype, they are not the entire story. At the molecular level, phenotype is determined by a person's ability—or inability—to make functional proteins.

 =

FIGURE 14.3 **Relationship between Genotype and Phenotype**

Genotype: genes coding for freckles

Phenotype: freckles

This includes pigment molecules, such as melanin, as well as enzymes, clotting factors, contractile proteins, and so on. For this reason, a person's phenotype for a trait may not be visible to the eyes.

By observing variations in genetic traits, such as freckles and eye color, we know that different versions of the same basic genes exist throughout the population. The term **allele** is used when referring to different versions of the same gene: the allele that codes for freckles, the allele that codes for no freckles, and so on. Let's say you inherited the freckles allele from both of your biological parents. Your genotype is considered **homozygous** for this gene, since your cells carry two copies of the same (*homo-*) allele. However, if your brother inherited two different (*hetero-*) version of the freckles gene—one that codes for the presence of freckles and one that codes for the absence of freckles—then his genotype is considered **heterozygous**.

Some traits and health conditions result from the activity of a single (*mono-*) pair of genes; this phenomenon is called **monogenic inheritance**. Freckles, dimples, sickle cell anemia, and cystic fibrosis all fall into this category. In the simplest form of inheritance—known as Mendelian inheritance or **simple inheritance**—the gene of interest has another distinguishing feature. Only two basic versions of this gene exist in the population, and one version is dominant to the other. A **dominant allele** will be physically expressed as long as a single copy is present. In contrast, a **recessive allele** will be masked if a dominant allele is present. Returning to the previous example, imagine that your brother's genotype is heterozygous for freckles. Even though he inherited an allele that codes for *no* freckles, this allele is not physically expressed because it is recessive. Since the freckles allele is dominant, your brother does indeed have freckles on his face.

EXERCISE 14.3 Examining Traits with Simple Inheritance Patterns

To begin this exercise, determine your phenotype for each trait listed in Table 14.1. To accurately assess these traits, you may need to work in pairs or examine yourself in the mirror. Once the class results have been pooled, you will then analyze the prevalence of dominant traits (example: presence of dimples) and recessive traits (example: absence of dimples) in the class.

Before you begin this experiment, create a hypothesis to test on the prevalence of dominant traits vs. recessive traits in your class.

_____

_____

TABLE 14.1 Examining Traits with Simple Inheritance Patterns

Trait	Dominant Phenotype	Recessive Phenotype	My Phenotype	Percentage of Class with Dominant Phenotype
Dimples (+ or −)	 Dimples	 No Dimples		
Freckles (+ or −)	 Freckles	 No Freckles		
Hairline (straight or Widow's peak)	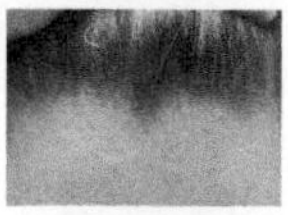 Widow's Peak	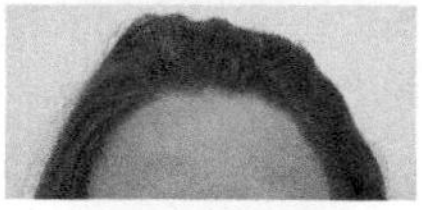 Straight Hairline		
Earlobes (attached or detached)	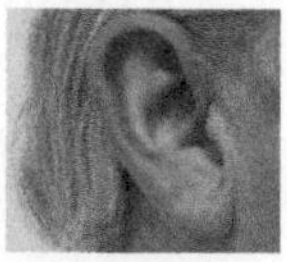 Unattached Earlobes	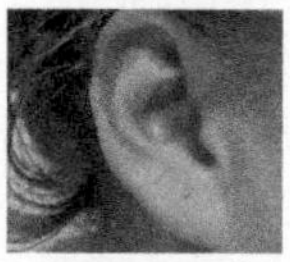 Attached Earlobes		
Cleft chin (+ or −)	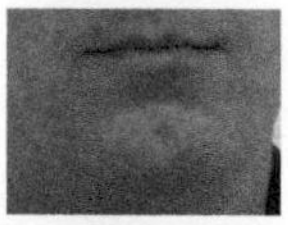 Presence of Cleft	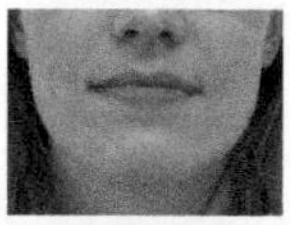 Absence of Cleft		
Tongue Rolling (+ or −)	 Ability to Roll Tongue	 Inability to Roll Tongue		
Mid-Digital Finger Hair (+ or −)	 Presence of Any Hair	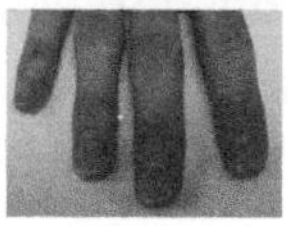 Absence of Hair		
Ability to taste phenylthiocarbamide (PTC) (+ or −)	Taster	Non-Taster		
Ability to taste sodium benzoate (+ or −)	Taster	Non-Taster		

Review Questions for Exercise 14.3

Do your class results suggest that dominant traits are always prevalent? Use specific examples from Table 14.1 to answer this question.

_____

_____

If you have a straight hairline, do you know your genotype for this trait? Explain your answer.

_____

_____

If you have a Widow's peak, do you know your genotype for this trait? Explain your answer.

_____

_____

The ability to taste PTC (phenylthiocarbamide) is one example of a dominant trait; some plants actually make bitter chemicals that are similar in taste to PTC. Using a variety of mechanisms, these chemicals deter animals from consuming the plant. Do you think PTC tasters may have a survival advantage over non-tasters, or perhaps vice versa? Explain your answer.

_____

_____

POLYGENIC INHERITANCE AND MULTIFACTORIAL INHERITANCE

When you look at the broader spectrum, it turns out that most traits and diseases are not regulated by a single gene. Just think of all the physical traits that were omitted from the previous exercise! **Polygenic traits**, such as skin color, are influenced by two or more (*poly-*) genes. Given the fact that several genes regulate the trait, as shown in Figure 14.4, the alleles can combine in a number of different ways. As a result, polygenic traits display a wider array of phenotypes than monogenic traits.

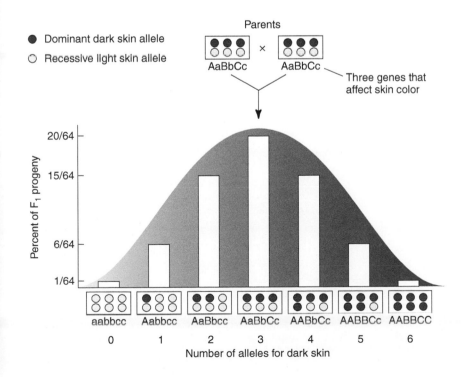

FIGURE 14.4 **Polygenic Inheritance**

To some extent, most genetic traits can also be influenced by environmental factors. This is certainly the case for skin color, given the fact that it can be altered by sun exposure. Can you think of a genetic trait that *cannot* be altered by the environment? Let's look at two examples: a person's height and the shape of a person's nose. Genetically, a child may be programmed to grow 6' 6" tall by adulthood. What happens to this child, though, if he or she suffers from a calcium or vitamin D deficiency? Without the proper building blocks for bone development, the genetic potential for a height of 6'6" cannot manifest. As strange as it may sound, rhinoplasty is a more extreme example of the environment influencing a person's phenotype. A person's genes may code for a particular nose shape, but with rhinoplasty, this phenotype can be surgically altered. Traits with **multifactorial inheritance** are influenced by genetics, as well as by traditional environmental factors, such as sun exposure, diet, activity levels, and exposure to various chemicals. The term *multifactorial* literally means that the trait is affected by multiple factors.

EXERCISE 14.4   Identifying Different Modes of Inheritance

During this exercise, you will determine whether the traits listed in Table 14.2 are governed by monogenic inheritance, polygenic inheritance, and/or multifactorial inheritance. Once a consensus has been reached for each trait, record the results in Table 14.2.

TABLE 14.2   **Identifying Modes of Inheritance**

Trait or Disorder	Mode(s) of Inheritance	How Do You Know?
Eye color		
Freckles		
Predisposition for Type II Diabetes		

TABLE 14.2   Identifying Modes of Inheritance (*continued*)

Trait or Disorder	Mode(s) of Inheritance	How Do You Know?
Height		
Cystic Fibrosis Normal CFTR channel · Mutant CFTR channel		
Hair Color		
Select an Additional Trait or Disorder:		

CODOMINANCE AND MULTIPLE ALLELE INHERITANCE

Have you ever donated blood at work or at school during a blood drive? If so, you probably received a donor card that listed your blood type for future reference. ABO blood types are another example of monogenic inheritance, since they are determined by one pair of genes. Given the nature of the ABO gene, though, additional modes of inheritance are also involved. Figure 14.5 illustrates the phenotypes associated with Type A, Type B, Type AB, and Type O blood. The ABO gene determines which antigens—if any—are present on your red blood cells. (An **antigen** is a chemical that can stimulate an immune response, assuming that the proper conditions are present.) The A allele codes for antigen A, the B allele codes for antigen B, and the O allele codes for no antigens. Since multiple versions of this gene exist throughout the population, the ABO gene demonstrates **multiple allele inheritance**. Keep in mind, though, that each individual only inherits two copies of the ABO gene: one from each parent.

FIGURE 14.5   ABO Blood Types

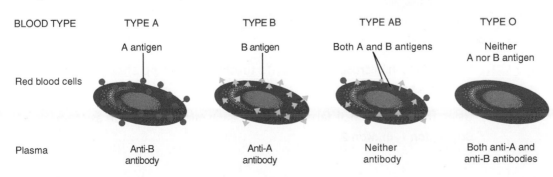

BLOOD TYPE	TYPE A	TYPE B	TYPE AB	TYPE O
	A antigen	B antigen	Both A and B antigens	Neither A nor B antigen
Red blood cells				
Plasma	Anti-B antibody	Anti-A antibody	Neither antibody	Both anti-A and anti-B antibodies

Type AB blood is also an example of **codominance**, since the A and B alleles are both expressed. Co-creators work together and share in the creative process; alleles with codominance share the spotlight, too, since one is not dominant to the other. You may wonder how both alleles get expressed in the case of Type AB blood. Why isn't one allele dominant to the other? Usually, dominant alleles code for the *production* of a substance, while recessive alleles code for the *absence* of a substance. Codominance occurs when two different alleles (example: the A and B alleles) code for two different substances (example: proteins with different sugars attached to them).

Although we tend to focus on the ABO gene when discussing blood types, in reality, red blood cells may express additional antigens that have clinical significance. Do you know what it means to have a *positive* or *negative* blood type? These qualifiers refer to the presence (positive) or absence (negative) of another surface antigen: the Rh factor, which is also called the D antigen. Although your ABO blood type and Rh status are both genetically determined, these traits involve two entirely different genes. These genes are not linked to one another, either, since they reside on different chromosomes.

STEP BY STEP 14.5   Blood Typing

During this exercise, you will analyze synthetic blood samples from a mother, a child, and two potential fathers. To determine each person's blood type, you will look for **agglutination** (clumping) after antibodies have been added to each sample. Figure 14.6 illustrates how agglutination occurs between specific antigens and specific antibodies. For instance, if a sample agglutinates following the addition of anti-A serum, this indicates that antigen A is present in the sample.

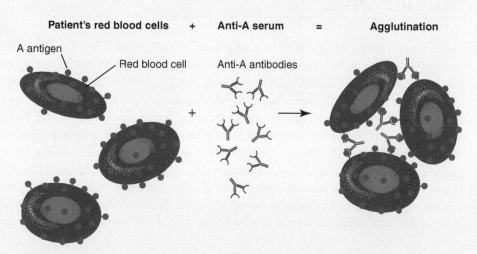

FIGURE 14.6   **Blood Agglutination**

Procedure

1. Collect 4 clean microscope slides, 12 toothpicks, and a grease pencil from the materials bench.

2. Using the grease pencil, label the top of each microscope slide as *Mom, Baby, Man 1,* or *Man 2.*

3. Draw three circles on each microscope slide, and label the circles as *A*, *B*, and *Rh*. Draw a line in between each circle on the slide; this will prevent overflow from one section to another.

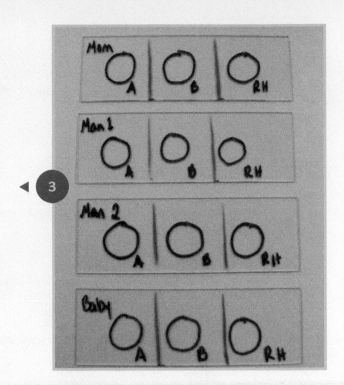

4. Place one drop of the appropriate blood sample into each circle.

5. Place one drop of the appropriate antiserum into each circle. **Note: Do not allow the tips of the antiserum bottles to touch the blood samples.**

   - Anti-A serum goes in circles labeled A
   - Anti-B serum goes in circles labeled B
   - Anti-Rh (Anti-D) serum goes in circles labeled Rh

6. Using a toothpick, mix the blood and antiserum together in one circle.

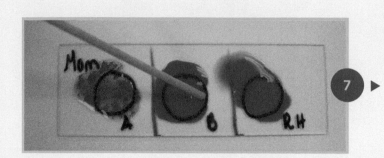

7. Repeat step 6 for each circle, using a clean toothpick every time.

8. Examine each circle for the presence or absence of agglutination. Record your results in Table 14.3.

Agglutination   No Agglutination

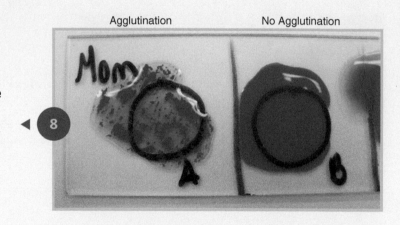

TABLE 14.3 Blood Typing Results

Blood Sample	Is antigen A present?	Is antigen B present?	Is the Rh factor present? (positive or negative)	Blood Type (example: AB positive)
Mother				
Baby				
Man 1				
Man 2				

Based on their blood types, could either man be excluded as the baby's potential father? Explain your answer.

_____

_____

Based on their blood types, could either man _potentially_ be the baby's father? Explain your answer.

_____

_____

According to the American Red Cross, "ABO blood typing is not sufficient to prove or disprove paternity or maternity."* Why do you think this is the case?

_____

_____

---

*(Source: http://www.redcrossblood.org/learn-about-blood/blood-types, accessed 10-18-10)

A CLOSER LOOK AT GENETIC PROBABILITY

Up to this point, we have focused on genetic variations that have no significant bearing on survival, such as eye color, dimples, and blood types. Although the ABO alleles are clinically significant when it comes to blood transfusions, other alleles are clinically significant for their roles in genetic disorders. Similar to freckles and dimples, some genetic disorders also follow simple inheritance patterns. These include sickle cell anemia, phenylketonuria (PKU), and Huntington's disease. Other genetic disorders are polygenic and/or multifactorial in nature, such as Type II diabetes, cardiovascular disease, and a person's susceptibility to certain types of cancers.

For traits _and_ diseases with simple inheritance patterns, it is relatively easy to calculate the probability of children inheriting these traits. To do so, the parents' genotypes must first be determined for the gene of interest. This may or may not require genetic testing, depending on the person's phenotype, as well as the phenotypes of his or her immediate relatives.

Let's say a couple decides to visit a genetic counselor before they start a family. Olivia and Byron want to know whether their children could inherit sickle cell anemia. Byron displays the phenotype for sickle cell anemia, but Olivia does not. Use the information provided on the next page to address Olivia and Byron's question.

(continued)

A specific mutation in the hemoglobin beta gene (Hbb) leads to the production of abnormal hemoglobin molecules. The alleles for this gene are abbreviated A and S:

- The A allele acts as a dominant allele, and it codes for functional hemoglobin molecules.

- The S allele acts as a recessive allele, and it codes for abnormal hemoglobin molecules.

Byron's genotype is homozygous recessive for the Hbb gene. Is Byron's genotype abbreviated as AA, AS, or SS? _____

Write Byron's genotype on the Punnett square below. Be sure to place each allele on a separate line.

Although Byron inherited two copies of the Hbb gene—one from each parent—his sperm cells randomly receive one copy or the other. What is the probability that each sperm will receive the S allele? _____

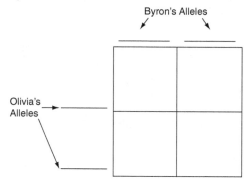

Byron's Alleles

Olivia's Alleles

Olivia's genotype is heterozygous for the Hbb gene. Is Olivia's genotype abbreviated as AA, AS, or SS? _____

Write Olivia's genotype on the Punnett square above. Be sure to place each allele on a separate line.

Even though Olivia inherited two copies of the Hbb gene—one from each parent—her eggs randomly received one copy or the other. What is the probability that each of her eggs carries the S allele? _____

The quadrants of the Punnett square represent the four possible genotypes that each child may inherit from Byron and Olivia. Since there are four possibilities—and each one has an equal probability of occurring—there is a 1 in 4 (25%) chance that each genotype will result during fertilization. To see the four possibilities, drag Byron's alleles to the boxes beneath them. Next, drag Olivia's alleles to the boxes directly on the right.

What percentage of their children (0%, 25%, 50%, 75%, or 100%) are expected to inherit sickle cell anemia? (Hint: Sickle cell anemia is a recessive disease.) _____

Let's say Olivia and Byron's first three children do not inherit sickle cell anemia. What is the probability that their fourth child will inherit this disease? Explain your answer. _____

_____

KARYOTYPE ANALYSIS

Did you know that some genetic syndromes, such as Down syndrome, can be diagnosed by examining a person's chromosomes? After the chromosomes are harvested from a patient's cell, they are organized into a profile called a **karyotype** for analysis. (The prefix *karyo-* refers to the *nucleus* of a cell.) Figure 14.7 shows a typical human karyotype. As you can see, 46 chromosomes are present per cell, and each chromosome belongs to a pair. Since two X chromosomes are present—*and the Y chromosome is absent*—this karyotype belongs to a female. With the exception of the sex chromosomes (X and Y), each chromosome in the human genome is numbered based on its relative size. Chromosome 1 is the largest chromosome, and chromosome 22 is the smallest chromosome. While analyzing karyotypes, geneticists ask themselves the following questions:

- How many chromosomes are present in each cell?

- If an abnormal number of chromosomes is present, which chromosome (1-22, X, or Y) is involved?

- Do any chromosomes have structural abnormalities, such as missing parts or additional parts?

- If a structural abnormality is identified, which chromosome (1-22, X, or Y) is involved?

Typically, somatic cells—cells other than sperm and eggs—carry two copies of each chromosome in the genome. Serious complications may arise during development if cells inherit an abnormal number of chromosomes. Some chromosomal abnormalities cannot lead to a viable pregnancy, while others may lead to genetic syndromes with severe birth

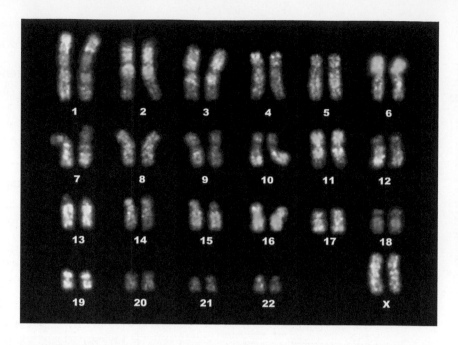

defects. If a particular chromosome is present in triplicate, then this condition is referred to as a **trisomy**. A person with Trisomy 21, for instance, carries three copies of chromosome 21 per cell. Although most trisomies do not lead to viable births, several trisomies have been identified and characterized in humans: Trisomy X (also known as Triple X syndrome), Trisomy 21 (also known as Down syndrome), Trisomy 18 (also known as Edwards syndrome), Trisomy 16, and Trisomy 13 (also known as Patau syndrome). If a single copy of a particular chromosome is present, then this condition is referred to as a **monosomy**. With the exception of the sex chromosomes in males, the only viable monosomy in humans is Monosomy X (also known as Turner syndrome).

EXERCISE 14.6   Two Sides of the Story: Living with a Chromosomal Abnormality and Diagnosing a Chromosomal Abnormality

To begin this exercise, you will read personal stories from individuals who are either living with a chromosomal abnormality or caring for a loved one with a chromosomal abnormality. Using the information provided, match each karyotype at the end of this exercise with the appropriate genetic syndrome: Turner syndrome, Patau syndrome, Down syndrome, or Klinefelter syndrome.

Terminology Used in Meriel's Story

Chromosomal Mosaicism—Certain cells in the body contain chromosomal abnormalities, while others do not.

Meriel's Story: Having a Child with Down syndrome

Having a child that has Down Syndrome (DS) was a shock.... I had no problems during my pregnancy and the baby was very active. My daughter was born full term.

It was a surprisingly quick labour, and due to drugs given late on, I was not very aware of what was going on. When she was born, I just remember the room being surprisingly silent. When I first saw my daughter I commented that her eyes looked slightly oriental in shape. The medical staff said nothing. Looking back this was as it should have been. I was able to meet my daughter, and spend the night with her. It was not till mid morning the next day that a nurse came and told me what they suspected. I will always be grateful that I had the chance to know my daughter, before being told that there was something different.

My daughter has changed my life completely—as any child would have. I have no other child to compare her with. The first couple of months were very difficult. It all seems a daze now. We were over come with love for her, but also distraught that she had a disorder that it seemed we could do little about, and with not much hope for the future. In those early days, we relied on what the medical profession told us – which was not much and not very positive. We were told she was at greater risk of heart defects, hearing problems, eye problems, learning difficulties, would probably be short and delayed in her development. Rather dismal! We asked them how she would personally be affected, and they told us they really could not say and we would have some idea at 1 year old. The uncertainty has been there from the start and still remains....

...I feel that conventional medicine offers very little at present. Unfortunately, very little money has been spent on researching how to treat DS. Until recently it was seen as too big a problem. Now, in part due to the human genome project and the mapping of chromosomes, research has begun to focus on which specific genes cause the learning difficulties. It may then be possible to turn down the effect. This exciting research gives us hope for the future.

...Our experiences of the NHS [National Health Service in the United Kingdom] are mixed. When I was told the news by a nurse, it was in the ward with another woman one bed away. We really should have been told in private. We don't have a lot of support from the special needs services. Unfortunately, we haven't found the services of much use. We went to a play and feeding therapy session and we were actually asked after a while what was wrong with our daughter. The problems we, and she may face, are of a slightly different nature to those some might expect with a child with DS. With no obvious identifying facial features and no sign of delayed

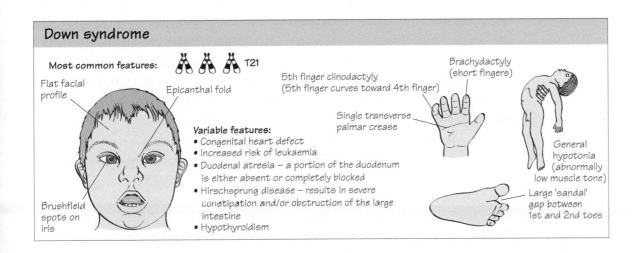

development we are looked upon as not really having many problems, but she has a label she cannot remove. I know it will affect the way people treat her and what they expect of her. It has now been suggested by a number of health professionals that our daughter may have a rare form of DS—mosaic. This means she may have less extra genetic material, and may be less affected.

... We've found out that there is a lot we can do to help our daughter reach her full potential, as you would with any other child, and there is hope for the future. The debate between nature and nurture has not been resolved. Yes the extra genetic material brings extra problems, but it's how you react that makes the difference....

Sian's Story: Living with Turner Syndrome

My family returned from Ireland when I was 10/11 [years old]. My sister who is four years younger than me was already taller than me at this point. My mother became aware of growth disorders and took up her concerns with the school nurse. I was referred to a hospital in London and was diagnosed with Turner Syndrome.

One thing that was important was the day I was diagnosed my mother asked the doctor if there was anything I should not be told—at least at that time. I was sent from the room while my mother was told the full extent of the condition, including the fact it meant I am infertile. Even though she told me several months later, it has meant that I have never really ever felt fully able to trust my parents or feel respected by them to this day, although I know they did what they thought was best at the time.

On a health level I have been on HRT [hormone replacement therapy] since I was 13 and will probably remain on it for another 15-20 years.... I also have some related health issues such as Hypothyrodism. Practically I am five foot tall (my sisters are 5'10") which can affect people's perceptions of me at times. Knowing I am infertile has meant I have found fulfilment in other areas of my life.

I also have felt not a desirable woman to men, so have not started dating until recently. Emotionally I have felt since my diagnosis that I don't have full control of my body. This is due in some part to the fact I have had to be on various medications for so long but also because I had heavy intervention from doctors at an important stage of my life (see below).

On the positive side I have made friends with many other women with TS who have become some of my dearest friends. I also appreciate the insights into perceptions of gender that having the condition has given me. When I was a teenager I was put on HRT to put me properly through puberty. The doctor who treated me used to examine how I was "coming along" in front of a group of medical students, and this made me feel a big distance from my sexuality (thus not feeling like a "proper" woman).

... About 18 months ago I attended my TS clinic and was ambushed by the clinician into answering very sensitive questions around relationships and sexuality,

This material has been reprinted courtesy of Telling Stories: Understanding Real Life Genetics (www.tellingstories.nhs.uk).

which while well intentioned, should not have been asked in the particular context and should have been given more time for discussion.

How to improve genetic education? Educate GPs [general practioners] about conditions! Also increase sensitivity around gender issues with this condition with TS clinics and keep continuity of care.

One thing I would add to my diagnosis story is that it was harder for my parents finding out than it was for me, which is often the case because of their concern as parents for their child (my Mum was in tears when she told me, whereas I was relatively unemotional). This may have added to my later perceptions of TS as something that others find tragic and hard to deal with. Also in my experience of talking to my friends, children are often able to take in things that doctors would not assume they would. . . .

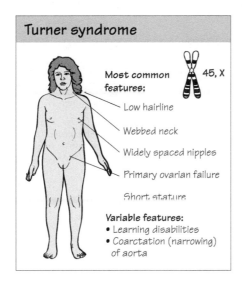

Terminology Used in Emily's Story

Tetrology of Fallot—combination of four specific heart defects

Pyloric Stenosis—narrowing of the portion of the stomach that connects to the small intestine

Hydronephrosis—accumulation of urine in the kidney(s) due to a blockage in the urinary tract

Emily's Story: Living with Trisomy 13 (as Told by her Mom)

Emily does not have "full" Trisomy 13. Her diagnosis is Partial Trisomy 13—specifically a duplication and inversion on her 13th chromosome. To date, we have never met anyone whose child has the same karyotype, although there are some that come close. Emily was born with Tetrology of Fallot, Pyloric Stenosis, and Hydronephrosis (all corrected surgically). She has a seizure disorder that presented at age 8 and is well controlled by medication. She has developmental delays across the board and many behavioral issues, mostly stemming from her inability to communicate fully. Her expressive language is unintelligible (she uses a touch-screen computer device), but her receptive language is excellent. Emily can walk, run, toilet herself, read at a 3rd grade level, do functional math and is learning social skills at school. At almost 14 years old, she is like a typical teenager in many ways—she loves her friends and music and dancing. However, her challenges are many and we predict that she will need assistance for the rest of her life. Our long term hope is that she will be able to live comfortably at a residential facility that serves the needs of the disabled population. Emily attends 8th grade at the local middle school; she is enrolled in a program called Academic Life Skills, which is a self-contained class-room where she receives remediation as necessary. She is a loving and very special young lady.

We also have 2 other children, both typically developing. Emily loves her siblings and enjoys family events very much.

Emily's story, and that of her family, are chronicled in the memoir Out Came the Sun: A Family's Triumph over a Rare Genetic Syndrome, *published by Academy Chicago Publishers in November 2008.*

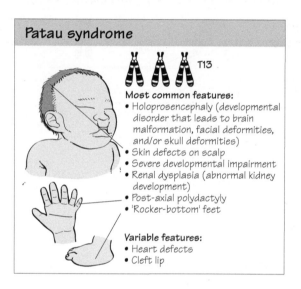

Terminology Used in Andrew's Story

Endocrinologist—a physician who specializes in the diagnosis and treatment of hormonal disorders

Cervical Spondylosis—degenerative disorder that affects the cartilage and bones in the neck

Andrew's Story: "" ...The buck stops with me..."

First of all let me explain, the genetic anomaly I have is not hereditary; it cannot be passed to further generations of my family—the main reason being I am sterile, so as to say, the buck stops with me! I was born on 5th March 1953, received all the relevant medical checks done at that time, and was pronounced a healthy baby boy. I married at the age of 24 years of age and after a few years, we decided on a family. Unsuccessful after a time, I was sent for a fertility test and it came back as a low sperm count. Still they said keep trying. Several years later and nine failed donor inseminations, IVF was suggested. Again, this failed, and it was not until forty years later that a freak accident i.e. the car door hit me in the chest and my doctor sent me to the hospital to be on the safe side.

Blood tests were carried out and as other tests were also, I was sent to see an endocrinologist who informed me that I had Klinefelter Syndrome.... After seeing yet another specialist, it was decided that I would benefit from testosterone treatment.... my world caved in on me: no family, failed IVF, failed adoptions—the only support I have ever had was from my wife, my parent's attitude towards me was that it wasn't their fault....

... How has it changed my life? What an understatement that is! I'm very bitter that this syndrome was not found out sooner than it was. It explains several reasons for the way I am. Osteoarthritis, left knee surgically stiffened. Right knee—unstable where I wear a full length calliper just so that I can stay on my feet! Osteoporosis. Cervical spondylosis. Peripheral vascular disease. Diabetes type 2. Severe mobility problems. Erectile dysfunction. The list is endless—high blood pressure. I worked from the age of fifteen until I was thirty two years old, then my health became worse and I was retired on the grounds of medically disabled and classed by the government medical officer as unfit for work—so ask yourself—how would you feel?...

... One experience that I remember: There are too many. But the one main thing is not being believed by consultants—mainly orthopedics—that there was something wrong with me. I was often told that it was impossible for me to have certain problems as they were more linked to women than men. They never bothered to check or read my notes—then they would have seen that my DNA or karyotype is 47XXY and that I carry an extra female chromosome. I was never believed and that pains me!...

How would you improve care for myself and family? By taking the time to understand what we feel. By making sure that the majority of hospitals have a genetics counsellor or at least someone in the field of genetics that knows what they are talking about! ...

_____

This material has been reprinted courtesy of *Telling Stories: Understanding Real Life Genetics* (www.tellingstories. nhs.uk).

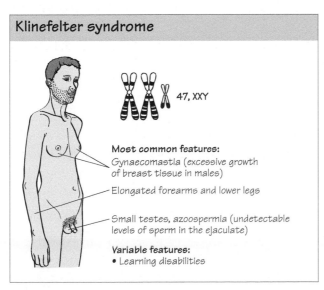

Klinefelter syndrome

47, XXY

Most common features:
Gynaecomastia (excessive growth of breast tissue in males)

Elongated forearms and lower legs

Small testes, azoospermia (undetectable levels of sperm in the ejaculate)

Variable features:
• Learning disabilities

KARYOTYPE 1

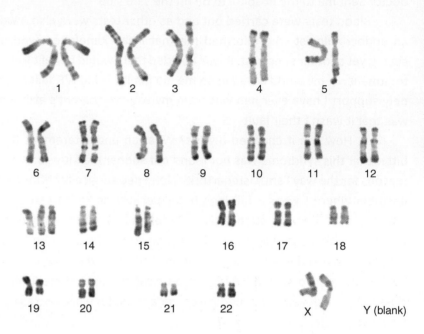

Examine the karyotype shown above. Does this karyotype belong to a male or a female?

_____

Are any chromosomal abnormalities present on this karyotype? If so, describe the nature of this chromosomal abnormality.

• Is a monosomy or trisomy present?

• Which chromosome is involved?

_____

_____

Does Karyotype 1 belong to a person with Turner syndrome, Patau syndrome, Down syndrome, or Klinefelter syndrome? _____

Name two common features of this genetic syndrome.

_____

_____

KARYOTYPE 2

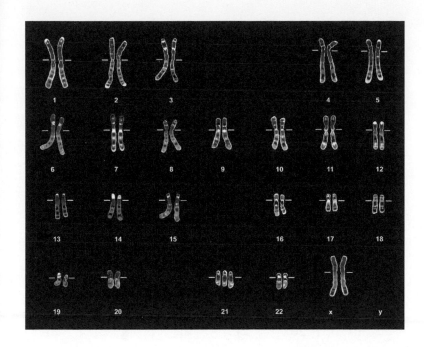

Examine the karyotype shown above. Does this karyotype belong to a male or a female?

_____

Are any chromosomal abnormalities present on this karyotype? If so, describe the nature of this chromosomal abnormality.

- Is a monosomy or trisomy present?
- Which chromosome is involved?

_____

_____

Does Karyotype 2 belong to a person with Turner syndrome, Patau syndrome, Down syndrome, or Klinefelter syndrome? _____

Name two common features of this genetic syndrome.

_____

_____

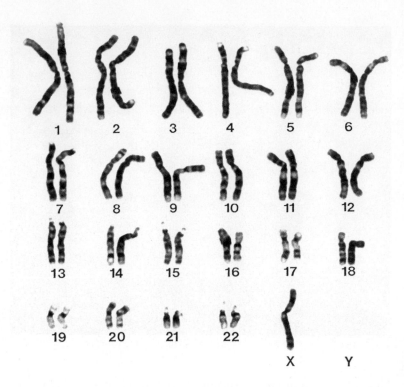

Examine the karyotype shown above. Does this karyotype belong to a male or a female?

_____

Are any chromosomal abnormalities present on this karyotype? If so, describe the nature of this chromosomal abnormality.

- Is a monosomy or trisomy present?
- Which chromosome is involved?

_____

_____

Does Karyotype 3 belong to a person with Turner syndrome, Patau syndrome, Down syndrome, or Klinefelter syndrome? _____

Name two common features of this genetic syndrome.

_____

_____

KARYOTYPE 4

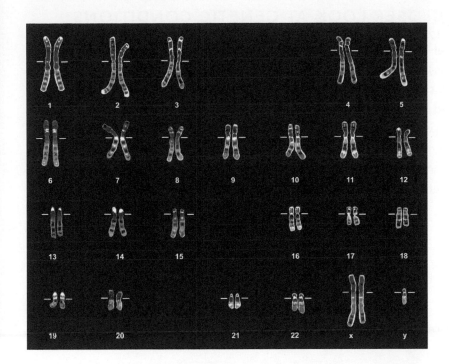

Examine the karyotype shown above. Does this karyotype belong to a male or a female?

_____

Are any chromosomal abnormalities present on this karyotype? If so, describe the nature of this chromosomal abnormality.

- Is a monosomy or trisomy present?
- Which chromosome is involved?

_____

_____

Does Karyotype 4 belong to a person with Turner syndrome, Patau syndrome, Down Syndrome, or Klinefelter syndrome? _____

Name two common features of this genetic syndrome.

_____

_____

A CLOSER LOOK AT GENETIC CONDITIONS

The National Library of Medicine (NLM), which belongs to the National Institutes of Health, "is the world's largest biomedical library and the developer of electronic information services." NLM's online library is accessed by health professionals, patients and their family members, educators, and researchers throughout the world. If you would like to learn more about the features, diagnosis, management, and pattern of inheritance for a specific genetic condition, you can visit the NLM Genetics Home Reference at **http://ghr.nlm.nih.gov/**. Information on genetic testing, genetic counseling, and gene therapy is also available on this website.

REVIEW QUESTIONS FOR LAB 14

1. Match each term on the left with the appropriate definition on the right.

   _____ Gene   a. The basic subunits of DNA and RNA molecules

   _____ Genome   b. Region of DNA that codes for a particular trait or protein

   _____ Nucleotide   c. Alternate versions of the same gene

   _____ Allele   d. A complete set of chromosomes

2. Why is Type II diabetes considered a multifactorial disease?

   _____

   _____

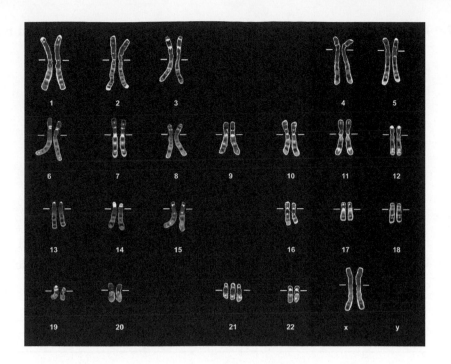

3. A _____ is shown above.
   a. Genotype   c. Phenotype
   b. Karyotype   d. Nucleotide

4. Do the chromosomes above belong to a male or female? How can you tell?
   _____

   _____

5. The chromosomes above belong to a person with _____.
   a. Turner syndrome (Monosomy X)
   b. Monosomy 21
   c. Patau syndrome (Trisomy 13)
   d. Down syndrome (Trisomy 21)

6. Match each term on the left with the appropriate choice on the right.
   _____ Genotype   a. Example = Skin Color
   _____ Phenotype   b. A person's physical traits
   _____ Homozygous   c. Inheriting two different alleles for
                                          the gene of interest
   _____ Heterozygous   d. A person's genetic makeup
   _____ Codominance   e. Example = Type AB Blood
   _____ Multifactorial Trait f. Inheriting two identical alleles for
                                          the gene of interest

7. Based on the agglutination results shown below, the patient's blood type is
   _____.

   a. AB positive
   c. AB negative
   b. B positive
   d. O negative

8. What is the difference between a dominant allele and a recessive allele? Are dominant traits always prevalent at the population level?

   _____

   _____

   _____

9. Why are height, skin color, and eye color considered polygenic traits?

   _____

   _____

10. Crystal and Brennan want to know how likely they are to have a child with sickle cell anemia. Neither of them displays the phenotype for this condition, but they are both heterozygous carriers. The alleles for the hemoglobin beta (Hbb) gene are abbreviated A and S:

    • The A allele codes for functional hemoglobin molecules.
    • The S allele codes for abnormal hemoglobin molecules.

    a. What is Crystal's genotype? (AA, AS, or SS) _____
    b. What is Brennan's genotype? (AA, AS, or SS) _____
    c. Based on Crystal's and Brennan's phenotypes, is sickle cell anemia a dominant or recessive disease? _____
    d. Calculate the probability that each of Crystal and Brennan's children will inherit sickle cell anemia.

LAB 15:
DNA Fingerprinting

By the End of This Lab, You Should Be Able to Answer the Following Questions:

- What are some practical applications of DNA fingerprinting?

- How do restriction enzymes physically alter DNA molecules?

- Why are digested DNA samples subjected to gel electrophoresis?

- Why must an agarose gel be stained after electrophoresis?

- When analyzing DNA fingerprints, how is a suspect matched to a DNA sample from the crime scene? What factors affect the statistical certainty of this match?

- How are suspects excluded as the source of crime scene DNA?

- What problems or potential errors may affect the reliability of DNA fingerprints?

INTRODUCTION

Crime scene investigations evolved dramatically throughout the course of the 20th century. During the 1950s, for instance, forensics revolved around analyzing imprints (fingerprints, footprints, and tire tracks), trace materials (cloth fibers, hairs, and paint chips), blood samples (splatter patterns and blood types), wounds, and weapons. Zoom ahead to the year 1985, when forensic science gained a priceless tool from Sir Alec John Jeffreys: DNA fingerprinting. Suddenly, a trace amount of biological evidence—blood, saliva, skin cells, or even a single hair strand—could become the cornerstone of a criminal investigation. The applications of DNA fingerprinting are not just confined to criminal cases, though. DNA fingerprints are also used to establish maternity and/or paternity, determine a patient's predisposition to genetic diseases, and even study evolutionary relationships among organisms.

During this lab period, you will participate in a simulated crime scene investigation. As a forensic geneticist, your task is to analyze DNA samples from the crime scene and five potential suspects. As shown in Figure 15.1, the DNA samples will first get cut into fragments by restriction enzymes. Using a technique called gel electrophoresis, these fragments will then get separated based on size, thereby creating the DNA fingerprints. Once the gel has been stained, you will see the DNA fingerprint associated with each sample. By comparing and analyzing these DNA fingerprints, you will determine the likelihood that the crime scene DNA originated from each suspect's body.

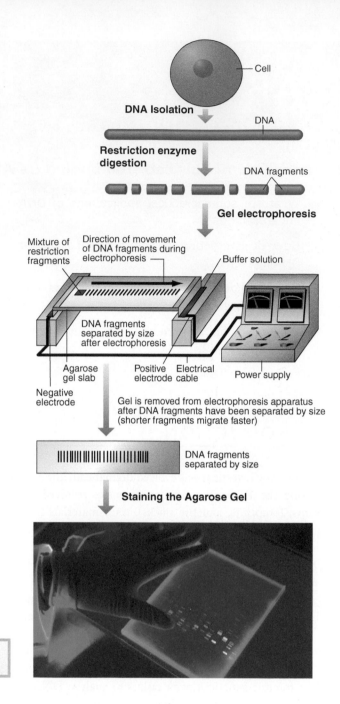

FIGURE 15.1 Flowchart for DNA Fingerprinting

RESTRICTION ENZYMES

Although human beings are 99.9% genetically identical*, variations do exist in certain parts of the human genome. (Typically, identical twins are the exception to this rule.) To produce DNA fingerprints, the DNA samples must be cut—or fragmented—at sites that tend to show variations between individuals. To do this, we rely on the activity of restriction enzymes. As shown in Figure 15.2, a **restriction enzyme** only cuts specific DNA sequences. These cut sites are commonly called **restriction sites** or **recognition sequences**. If two DNA samples differ in sequence at a potential cut site, then during restriction digestion, the samples are cut into different sized fragments. The more two DNA samples differ in sequence from one another, the more their DNA fingerprints will differ from one another. Table 15.1 lists the sources of several restriction enzymes, as well as their restriction sites.

*Source: *Frequently Asked Questions About Genetic and Genomic Science.* National Human Genome Research Institute. Accessed 11-28-10 from http://www.genome.gov/19016904.

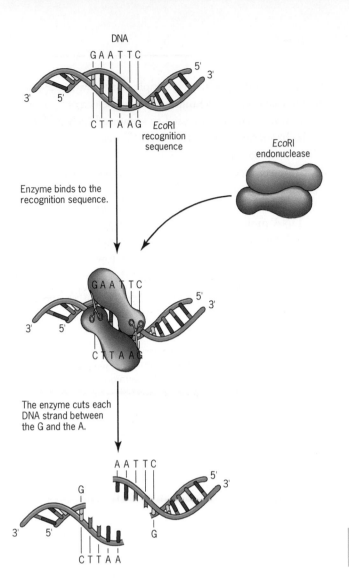

DNA

GAATTC

5'
3'

3'
5'

CTTAAG   *Eco*RI
recognition
sequence

*Eco*RI
endonuclease

Enzyme binds to the
recognition sequence.

GAATTC

5'
3'

3'
5'

CTTAAG

The enzyme cuts each
DNA strand between
the G and the A.

AATTC

5'
3'

G

3'
5'

G

CTTAA

FIGURE 15.2 **Activity of
Restriction Enzymes**

TABLE 15.1   **Recognition Sites for Selected Restriction Enzymes**

Enzyme	Source	Recognition Sequence	Cut
EcoRI	*Escherichia coli*	5'—GAATTC—3' 3'—CTTAAG—5'	5'—G AATTC—3' 3'—CTTAA G—5'
BamHI	*Bacillus amyloliquefaciens*	5'—GGATCC—3' 3'—CCTAGG—5'	5'—G GATCC—3' 3'—CCTAG G—5'
HindIII	*Haemophilus influenzae*	5'—AAGCTT—3' 3'—TTCGAA—5'	5'—A AGCTT—3' 3'—TTCGA A—5'
HaeIII	*Haemophilus aegyptius*	5'—GGCC—3' 3'—CCGG—5'	5'—GG CC—3' 3'—CC GG—5'
AluI	*Arthrobacter luteus*	5'—AGCT—3' 3'—TCGA—5'	5'—AG CT—3' 3'—TC GA—5'
PstI	*Providencia stuartii*	5'—CTGCAG—3' 3'—GACGTC—5'	5'—CTGCA G—3' 3'—G ACGTC—5'

Note: Most restriction digestions require a lengthy incubation period. Unless otherwise specified by your instructor, the DNA samples have been predigested for DNA
fingerprinting.

GEL ELECTROPHORESIS

To produce DNA fingerprints, the DNA fragments in each sample must now be separated and identified. To do this, we rely on a laboratory technique called gel electrophoresis. In molecular biology labs, gel electrophoresis is commonly used to separate DNA molecules, RNA molecules, or protein molecules based on their relative sizes. First, a gel matrix is created by combining a buffer solution with a gelatinous ingredient, such as agarose powder. (Agarose is a carbohydrate produced by seaweed.) Small pores are created throughout the gel matrix as it solidifies; these pores will serve as passageways for DNA fragments during electrophoresis. Figure 15.3 illustrates how DNA fragments move through the gel during electrophoresis. Keep in mind that DNA molecules have a uniform negative charge, since each nucleotide—the subunits of DNA molecules—contains a phosphate ion. When an electrical current is present, these charged DNA molecules migrate toward the positive pole (anode) of the gel box. Smaller DNA fragments squeeze through the pores of the gel much more easily than larger DNA fragments. As a result, the smaller DNA fragments travel faster and farther during electrophoresis.

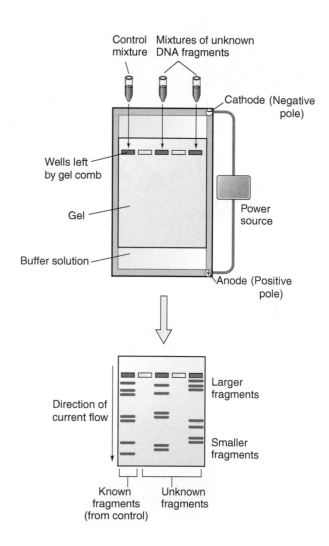

FIGURE 15.3 **Gel Electrophoresis**

Visualizing
THE LAB

Note: If pre-cast gels are provided by your instructor, then skip to Step-by-Step 15.2: Gel Electrophoresis.

The recipe below is suitable for pouring one 0.8% agarose mini gel.

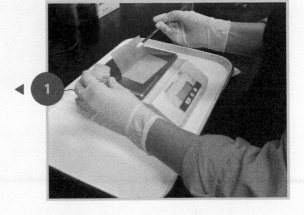

1. Weigh out 0.4 g of agarose powder on an analytical scale, and place it into a 250 mL Erlenmeyer flask.

2. Using a graduated cylinder, add 50 mL of 1x TAE (Tris Acetate EDTA) buffer to the flask.

● ● ● **SAFETY NOTE**

To dissolve the agarose powder, the agarose solution must be heated to approximately 85°C. At this temperature, the solution *and* the flask must be handled with great care. Be sure to wear safety goggles, heat-resistant gloves, and a lab coat to avoid the risk of severe burns.

3. Using a hot plate or a microwave, heat the solution until the agarose granules completely dissolve. To prevent the solution from boiling over, you must be vigilant during this step. Remove the flask from the heat source intermittently, and gently swirl its contents.

4. Once the agarose has melted, remove the flask from the heat source. Before the gel is poured, the solution must cool to approximately 55°C. The solution is still quite warm when it approaches 55°C, but the flask is not uncomfortable to hold.

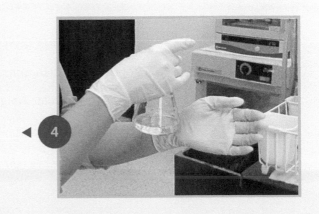

5. Pour the agarose solution into the gel casting tray. **Make sure that the gel comb is in place - and the ends of the casting tray are taped or closed— before pouring the gel.**

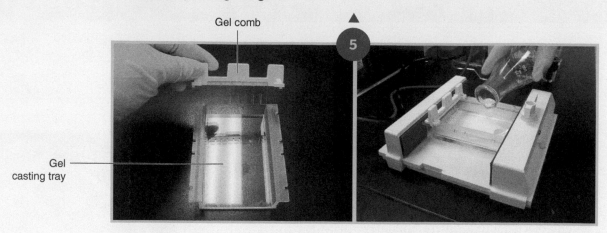

Gel comb

5

Gel casting tray

6. Allow the gel to solidify without movement or disruption. This process typically takes 15–20 minutes. As it cools and solidifies, the clear gel solution becomes cloudy.

Visualizing THE LAB

STEP BY STEP 15.2 Gel Electrophoresis

Procedure

1. Collect one set of microcentrifuge tubes, which have been pre-filled and labeled as follows:

   CS—contains 20 microliters (µL) of Digested Crime Scene DNA

   S1—contains 20 µL of Digested DNA from Suspect 1

   S2—contains 20 µL of Digested DNA from Suspect 2

   S3—contains 20 µL of Digested DNA from Suspect 3

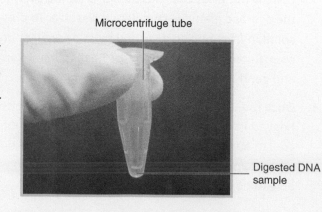

Microcentrifuge tube

Digested DNA sample

S4—contains 20 µL of Digested DNA from Suspect 4

S5—contains 20 µL of Digested DNA from Suspect 5

M—contains 10 µL of DNA Molecular Weight Ladder

LD—contains 40 µL of Loading Dye

2. Gently tap each sealed tube on the lab bench. (This step ensures that the sample is collected at the bottom of the tube.)

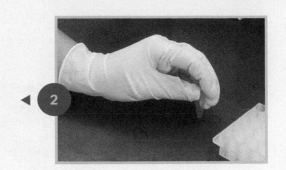

◄ 2

3. Using a micropipettor, add 5 µl of loading dye (LD) to each of the following tubes. **To prevent cross-contamination, use a new micropipette tip with each sample.**

> Crime Scene DNA (CS)
>
> Suspect 1 (S1)
>
> Suspect 2 (S2)
>
> Suspect 3 (S3)
>
> Suspect 4 (S4)
>
> Suspect 5 (S5)
>
> DNA Molecular Weight Ladder (M)

◄ 3

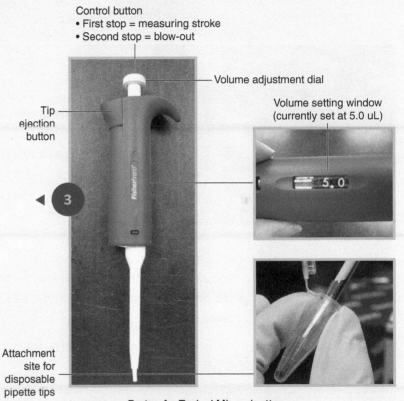

Control button
• First stop = measuring stroke
• Second stop = blow-out

Volume adjustment dial

Tip ejection button

Volume setting window
(currently set at 5.0 uL)

5.0

Attachment site for disposable pipette tips

Parts of a Typical Micropipettor

4. To mix the dye and the DNA sample together, flick the bottom of each sealed tube with your finger. Afterwards, tap each sealed tube on the lab bench again.

5. Place the agarose gel into the electrophoresis chamber. **Make sure the wells of the gel are situated near the black (–) electrode.**

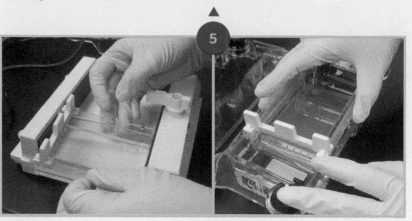

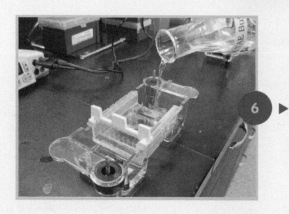

6. Fill the electrophoresis chamber with 1x TAE buffer until the gel is completely submerged.

7. Using a micropipettor, carefully load each empty lane (well) with the sample volume specified below. **To prevent cross-contamination, use a new micropipette tip with each sample.**

Lane 1: 10 μl of DNA Molecular Weight Ladder (M)

Lane 2: 20 μl of Crime Scene DNA (CS)

Lane 3: 20 μl of Suspect 1 (S1)

Lane 4: 20 μl of Suspect 2 (S2)

Lane 5: 20 μl of Suspect 3 (S3)

Lane 6: 20 μl of Suspect 4 (S4)

Lane 7: 20 μl of Suspect 5 (S5)

Lane 8: Empty

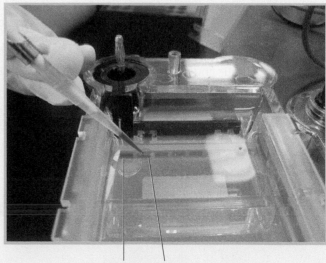

Lane 1   Lane 2

8. Secure the lid onto the electrophoresis chamber. Be sure to line up the black (–) jack on the lid with the black (–) plug on the base. If this is done properly, the red (+) jack will also line up with the red plug. ◄ 8

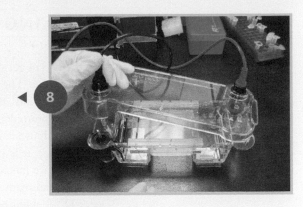

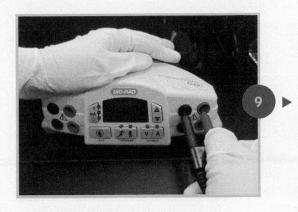

9 ▶ 9. Plug the leads into the power supply. The black (–) plug connects to the black jack and the red (+) plug connects to the red jack.

Note: The voltage and length of time specified below are suitable for most mini gel electrophoresis systems. These parameters may need to be adjusted, depending on equipment specifications.

10. Set the power supply to 110 Volts (V), and turn on the power ◄ 10 source.

11. Allow the gel to run for approximately 30 minutes. Turn off the power supply sooner if the tracking dye approaches the end of the gel.

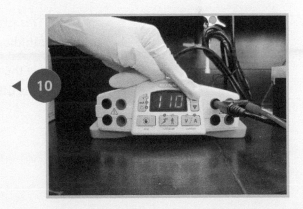

STAINING AN AGAROSE GEL

By the end of gel electrophoresis, the DNA fragments in each sample have been separated based on size. The DNA fingerprints remain invisible, though, until they are stained with a dye that binds to DNA. Figure 15.4 compares two DNA stains: methylene blue and ethidium bromide. Methylene blue is visible to the eyes under natural light, and it is much less hazardous than ethidium bromide. With that said, though, methylene blue only detects DNA when it is present in high quantities. Fluorescent DNA stains, such as ethidium bromide, are much more sensitive to DNA. These stains are only visible under UV light, though, and many of them contain chemicals that mutate DNA. For this reason, fluorescent DNA stains must be used with great care.

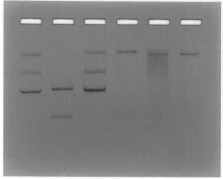 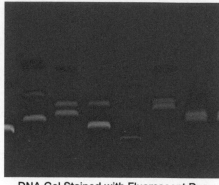

FIGURE 15.4   DNA Stains

DNA Gel Stained with Methylene Blue   DNA Gel Stained with Fluorescent Dye

Note: This procedure may be modified for use of a fluorescent stain, such as 1x ethidium bromide, or overnight staining with 1x methylene blue.

1. Remove the lid from the electrophoresis chamber. **Make sure that the power supply is off before you remove the lid.**

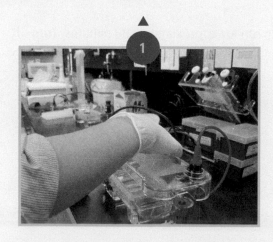

2. Carefully transfer the gel into a staining tray.

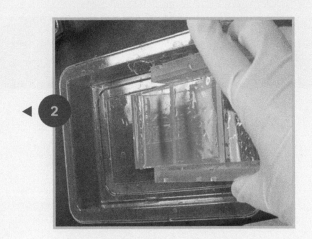

3. Submerge the gel in 100 x methylene blue stain for 2 minutes.

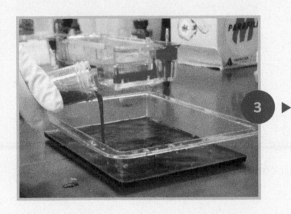

4. Carefully transfer the gel into a rinsing tray. Rinse the gel with warm tap water for 10 seconds.

5. Immerse the gel in fresh tap water and gently agitate the tray for five minutes. Repeat this step, as needed, until the DNA bands are distinguishable.

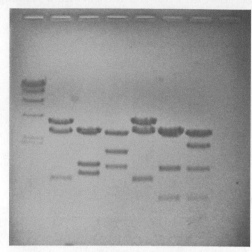

FIGURE 15.5   Results of
DNA Fingerprinting

Courtesy of Bio-Rad Laboratories, Inc.

1. Identify the lanes on your gel that contain each DNA sample. Label these lanes
    on Figure 15.5.

    Lane 1: DNA Molecular Weight Ladder   Lane 5: Suspect 3

    Lane 2: Crime scene DNA   Lane 6: Suspect 4

    Lane 3: Suspect 1   Lane 7: Suspect 5

    Lane 4: Suspect 2   Lane 8: Empty

2. Label each band in the DNA ladder according to its fragment length, which is
    stated in units of base pairs (bp). If you used a Lambda/HindIII DNA ladder, then
    the fragment lengths are as follows:

    23,130 bp

    9,416 bp

    6,557 bp

    4,361 bp

    2,322 bp

    2,027 bp

    564 bp (not visible on most gels stained with methylene blue)

    125 bp (not visible on most gels stained with methylene blue)

3. Using graph paper or a computer program, the DNA ladder can be used to
    generate a standard curve for analysis. This graph relates the size of each DNA
    molecule to the distance it migrated during electrophoresis. A computer pro-
    gram determines that the crime scene DNA lane contains 3,354 bp fragments,
    2,548 bp fragments, and 591 bp fragments. Label each band in the crime scene
    DNA lane accordingly.

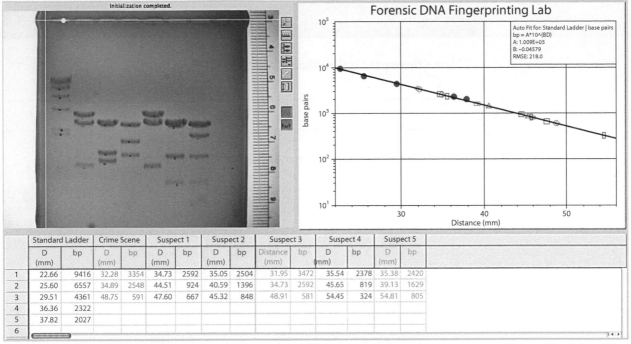

Forensic DNA fingerprinting gel stained with Fast Blast DNA stain. The Image was captured using the ProScope HR high resolution camera, and base pair determination and analysis were accomplished with *Logger Pro* software.

Courtesy of Bio–Rad Laboratories, Inc.

4. Compare the crime scene DNA fingerprint to each suspect's DNA fingerprint.

   a. Which suspect(s) can be excluded as the source of the crime scene DNA? How did you reach this conclusion?

   _____

   _____

   b. Which suspect(s) *cannot* be excluded as the source of the crime scene DNA? How did you reach this conclusion?

   _____

   _____

5. Imagine there is a 1 in 50 chance that each DNA band would match a suspect and the crime scene by random chance. Statistically, what is the likelihood that three DNA bands would match by random chance? _____

6. How can the likelihood of a random match be minimized during forensic investigations?

   _____

   _____

7. What problems or potential errors may affect the reliability of DNA finger-prints? How can these factors be minimized?

   _____

   _____

1. Name three practical applications of DNA fingerprinting.

   _____

   _____

   Use the diagram below to answer questions 2–4.

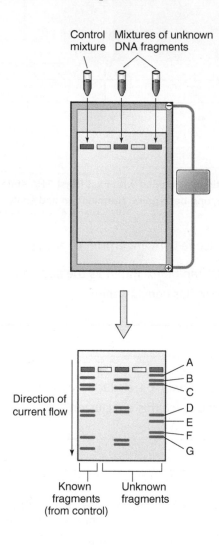

2. Label the wells on the gel shown above. Which piece of equipment is used to transfer samples into the wells? _____

3. Which DNA band (A-G) contains the smallest DNA fragments?

   _____

4. Why does DNA migrate toward the anode (positive electrode) during electrophoresis?

   _____

   _____

5. How do restriction enzymes physically alter DNA molecules? Is this a random process or a specific process? Explain your answer.

_____

_____

_____

Use the image below to answer questions 6–7.

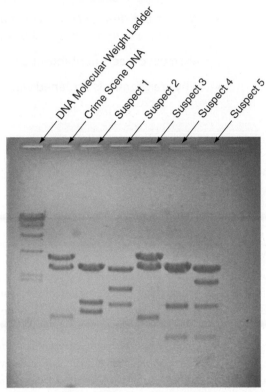

6. Which suspect(s) can be excluded as the source of the crime scene DNA?

_____

7. Which suspects *cannot* be excluded as the source of the crime scene DNA?

_____

8. What is the purpose of running a DNA ladder on a gel?

_____

_____

9. What problems or potential errors may affect the reliability of DNA fingerprints?

_____

_____

_____

10. Gel electrophoresis separates DNA fragments based on _____.
    a. their relative charges
    b. their relative sizes
    c. the sequence of DNA bases (A, C, T, G) within them
    d. all of the above

LAB 16:
Cancer

By the End of This Lab, You Should Be Able to Answer the Following Questions:

- What is the difference between a benign tumor and a malignant tumor?

- How do cancer cells metastasize to different parts of the body?

- What is the purpose of running an invasion assay?

- How can you tell whether a tissue biopsy appears normal or abnormal?

- What is the difference between a carcinoma, a sarcoma, a lymphoma, and leukemia?

- Are most cancers heritable or sporadic in nature?

- Why are chemicals screened with the Ames test?

INTRODUCTION

Upon hearing the word *cancer*, most people experience a visceral feeling of fear or sadness. Almost everyone's life has been touched by cancer, either directly or indirectly through the life a loved one. But what is cancer? What causes cancer? How do I know if I will get cancer?

Each day, a delicate balancing act occurs inside of your body to replace dead cells with newly formed ones. Cells perish for a number of different reasons: injury, age, disease, the onset of structural problems, and in some cases, because they are no longer needed by the body. How on earth does your body know *when, where,* and *how many* new cells are needed? Similar to solving problems in our own society, the answer lies in communication, although cells exchange messages that are chemical in nature. Let's say you lose a patch of skin cells due to a nasty paper cut. The cells that survive this trauma physically sense that neighbors are missing on certain sides. By releasing small molecules called **growth factors**, these survivors can alert one another that it's time to start dividing. Within a matter of days, or perhaps even hours, growth factors will stimulate enough cell division to completely fill in the gap. Not only does this halt the release of additional growth factors, but in addition, the same cells may start releasing **anti-growth factors** to prevent cell division at inappropriate times. By communicating with one another, like good neighbors do, the problem has been resolved for the entire body.

Similar to bad neighbors, tumor cells disregard communication efforts from surrounding cells. Regardless of the effects on the body as a whole—not to mention the incoming messages that urge them to stop—tumor cells continue dividing at inappropriate times. It is important to keep in mind, though, that not all tumor cells are cancerous in nature. Figure 16.1 illustrates the difference between benign tumor cells and malignant (cancerous) tumor cells. If a tumor is confined to one location—and its cells are incapable of invading other tissues—then the tumor is considered **benign**. Benign tumors are generally not harmful to a person's health, unless they compress delicate organs such as the brain or blood vessels. The cells in **malignant** tumors, on the other hand, can invade surrounding tissues. When people use the word *cancer*, they are referring to the presence of malignant tumor cells.

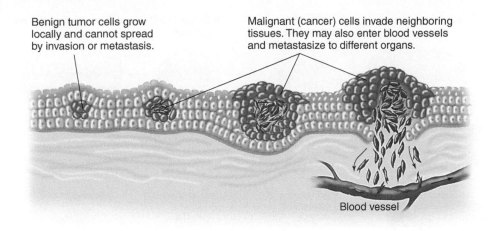

Benign tumor cells grow locally and cannot spread by invasion or metastasis.

Malignant (cancer) cells invade neighboring tissues. They may also enter blood vessels and metastasize to different organs.

Blood vessel

FIGURE 16.1 **Characteristics of Benign and Malignant Tumor Cells**

METASTASIS

If it is apparent that tumor cells are malignant in nature, then the question becomes whether or not some have traveled to distant parts of the body. The term **metastasis** refers to the migration of cancer cells from the primary tumor, or the primary site of development, to secondary sites within the body. As shown in Figure 16.2, tumor cells must overcome ten essential obstacles to metastasize:

1. Tumor cells must **proliferate** (divide rapidly) and stimulate **angiogenesis** (the formation of new blood vessels).

2. Certain cells must detach from the tumor, invade surrounding tissues, and ultimately enter blood vessels and/or lymphatic vessels.

3. Tumor cells must group together with circulating blood cells, thereby forming an **embolus**.

4. The embolus must migrate through the bloodstream.

5. The embolus must arrest in the vasculature (blood vessels) of a distant organ.

6. The embolus must adhere to the walls of a blood vessel.

7. Tumor cells must move out of the blood vessel and into the tissues of this secondary organ. This process is called **extravasation**.

8. Tumor cells must establish a microenvironment within this secondary organ.

9. Proliferation and angiogenesis must occur at this secondary site.

10. METASTASIS (process complete)

Unless tumor cells complete *and* survive each one of these steps, metastasis will fail to occur. Furthermore, each step serves as a significant obstacle for tumor cells to overcome. Laboratory studies demonstrate that within 24 hours of entering the bloodstream, less than 0.1% of tumor cells are still viable (living), and less than 0.01% of these viable cells go on to form metastases.* In other words, although many cancer cells might gain the ability to move away from the primary tumor, very few cells actually survive the process of metastasis. Metastasis remains a significant concern in the field of **oncology**—the branch of medicine devoted to the diagnosis, treatment, and prevention of cancer—since over 90% of all cancer-related mortalities result from disease spread.

---

*Fidler, I.J. Metastasis: quantitative analysis of distribution and fate of tumour emboli labeled with ^{125}I-5-iodo-2'-deoxyuridine *J. Natl Cancer Inst.* **45**, 773–782 (1970).

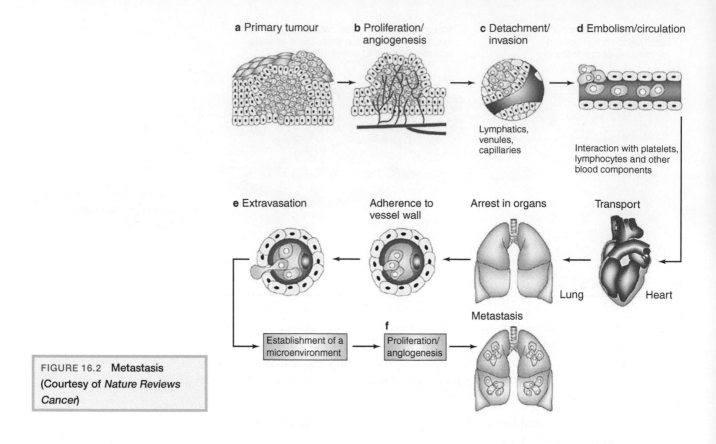

FIGURE 16.2 **Metastasis** (Courtesy of *Nature Reviews Cancer*)

Visualizing THE LAB

STEP BY STEP 16.1 Performing an Invasion Assay

In order to metastasize, cancer cells must acquire the ability to invade surrounding tissues. (The exception to this rule, of course, would be a cancer of the blood cells.) Typically, invasion relies on **protease** enzymes, which digest proteins (example: collagen fibers) in the extracellular matrix of connective tissue. As shown in Figure 16.3, invasion can be readily observed *in vitro* – outside of a living organism— using a Boyden Chamber. Invasion assays are used by cancer researchers across the globe to test strategies for inhibiting metastasis. During this exercise, you will conduct your own invasion assay on a cancer cell line that is frequently used by researchers.

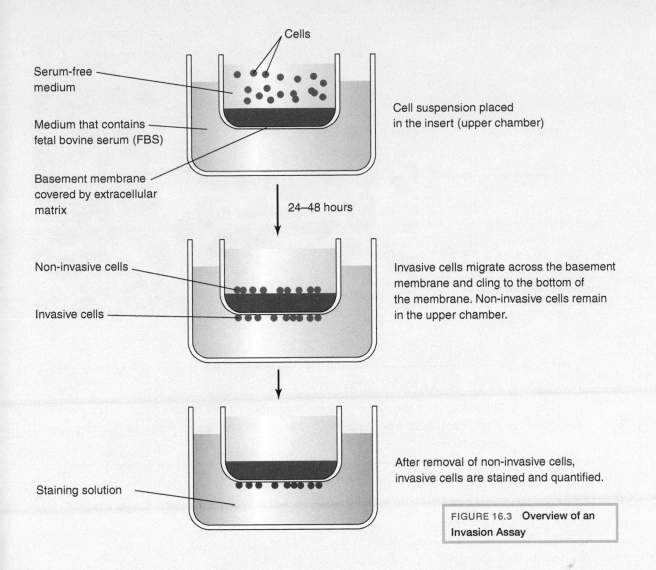

Cells

Serum-free medium

Medium that contains fetal bovine serum (FBS)

Basement membrane covered by extracellular matrix

Cell suspension placed in the insert (upper chamber)

24–48 hours

Non-invasive cells

Invasive cells

Invasive cells migrate across the basement membrane and cling to the bottom of the membrane. Non-invasive cells remain in the upper chamber.

Staining solution

After removal of non-invasive cells, invasive cells are stained and quantified.

FIGURE 16.3 **Overview of an Invasion Assay**

The AN3CA cancer cell line was derived from a 55-year-old woman with metastatic endometrial cancer. The spindle morphology, or spindle shape, of AN3CA cells is shown in Figure 16.4. Spindle morphology is a classic hallmark of tumor cells that have an enhanced ability to become motile, and therefore, metastasize to distant tissues.

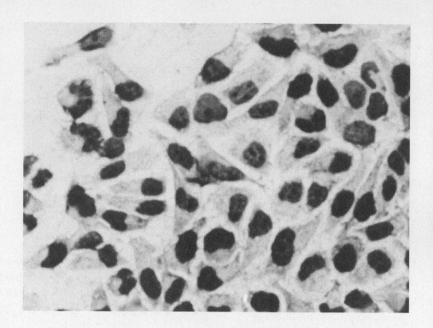

FIGURE 16.4 **Morphology of AN3CA Cells**

●●● **SAFETY NOTE**

This procedure should be performed inside of a cell culture hood. Latex gloves, a lab coat, and eye protection should also be worn while performing this procedure.

Procedure

1. Remove a cell culture plate that contains AN3CA cells from the 37°C incubator.

2. Using a sterile pipette, remove the liquid medium from the AN3CA cells. Place the medium into a hazardous waste container for disposal.

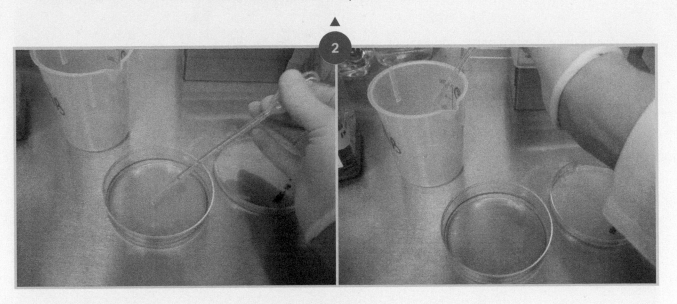

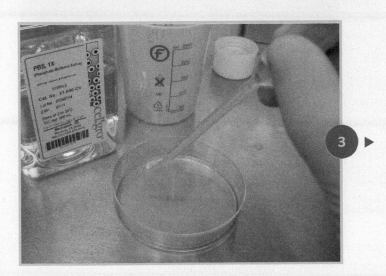

3. Wash the cells three times with 5 mL aliquots of 1x phosphate buffered saline (PBS). Following each wash, transfer the PBS from the plate to the hazardous waste container.

4. Add 1 mL of 0.05% Trypsin EDTA solution to the culture plate.

5. Place the culture plate into a 37°C incubator that contains 5% carbon dioxide for 3–5 minutes. During this incubation period, the enzyme trypsin will detach AN3CA cells from the bottom of the plate.

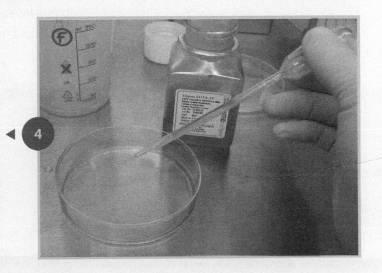

6. Using a sterile pipette, add 5 mL of warm (37°C) *serum free* Minimum Essential Medium (MEM) to the plate.

◄ **6**

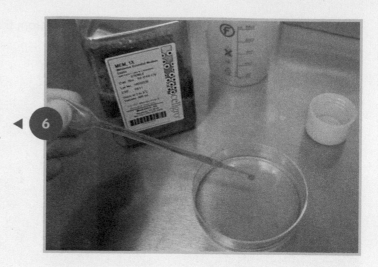

7. Using a sterile pipette, transfer the AN3CA cell suspension to a sterile 15 mL tube.

▲ **7**

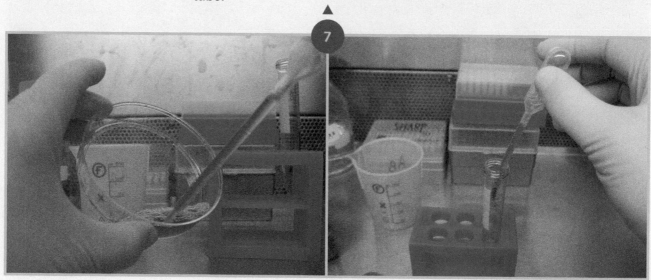

8. Pipette the cell suspension up and down 10–15 times in the tube; this will evenly distribute cells throughout the liquid medium.

9. Collect a preincubated invasion chamber from the 37°C incubator. This chamber already contains liquid medium and an invasion insert, but it does not contain tumor cells.

▲ **9**

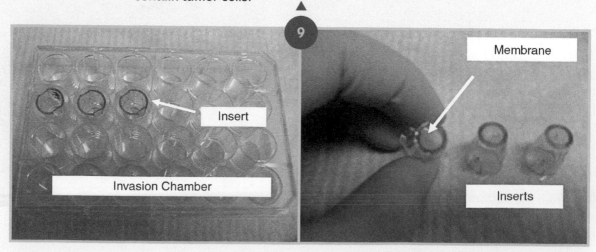

Insert

Invasion Chamber

Membrane

Inserts

10. Using a sterile pipette, remove the liquid medium from the insert *and* the lower chamber. Place the medium in a hazardous waste container for disposal.

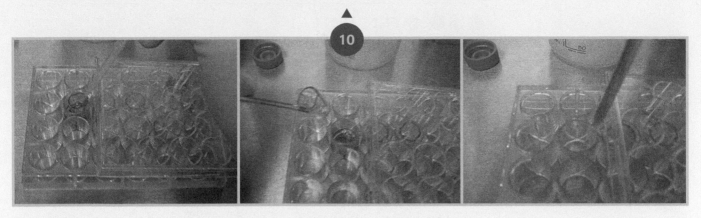

11. Place the insert back into the invasion chamber.

12. Using a sterile pipette, transfer 0.5 mL of the AN3CA cell suspension into the insert.

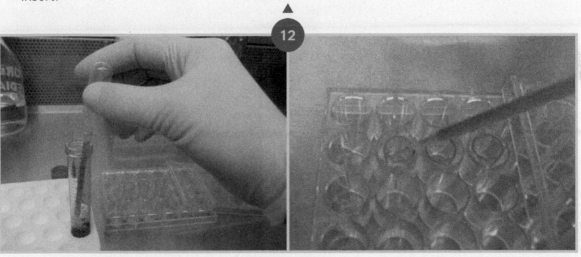

13. Using a sterile pipette, transfer 0.75 mL of MEM medium *with serum* to the lower chamber. **Perform this step carefully so the insert is not disrupted!** The serum serves as a chemoattractant for the tumor cells.

14. Carefully place the invasion chamber (containing the insert) into a 37°C incubator for 24 hours.

Next Lab Period

> **SAFETY NOTE**
>
> Unless otherwise stated by your instructor, this procedure may be performed outside of a cell culture hood. Latex gloves, a lab coat, and eye protection should be worn while performing this procedure.

1. Remove the invasion chamber from the 37°C incubator.

2. Remove the insert from the invasion chamber.

   Note: Perform steps 3–5 in a timely manner so that the membrane (located at the bottom of the insert) does not dry out. Use caution so as not to scrape, touch, or disrupt the bottom of the membrane; this region contains invasive cells.

Membrane

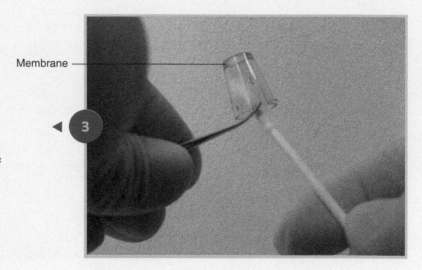

3. Insert a cotton-tipped swab into the insert. Apply gentle but firm pressure while moving the swab over the surface of the membrane. (This "scrubbing" step removes noninvasive cells from the top of the membrane.)

4. Moisten a second cotton swab with 1x PBS and repeat the scrubbing procedure.

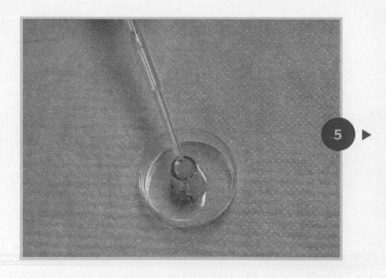

5. Collect a 35 mm cell culture plate and set the insert **upside down** in this plate. Using a pipette, add 2–3 drops of 100% ethanol to the bottom of the insert. There should be enough ethanol to cover the entire membrane.

6. Allow the ethanol to sit on top of the membrane for 5 minutes.

7. Grasp the insert with a pair of forceps and gently dunk it in a beaker of distilled water.

8. Set the insert **upside down** in the 35 mm cell culture plate.

> ●●● SAFETY NOTE
>
> Hematoxylin is an acidic compound that may cause eye, skin, and respiratory tract irritation. Wear eye protection, gloves, and a lab coat while using this dye. If hematoxylin comes in direct contact with the skin or eyes, flush the area with water for at least 15 minutes. Notify your lab instructor about the situation to determine whether medical attention is necessary.

9. Add several (2–3) drops of hematoxylin stain to the bottom of the insert; there should be enough hematoxylin stain to cover the entire membrane.

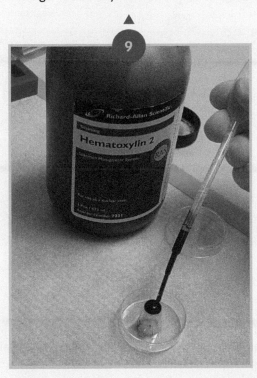

10. Grasp the insert with a pair of forceps. To rinse off the excess stain, dunk the insert into a beaker of distilled water.

11. Allow the insert to air dry **upside down** for 10–15 minutes. ◀ 11

12. Carefully insert the tip of a scalpel blade at the edge of the insert-membrane junction. Rotate the insert while carefully cutting the edge of the membrane, but **do not fully release the membrane!** Leave a very small point of attachment during this step.

▲ 12

Membrane

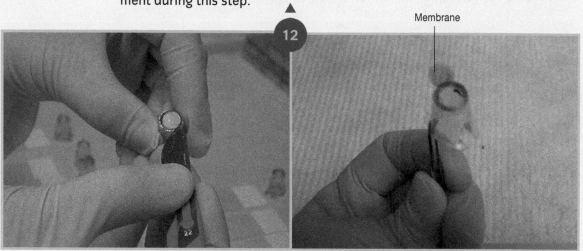

●●● SAFETY NOTE

Permount contains toluene: a flammable chemical that is found in most paint thinners. Use this product in a well-ventilated area; its vapors may irritate the respiratory tract and/or cause dizziness. Gloves, safety goggles, and a lab coat should also be worn while using this product. If Permount comes in direct contact with your skin or eyes, flush the area with water for at least 15 minutes. Notify your lab instructor of the situation immediately to determine whether medical attention is necessary.

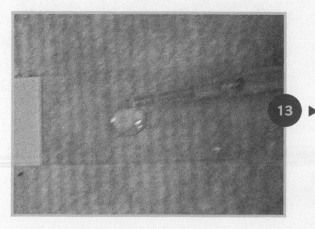

13 ▶

13. Place a small drop of Permount on the center of a clean microscope slide.

14. Peel the membrane from the remaining point of attachment with a pair of forceps.

15. Place the membrane **bottom side down** on the microscope slide. (The bottom side of the membrane was stained, and it originally faced the lower chamber.)

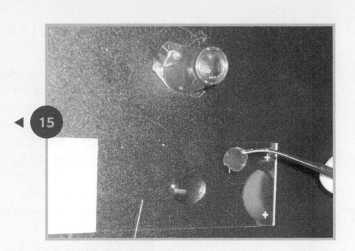

16

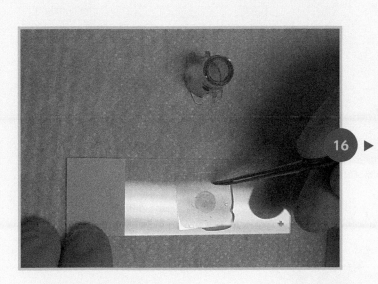

16. Cover the membrane with a cover slip. Using a pair of forceps, apply gentle pressure to the cover slip. This pressure will disperse the Permount and expel any air bubbles.

17. View your microscope slide on a compound microscope, and determine whether the AN3CA cells invaded the extracellular matrix.

17

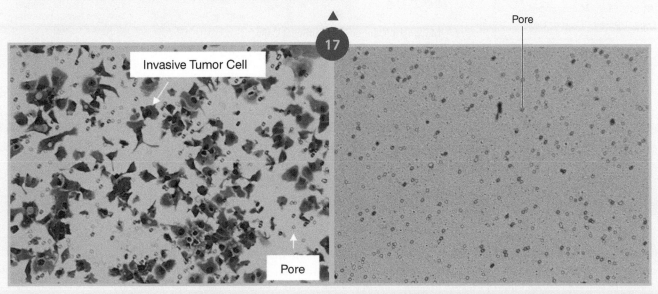

Invasive Tumor Cell

Pore

Pore

Positive Result for Invasion Assay

Negative Result for Invasion Assay

CANCER DIAGNOSIS AND CLASSIFICATION

Imagine that you work in a clinical laboratory as a pathologist. Each day at work, you examine patient **biopsies** (tissue specimens), such as the ones collected during Pap smears. How do you determine whether each biopsy appears normal or abnormal? Due to your education and training in this field, you understand that a cell's structure is related to its function. When a cell is traveling down the road toward malignancy, its function *and* its physical structure change along the way. Figure 16.5 summarizes the physical characteristics that distinguish normal cells from cancerous or precancerous cells. Features such as cell shape, cell arrangement, nucleus size, and the number of nuclei present per cell give health professionals clues about the presence or absence of malignancy.

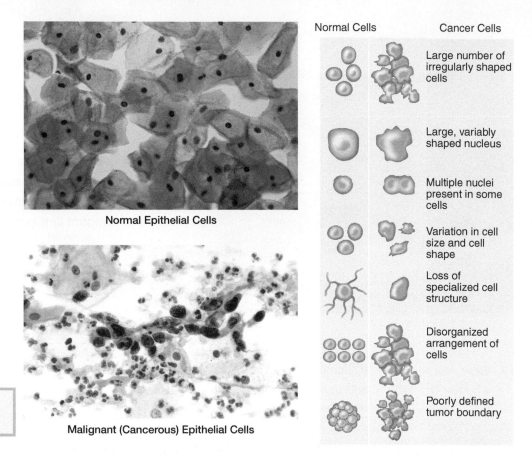

Normal Epithelial Cells

Malignant (Cancerous) Epithelial Cells

Normal Cells | Cancer Cells

Large number of irregularly shaped cells

Large, variably shaped nucleus

Multiple nuclei present in some cells

Variation in cell size and cell shape

Loss of specialized cell structure

Disorganized arrangement of cells

Poorly defined tumor boundary

FIGURE 16.5 **Microscopic Appearance of Cancer Cells**

In this day and age, additional diagnostic tests may be used in conjunction with biopsy analysis, such as imaging procedures (examples: X-rays, endoscopy, MRIs, and CT scans), analysis of bodily fluids (examples: blood tests for prostate-specific antigen and fecal occult tests), and molecular tests (example: evaluating gene expression and/or protein expression in the cells of interest). If a patient is diagnosed with cancer, as shown in Figure 16.6, the name ascribed to the cancer indicates its tissue of origin:

- **Carcinomas** arise from epithelial cells.

- **Sarcomas** arise from connective tissue or muscle tissue.

- **Leukemias** arise from bone marrow or any tissue that produces new blood cells. Malignant cells ultimately circulate in the bloodstream.

- **Lymphomas** arise from lymphatic cells, and malignant cells form solid tumors in organs such as the lymph nodes.

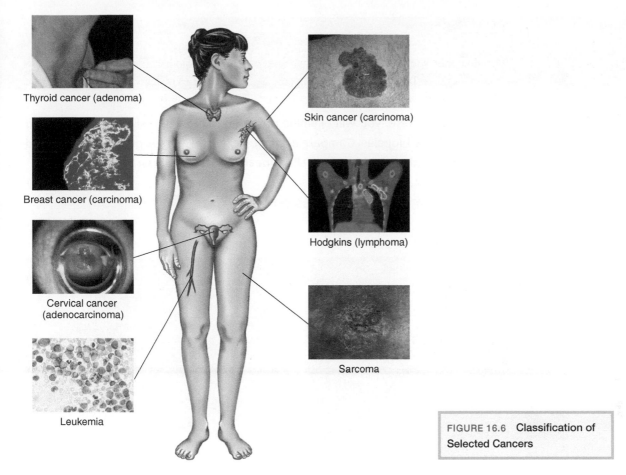

Thyroid cancer (adenoma)

Breast cancer (carcinoma)

Cervical cancer (adenocarcinoma)

Leukemia

Skin cancer (carcinoma)

Hodgkins (lymphoma)

Sarcoma

FIGURE 16.6 **Classification of Selected Cancers**

The name ascribed to a cancer also indicates the specific tissue type (or organ) that carries (or carried) the primary tumor:

- The prefix **adeno-** is used when referring to a gland (example: breast adenocarcinoma).

- The prefix **lipo-** is used when referring to adipose tissue (example: liposarcoma of the thigh).

- The prefix **myo-** is used when referring to muscle tissue (example: uterine leiomyosarcoma).

- The prefix **myelo-** is used when referring to white blood cells that are *not* B lymphocytes or T lymphocytes (example: acute myeloid leukemia)

To determine the best course of treatment, physicians must determine how far cancer cells have spread throughout the body. Typically, cancers are assigned stages based on their progression, or spread, through a patient's body. Table 16.1 summarizes the distinguishing features of each cancer stage.

TABLE 16.1   Assigning Cancer Stages*

Stage	Definition
Stage 0	Carcinoma in situ: abnormal cells remain in the tissue layer where they first formed.
Stage I, Stage II, and Stage III	Higher numbers indicate more extensive disease: larger tumor size and/or spread of the cancer to nearby lymph nodes and/or organs adjacent to the primary tumor.
Stage IV	The cancer has spread to another organ.

* Source: Cancer Staging Fact Sheet from the National Cancer Institute. Accessed 12-4-10 at http://www.cancer.gov/cancertopics/factsheet/Detection/staging.

During this exercise, you will examine prepared microscope slides of normal and abnormal tissue biopsies. Compare each biopsy to the images provided below, and determine whether the biopsied cells appears normal or abnormal.

Biopsies of Epithelial Tissue

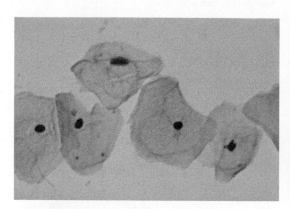 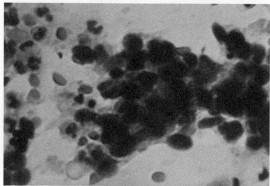

Label each biopsy as either *normal* or *abnormal*. How did you reach this conclusion?

_____

_____

Is cancer that arises from epithelial tissue considered a sarcoma, lymphoma, carcinoma, or leukemia?

_____

The overwhelming majority of cancers actually originate from epithelial tissue. What is a logical explanations for this? (Hint: Think about the functions and locations of epithelial tissue throughout the body.)

_____

_____

Blood Biopsies

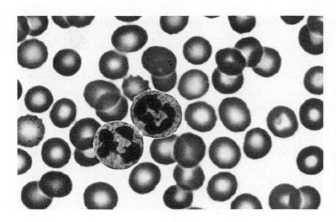 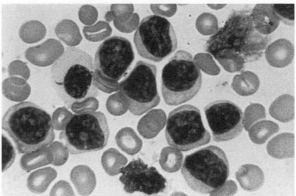

Label a red blood cell and a white blood cell on each biopsy.

Label each biopsy as either *normal* or *abnormal*. How did you reach this conclusion?

_____

_____

Is a cancer of the blood cells considered a sarcoma, lymphoma, carcinoma, or leukemia?

_____

Biopsies of Muscle Tissue

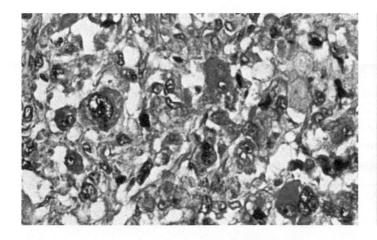 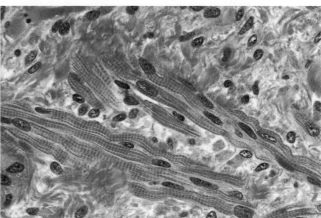

Label a muscle fiber (muscle cell) on each biopsy. Do both of these biopsies contain smooth muscle, skeletal muscle, or cardiac muscle? _____

Label each biopsy as either *normal* or *abnormal*. How did you reach this conclusion?

_____

_____

Is cancer that arises from muscle tissue considered a sarcoma, lymphoma, carcinoma or leukemia?

_____

Biopsies of Lymph Nodes

Label a lymphocyte on each biopsy.

Label each biopsy as either *normal* or *abnormal*. How did you reach this conclusion?

_____

_____

Is cancer that arises from lymphatic tissue considered a sarcoma, lymphoma, carcinoma, or leukemia? _____

Imagine that malignant cells are identified in a biopsy from a left axillary lymph node. (The term *axillary* refers to the armpit.) This patient currently has an adenocarcinoma in her left breast. Based solely on these results, which cancer stage would be assigned to the patient? _____

GENETIC MUTATIONS

Genetic mutations lie at the root of **carcinogenesis**, or cancer (*carcino-*) development (*-genesis*) in the body. In essence, a **genetic mutation** is a spelling error that permanently alters the information carried by DNA molecules. Genetic mutations come in a variety of shapes and sizes, as shown in Figure 16.7. **Point mutations** only change one letter (DNA base) in the entire sequence, while other mutational events affect larger segments of the molecule. Several possible outcomes exist for a mutated cell: DNA repair, cell death, or survival in the midst of the mutation. Most mutations are actually identified and corrected by DNA repair enzymes; these enzymes function as spell checkers while they scan DNA molecules for mutations. If a mutation is identified, but it is well beyond repair, cells normally take it upon themselves to undergo **apoptosis**: programmed cell death. Apoptosis prevents DNA mutations from passing on to future generations of cells. When safeguards like DNA repair and apoptosis fail, though, mutated cells can survive and pass on their mutations to future generations of cells.

Types of Mutations

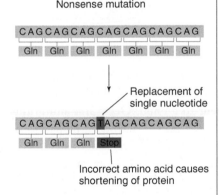

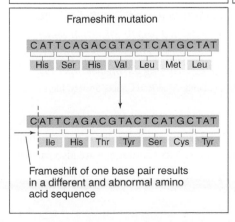

Structural Abnormalities in Chromosomes

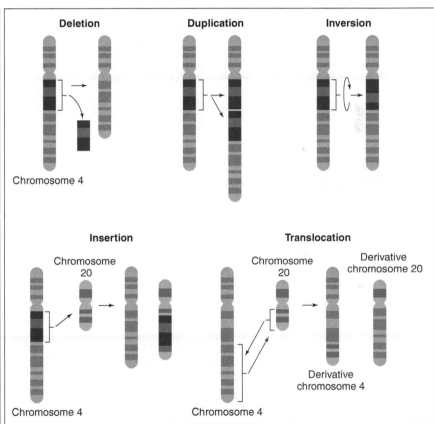

FIGURE 16.7 DNA Mutations and Chromosomal Aberrations

Not all mutations are linked to carcinogenesis; it just depends on what region of DNA is altered by a mutational event. Genes that regulate cell division—such as proto-oncogenes and tumor suppressor genes—are commonly mutated in tumor cells. **Proto-oncogenes** code for proteins that stimulate cell division at appropriate times, such as periods of growth and repair. (The prefix *onco-* is commonly used in medical terminology when referring to tumors.) A mutated proto-oncogene is called an **oncogene**; due to the presence of oncogenes, a cell may divide at inappropriate times. **Tumor suppressor genes** code for proteins that either prevent cell division at *inappropriate* times or stimulate apoptosis at *appropriate* times. Due to the presence of mutated tumor suppressor genes, a cell may divide at inappropriate times and/or fail to undergo apoptosis, in spite of the presence of DNA mutations. Figure 16.8 shows how oncogenes and mutated tumor suppressor genes affect cell division. Most metastatic cells have also accumulated genetic mutations that lead to angiogenesis and protease production.

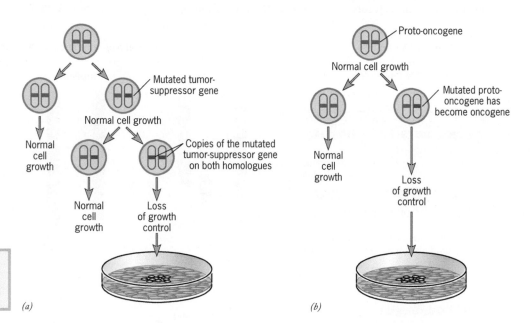

FIGURE 16.8 **Effects of Oncogenes and Mutated Tumor Suppressor Genes**

So what exactly prompts the formation of DNA mutations? Sometimes, spelling errors spontaneously occur while DNA is getting copied before cell division. Other times, DNA mutations are passed directly from a parent to a child. **Germ line mutations**—mutations carried by chromosomes in sperm and egg cells—can indeed be passed from one generation to another. However, most mutations arise sporadically and cannot be attributed to heredity. As discussed in *Cancer Facts & Figures 2010* from the American Cancer Society:

> *All cancers involve the malfunction of genes that control cell growth and division. About 5% of all cancers are strongly hereditary, in that an inherited genetic alteration confers a very high risk of developing one or more specific types of cancer. However, most cancers do not result from inherited genes but from damage to genes occurring during one's lifetime.*
> (American Cancer Society. Cancer Facts and Figures 2010. Atlanta: American Cancer Society, Inc.)

Exposure to certain environmental and industrial factors may also increase the likelihood of mutagenesis: the development of DNA mutations. A **mutagen** is any substance or agent that increases the frequency of DNA mutations. Most mutagens are also carcinogens by nature; a **carcinogen** is a cancer-causing agent or substance. Table 16.2 lists several environmental and industrial risk factors for cancer.

TABLE 16.2 Selected Environmental and Industrial Risk Factors for Cancer

Class	Examples	Associated Cancers	Illustration
Chemicals	Ethylmethane-sulfonate (EMS)		
	Carcinogens in Tobacco Smoke	Lung, Mouth, Throat, Bladder, Kidney, and Esophageal Cancers	
	Asbestos	Lung Cancer, Mesothelioma	
	Benzene	Leukemia	
Radiation			
	Ultraviolet Light	Skin Cancer	
	X-rays and Gamma Rays	Leukemia	
	Radon Gas	Lung Cancer	
Viruses	High-Risk Human Papillomaviruses (HR-HPVs)	Cervical Cancer	
	Hepatitis B and C	Liver Cancer	

During the 1970s, a scientist named Bruce Ames developed a quick and inexpensive way to identifying mutagens, which is now called the Ames test. During this test, potential mutagens are exposed to bacteria that cannot make the amino acid histidine. Unless these bacteria acquire the ability to make histidine through a mutational event, they will die and not produce visible growths on solid media. Figure 16.9 summarizes the steps involved in a traditional Ames test. Mutations can actually develop for one of two reasons during this test: as a result of mutagen activity *or* as a result of spontaneous mutations. To distinguish these events, a control plate must be run along with the test plate. The control plate sets a baseline for the frequency of spontaneous mutational events. Ultimately, the growth on each plate is compared to determine whether chemicals are mutagenic.

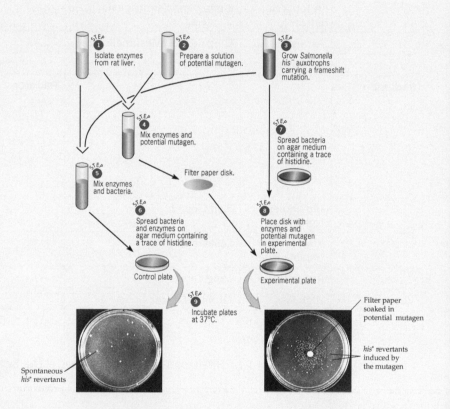

FIGURE   16.9 The Ames Test

Before you perform the Ames test, brainstorm about substances to test from everyday life. Several ideas are included below:

- Alcohol (What type? What concentration?)

  _____

  _____

- Weak acids (example: acetic acid in vinegar) or weak bases (example: sodium bicarbonate in baking soda)

  _____

  _____

- Tobacco (Cigarettes, cigars, or chewing tobacco? How will the chemicals in tobacco be extracted into liquid?)

  _____

  _____

- Over-the-counter pesticides, herbicides, or insecticides (What safety precautions are necessary? For instance, does this product need to be used in a fume hood?)

  _____

  _____

- Other ideas:

  _____

  _____

Create a hypothesis to test regarding the effects of _____ (your test substance) on DNA.

_____

_____

Procedure

1. Collect two Petri plates that contain sterile Ames Agar. Label the **bottom** of each plate as either *Control* or *Test*.

2. Collect two test tubes of sterile Ames liquid agar from the 55°C water bath.

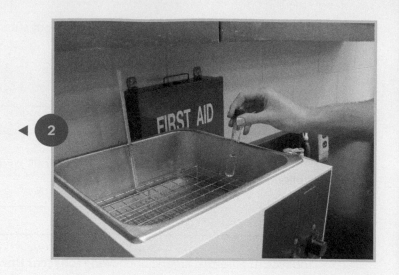

3. Aseptically transfer 0.1 mL of *Salmonella typhimurium* broth culture to each test tube. Refer to *A Closer Look at Aseptic Technique* in Lab 17 (Microorganisms and the Human Body), as needed, to review aseptic technique.

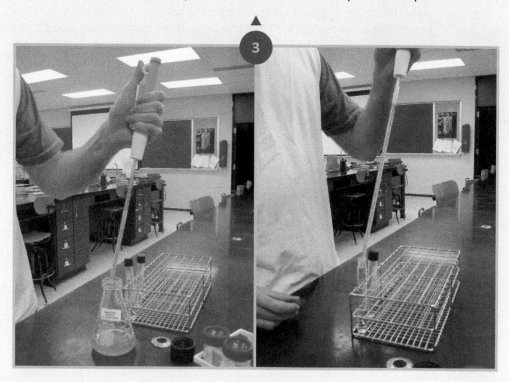

4. Add 0.1 mL of biotin–histidine solution to each test tube. This small amount of histidine gives *S. typhimurium* enough time to mutate. (If a small amount of histidine was not added, then all of the bacteria would immediately die.)

5. Mix the contents of one test tube by rolling it between your palms.

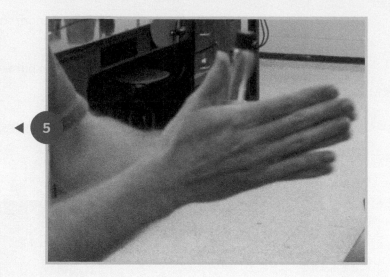

6. Open one Ames agar plate, and pour the contents of the mixed test tube over the solid agar.

7. Gently tilt the plate around to evenly distribute the liquid. Place the lid back onto the Petri plate, and let the liquid agar solidify for 5 minutes.

8. Repeat steps 5–7 with the other test tube and Ames agar plate.

9. Following the solidification period, sterilize a pair of forceps as described below.

   a. Light a Bunsen burner with a striker.

b. Insert the tips of the forceps into a container of ethanol. **Note: Do not get ethanol on anything other than the *tips* of the forceps!**

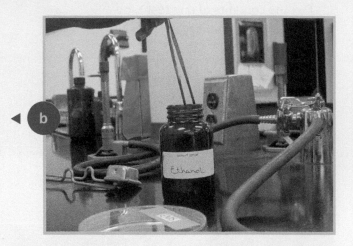

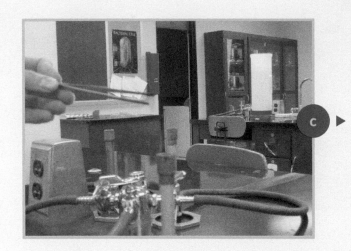

c. *Quickly* pass the tips of the forceps through the flame. The tips will get heat sterilized when they light on fire. **Note: Do not hold the forceps in the flame for a prolonged period of time.**

10. Using the sterile forceps, transfer one *Test* chemical disc to the center of the test plate. This disc is saturated with the liquid you are testing.

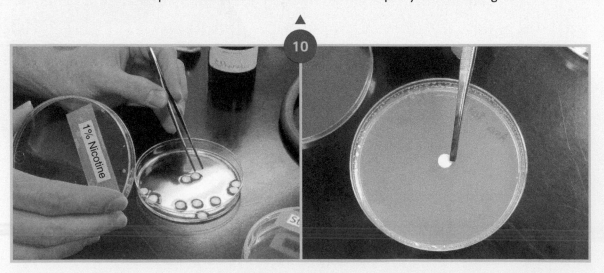

11. Re-sterilize your forceps. Using the sterile forceps, aseptically transfer one *Sterile Water* disc to the center of the control plate.

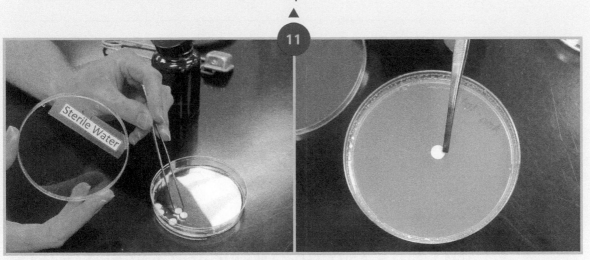

12. Incubate both plates in a 37°C incubator for 48 hours.

Next Lab Period

1. Remove your agar plates from the incubator or refrigerator.

2. Count the number of bacterial colonies growing on each plate, and record your results in the space provided below.

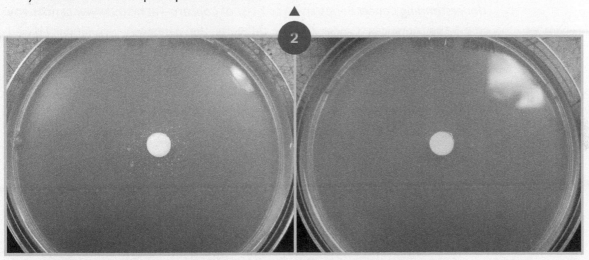

Test Substance: _____

Control Plate: _____ colonies

Test Plate: _____ colonies

Review Questions for Step by Step 16.3

Was your initial hypothesis supported or rejected by the results of the Ames test? Explain how you reached this conclusion.

_____

_____

_____

Did any other students test the chemical that you chose? If so, how do your results compare with those obtained by other students?

_____

_____

What are potential sources of error in this experiment? What improvements (if any) could be made to this experiment?

_____

_____

A CLOSER LOOK AT THE NATIONAL CANCER INSTITUTE AND THE AMERICAN CANCER SOCIETY

The National Cancer Institute (NCI) is one of 11 agencies that belong to the United States Department of Health and Human Services. The mission of NCI is to conduct and support "research, training, health information dissemination, and other programs with respect to the cause, diagnosis, prevention, and treatment of cancer, rehabilitation from cancer, and the continuing care of cancer patients and the families of cancer patients." You can access information on the NCI website—including the *Understanding Cancer Series* and *A to Z List of Cancers*—at **http://www.cancer.gov**.

The American Cancer Society (ACS) "is the nationwide, community-based, voluntary health organization dedicated to eliminating cancer as a major health problem by preventing cancer, saving lives, and diminishing suffering from cancer, through research, education, advocacy, and service." The ACS website contains a wealth of information for patients, survivors, family and friends, researchers, and medical professionals. You can access information from ACS online at **http://www.cancer.org**.

1. Why are chemicals screened with the Ames test?

    _____

    _____

2. Label each tissue biopsy as *normal* or *abnormal*. How did you reach this conclusion?

    _____

    _____

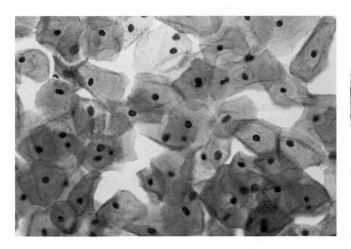 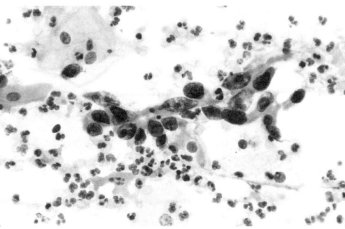

    _____   _____

3. A cancer that originates from epithelial tissue is called a _____.
    a. sarcoma
    b. leukemia
    c. carcinoma
    d. lymphoma

4. What is the purpose of running an invasion assay?

    _____

    _____

5. What is the difference between a benign tumor and a malignant tumor?

    _____

    _____

6. Which of the following processes is illustrated below?

   a. metastasis   c. remission

   b. angiogenesis   d. apoptosis

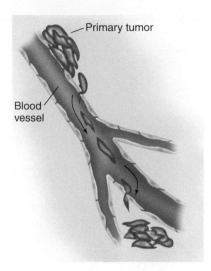

7. True or False? Most cancers are sporadic in nature, not heritable.

8. Based on the Ames test results shown below, does substance 1 appear to be a mutagen? Why or why not?

   _____

   _____

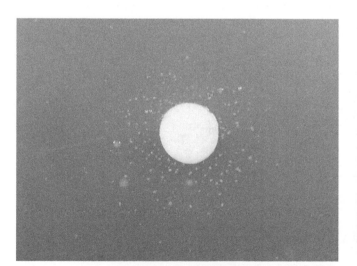

Test Plate — Substance 1

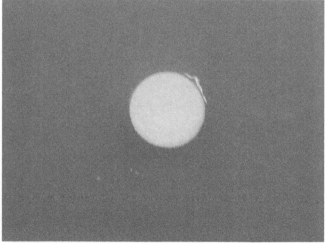

Control Plate

9. The biopsy below originated from a skeletal muscle. Do carcinomas, sarcomas, leukemias, or lymphomas originate from muscle tissue?

_____

_____

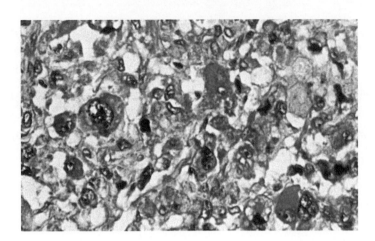

10. How do cancer cells metastasize to different parts of the body?

_____

_____

_____

LAB 17:

Microorganisms and the Human Body

By the End of This Lab, You Should Be Able to Answer the Following Questions:

- Are all bacteria harmful to humans?

- How can you distinguish bacterial cells from animal cells on a microscope slide?

- What is the purpose of using aseptic techniques?

- What characteristics allow us to differentiate bacterial species from one another?

- Why would a health professional need to identify the bacteria growing in a patient's specimen?

- What is a pathogen? What is epidemiology?

- How do epidemiologists determine the prevalence of a disease in a population?

- What is the purpose of running the Kirby-Bauer test?

INTRODUCTION

Did you know that bacteria are living inside of your body at this very moment? Although we tend to associate bacteria with illness, the truth is not all bacteria cause disease. In fact, some bacterial species are actually beneficial to our health. As shown in Figure 17.1, **normal bacterial floras** reside in the mouth, the nose, the intestines, and a number of additional organs. The normal flora aids in digestion and competes with disease-causing bacteria for space and nutrients. Certain yeast species, such as *Candida albicans*, also belong to the normal flora of the human body. Yeast are **unicellular**, or one-celled, members of the Kingdom Fungi.

Humans only inhabit certain regions of the planet, since our livelihood requires strict temperature conditions, oxygen levels, and pH conditions. Bacterial species are so diverse, however, that they have been identified in almost every habitat that exists on the planet: water, soil, vegetation, the Earth's crust, hot springs, animal bodies, the atmosphere, and so on. Some bacterial species are beneficial to humans, such as the ones that typically belong to the normal flora. Other bacterial species coexist with us on Earth in a neutral manner, since they neither benefit nor harm us. Certain bacterial species are considered **pathogenic**, since they can transmit infectious diseases to other organisms, such as humans.

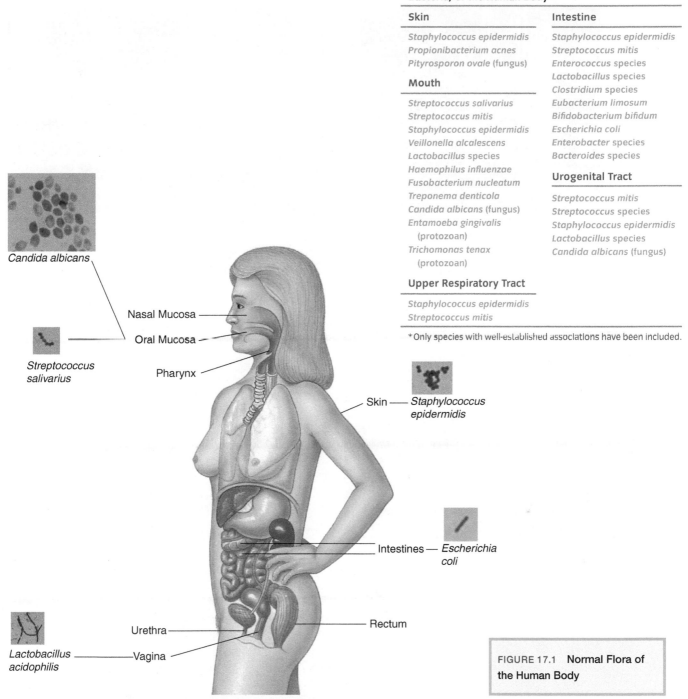

Major Normal Microflora (Unless Otherwise Noted, Bacteria) of the Human Body*

Skin	Intestine
Staphylococcus epidermidis	*Staphylococcus epidermidis*
Propionibacterium acnes	*Streptococcus mitis*
Pityrosporon ovale (fungus)	*Enterococcus* species
	Lactobacillus species
Mouth	*Clostridium* species
Streptococcus salivarius	*Eubacterium limosum*
Streptococcus mitis	*Bifidobacterium bifidum*
Staphylococcus epidermidis	*Escherichia coli*
Veillonella alcalescens	*Enterobacter* species
Lactobacillus species	*Bacteroides* species
Haemophilus influenzae	
Fusobacterium nucleatum	**Urogenital Tract**
Treponema denticola	*Streptococcus mitis*
Candida albicans (fungus)	*Streptococcus* species
Entamoeba gingivalis	*Staphylococcus epidermidis*
(protozoan)	*Lactobacillus* species
Trichomonas tenax	*Candida albicans* (fungus)
(protozoan)	

Upper Respiratory Tract

Staphylococcus epidermidis
Streptococcus mitis

*Only species with well-established associations have been included.

Candida albicans

Streptococcus salivarius

Nasal Mucosa
Oral Mucosa
Pharynx

Skin — *Staphylococcus epidermidis*

Intestines — *Escherichia coli*

Rectum

Urethra
Vagina

Lactobacillus acidophilis

FIGURE 17.1 **Normal Flora of the Human Body**

ASEPTIC TECHNIQUES

During this lab period, you will work with items that contain microorganisms, such as yeast and bacteria. By definition, a **microorganism** (or **microbe**) is too small to be seen with the naked eye. While working with microbes, measures must be taken to prevent contamination of your body, the surrounding environment, *and* the specimens that are being tested. **Aseptic techniques** are strategies used to maintain sterility, which is the absence of microbial contamination. Aseptic techniques are commonly used by microbiologists, health care professionals, and students who work with microorganisms in a laboratory setting. The following aseptic techniques should be utilized during this lab period:

1. Disinfect your lab bench at the beginning and end of the lab period.

2. Use sterile tools to collect and transfer specimens for testing: sterile cotton swabs, metal loops sterilized with heat or chemicals, and so on.

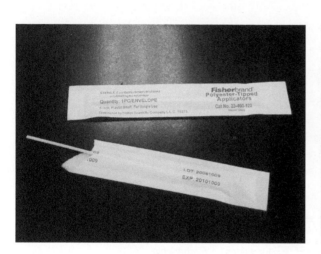

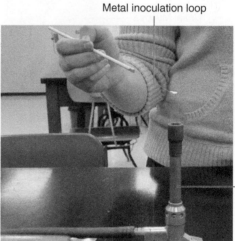

Metal inoculation loop

Bunsen burner

3. Place all contaminated materials (example: used cotton swabs) into a biohazard bag. These materials will be sterilized prior to disposal.

4. Be sure to wash your hands thoroughly at the beginning and end of the lab period.

During this exercise, you will detect microorganisms that are living on an object (example: a door handle) or inside of a substance (example: aquarium water) of your choice. Before you perform this exercise, brainstorm as the class about potential environments to test. Several ideas are included below.

- Your shoes
- Cell phones
- Pens, pencils, and papers
- Door handles
- Your hair
- Your skin
- Body orifices (throat, mouth, ears) *only if permitted by your instructor*

Create a hypothesis to test regarding the presence (or absence), abundance, and diversity of microbes in the environment you are testing.

Environment Tested: _____

Hypothesis:

_____

_____

Procedure

1. Collect one Petri plate filled with sterile nutrient agar. (This media contains the nutrients required for microbial survival and reproduction.)

    Note: Do not remove the lid until instructed to do so!

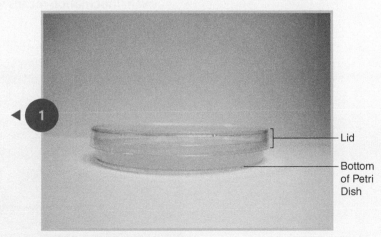

Petri Dish Filled with Nutrient Agar

Lid

Bottom of Petri Dish

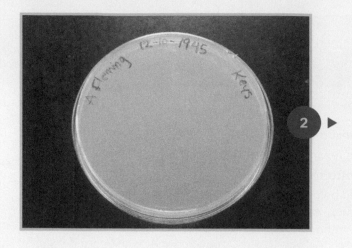

2. Label the bottom of the plate with your name, the date, and the environment you are testing.

3. Unwrap a sterile cotton swab and use it to swab the environment of your choice.

   Note: If you are swabbing a dry environment, moisten the cotton swab with sterile water prior to swabbing.

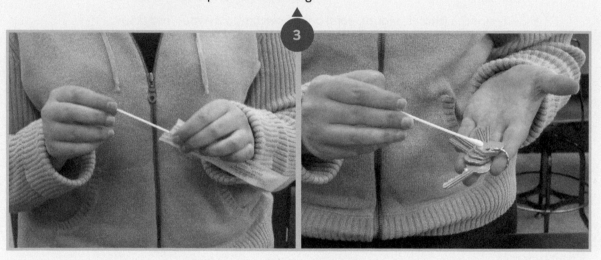

4. Inoculate the surface of the nutrient agar by creating a zigzag with the cotton swab. **Inoculation** means you are introducing microorganisms into the sterile growth media.

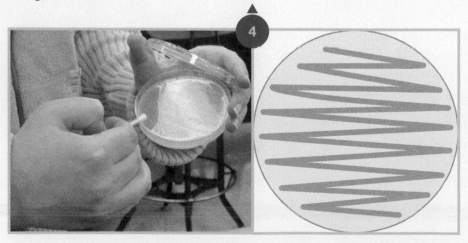

5. Place the lid back on the Petri plate.

6. Place your Petri plate upside down in the location specified by your instructor. To promote microbial growth, the Petri plate may be incubated inside of a 37°C incubator for 24–48 hours.

Next Lab Period

1. Examine your Petri plate for any signs of microbial growth.

As shown below, a colony is a visible cluster of microbes that all descended from one original cell. As a result, every microbe in the colony should be genetically identical. Do you see any isolated colonies growing on the surface of your media? If so, how would you describe the shape, texture, and color of these colonies?

_____

_____

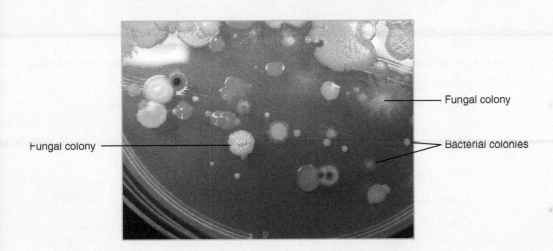

2. Compare your Petri plate with ones created by other students. Pay attention to the environment tested, the abundance of growth on each plate, and the diversity of growth on each plate.

Review Questions for Step by Step 17.1

Was your initial hypothesis supported or rejected by the results of this experiment? If necessary, revise your initial hypothesis in the space provided below.

_____

_____

Were bacteria, fungi, or a mixture of the two growing on your Petri plate? How can you distinguish bacterial colonies and fungal colonies?

_____

_____

In regards to microbial abundance and microbial diversity, how did your environment compare to those tested by other students?

_____

_____

What are potential sources of error in this experiment? What improvements (if any) could be made to this experiment?

_____

_____

DISTINGUISHING BACTERIAL CELLS FROM ANIMAL CELLS

It practically goes without saying that bacteria and humans are quite different from one another. With that said, though, it may be surprising to learn that we also share many similarities with bacteria. The following traits are shared by *all* living organisms, including bacteria, fungi, plants, protists, and animals:

- Cells are the basic building blocks, or structural units, of all living organisms.

- The same four "letters" (DNA bases) are found in every organism's DNA alphabet.

- ATP is the universal energy currency for cellular activities.

- A method of reproduction must exist to perpetuate the species.

Some differences are obvious between humans and bacteria, such as overall size. Bacteria cannot be seen with the naked eye since they are unicellular organisms. In addition to the size difference found at the organismal level, bacterial cells are approximately 100 times smaller than animal cells. For this reason, we must use the 100x (oil immersion) objective to view bacteria on a light microscope.

Humans are multicellular organisms with specialized tissues, organs, and organ systems. Each organ system focuses on a specific aspect of homeostasis for the good of the entire body. Skin, for instance, protects delicate internal organs from changes in the external environment. But how do bacteria withstand environmental changes, since they are only composed of a single cell? As shown in Figure 17.2, bacteria have an additional layer of armor that is not found on human cells: a rigid barrier is called the **cell wall**. A molecule called **peptidoglycan** is the main component of cell walls in bacteria. As the name implies, this molecule is composed of proteins (*peptido-*) and carbohydrates (*-glycan*). Animal cells also differ from bacterial cells based on the presence or absence of a nucleus. Human cells are considered **eukaryotic** since DNA is sequestered within a true (*eu-*) nucleus (*karyo-*). Bacteria, on the other hand, are considered **prokaryotic** organisms since they evolved before (*pro-*) organisms with nuclei (*karyo-*) existed on earth.

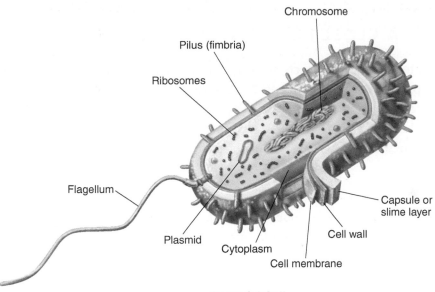

Bacterial Cell

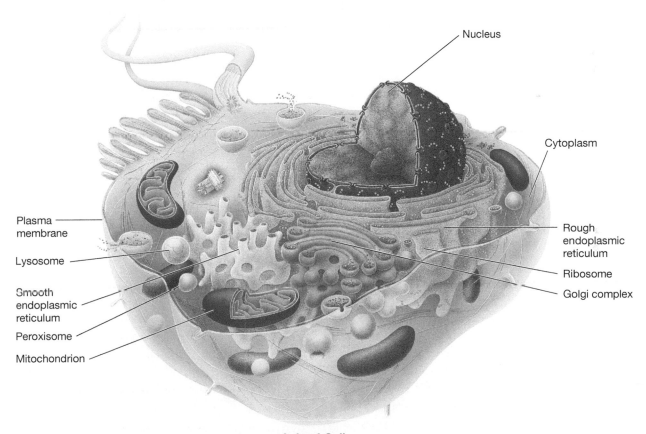

Animal Cell

FIGURE 17.2 **Structural Differences in Bacterial and Animal Cells**

A CLOSER LOOK AT OIL IMMERSION

Since bacterial cells are much smaller than animal cells, we will modify the microscopy techniques used in Lab 2 (Observing Cells with Light Microscopy) to view bacteria at higher levels of magnification. More specifically, two additional steps will be used to view bacteria: (1) application of oil to the microscope slide and (2) use of the 100x (oil immersion) objective.

Procedure

1. Focus the bacterial slide under the 4x, 10x, and 40x objectives. Refer to Lab 2, as needed, to review the steps for proper usage of a compound light microscope.

2. Once the image is focused using the 40x objective, turn the nosepiece so that it is balanced halfway between the 40x and 100x objectives. This creates a space for the application of immersion oil.

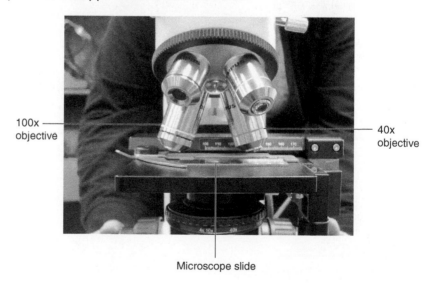

100x objective

40x objective

Microscope slide

3. Place one drop of immersion oil on top of the microscope slide; this drop should be placed directly over the light beam traveling through the slide.

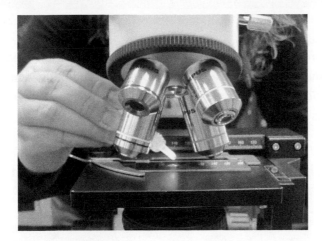

4. Turn the nosepiece and position the 100x objective directly above the light beam. The objective lens will be immersed in oil once this step is completed.

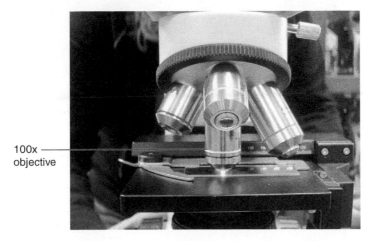

100x objective

5. Use the fine focus knob to bring the image into sharp focus.

6. When you are finished viewing the slide, use lens paper and lens cleaner (if available) to remove any residual oil from the 100x objective.

During this exercise, you will observe a prepared microscope slide that contains a mixture of animal cells and bacterial cells. Compare your focused microscope slide to the photomicrograph provided below, and answer the review questions that accompany this exercise.

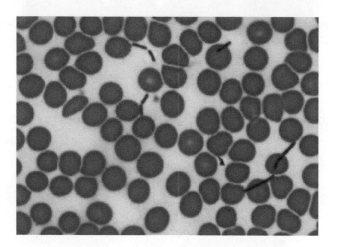

Label a red blood cell and a bacterial cell on the image above. How can you tell these cells apart?

_____

_____

How would you describe the shape of the bacteria on this microscope slide?

_____

_____

The presence of bacteria in blood indicates that a patient has a severe infection called bacteremia. What does the suffix –*emia* mean? Refer to Lab 6 (The Cardiovascular System), if needed, to help answer this question. What treatment options are available for patients with bacteremia?

_____

_____

_____

During this exercise, you will observe bacteria derived from the plaque in your own oral cavity. Good oral hygiene (brushing your teeth, flossing, etc.) helps reduce plaque buildup and the development of associated dental diseases, such as dental caries (cavities) and gingivitis (gum disease). Answer the following questions before you perform this experiment.

How long has it been since you consumed your last meal? Did you brush your teeth and/or floss after this meal? _____

How long has it been since you brushed your teeth? _____

Optional: Do you have a history of gum disease? _____

Procedure

1. Using aseptic technique, place a small drop of water on the center of a microscope slide.

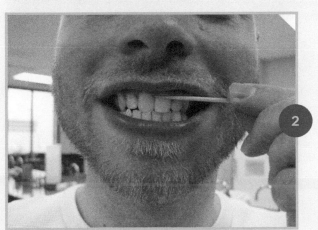

2. Unwrap a toothpick and gently scrape your teeth near the gum line.

3. Mix the scrapings from your teeth into the water on the microscope slide.

4. Place the toothpick into a biohazard bag for disposal.

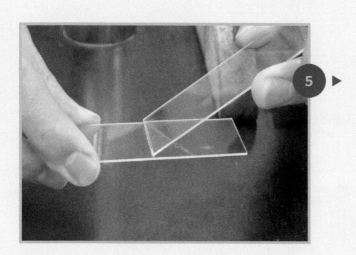

5. Spread the water into a thin film using another microscope slide. Place the slide you used for spreading in a biohazard bag.

6. Let your microscope slide air dry.

7. Secure one end of your microscope slide into a slide holder.

8. Quickly pass your slide through the flame on a Bunsen burner. *Repeat this step two additional times.* The heat will fix bacteria onto the glass microscope slide; bacteria are also killed by heat during this process.

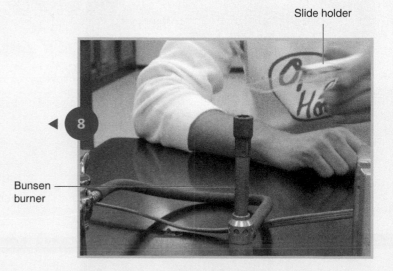

Slide holder

Bunsen burner

9. Set your microscope slide on top of a staining rack and flood its surface with methylene blue stain.

10. Let the stain sit on your microscope slide for one minute.

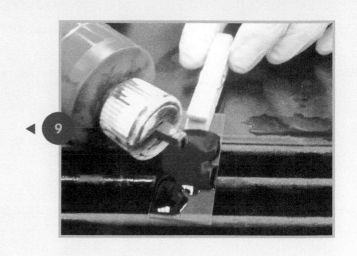

11. Gently rinse the excess stain off of your slide with water.

12. Gently blot the microscope slide with a paper towel.

13. Observe your microscope slide under the 100x (oil immersion) objective. Refer to *A Closer Look at Oil Immersion*, as needed, while performing microscopy.

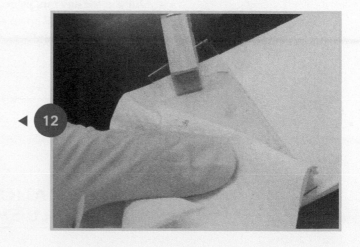

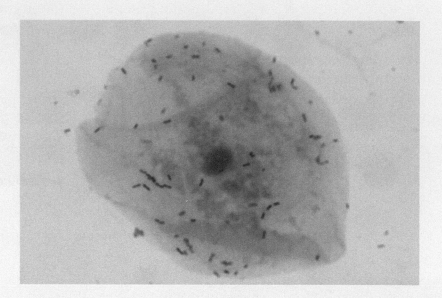

Label a cheek cell, its nucleus, and a bacterial cell on the picture above. How can you tell cheek cells apart from bacterial cells?

_____

_____

How would you describe the shapes, arrangements, and abundance of bacteria in your dental plaque?

_____

_____

Did you see any tan, jagged objects on your microscope slide? If so, how do you think these objects ended up on your slide? (_Hint_: They are not derived from your cells, bacterial cells, or the stain!)

_____

_____

CLASSIFICATION OF BACTERIA: THE GRAM STAIN, CELL SHAPE, AND CELL ARRANGEMENT

Bacteria are divided into two major groups based on differences in their cell wall composition: Gram-positive bacteria and Gram-negative bacteria. The Gram stain procedure is summarized in Figure 17.3, and the details of this technique are described in _A Closer Look at the Gram Stain_ later in this lab. **Gram-positive bacteria** display a purple color following the Gram stain, while **Gram-negative bacteria** display a pink color.

Gram-positive coccus

Application of Crystal Violet
(Primary Stain)

Application of Gram's Iodine
(Mordant)

Alcohol rinse (decolorizes
Gram-negative bacteria)

Application of Safranin
(counterstains Gram-negative
bacteria)

Gram-negative bacillus

FIGURE 17.3 **The Gram Stain**

Within a medical setting, this information can be used to select appropriate antibiotics for patients with bacterial infections. Infections caused by Gram-positive pathogens, such as *Staphylococcus aureus*, may be treatable with antibiotics such as penicillin. Penicillin cannot treat bacterial infections caused by Gram-negative pathogens, though, such as *Neisseria gonorrhoeae*.

In addition to their Gram reactions, bacteria are also distinguished based on differences in their cell shapes and cell arrangements:

Common Bacterial Shapes

- The term *cocci* refers to spherical bacteria (*coccus* = singular)

- The term *bacilli* refers to rod-shaped bacteria (*bacillus* = singular)

- The term *spirilli* refers to spiral-shaped bacteria (*spirillum* = singular)

Common Bacterial Arrangements

- The prefix *diplo-* indicates that bacterial cells are arranged in pairs (example: diplococcus)

- The prefix *strepto-* indicates that bacterial cells are arranged in chains (examples: streptococcus and streptobacillus)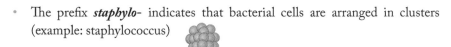

- The prefix *staphylo-* indicates that bacterial cells are arranged in clusters (example: staphylococcus)

The names of some bacterial species actually give us clues about their physical attributes, their location(s) in the host's body, and/or infectious diseases caused by the species. Examples include *Staphylococcus epidermidis*, *Streptococcus salivarius*, and *Streptococcus pneumoniae*. Cell shape, cell arrangement, and Gram reaction are all clues that help microbiologists and health professionals identify different bacterial species.

During this exercise, you will observe prepared microscope slides of several different bacterial species. Identify the cell shape, cell arrangement, and Gram reaction (Gram-positive or Gram-negative) for each bacterial species you observe. To test your knowledge after performing microscopy, fill out the chart provided below.

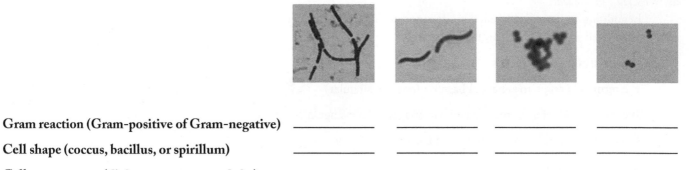

Gram reaction (Gram-positive of Gram-negative) _____ _____ _____ _____

Cell shape (coccus, bacillus, or spirillum) _____ _____ _____ _____

Cell arrangement (*diplo-, strepto-, or staphylo-*) _____ _____ _____ _____

A CLOSER LOOK AT THE GRAM STAIN

In 1884, the Gram stain was invented by a Danish microbiologist named Hans Christian Gram. Since this staining technique is quick and inexpensive to perform, it is still frequently used to distinguish bacterial species from one another. The Gram stain exploits the structural differences in the cell walls of Gram-positive and Gram-negative bacteria. As shown in Figure 17.4, the cell walls of Gram-positive bacteria contain a thick layer of peptidoglycan. In contrast, the cell walls of Gram-negative bacteria contain a thin layer of peptidoglycan surrounded by another lipid membrane. These structural differences are evident following the Gram stain due to differences in stain retention.

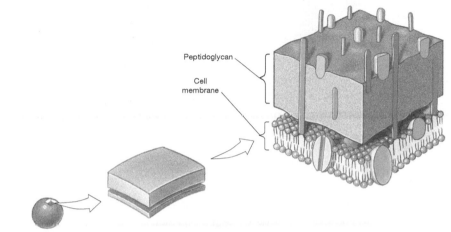

Peptidoglycan

Cell
membrane

(a) Gram-positive bacteria

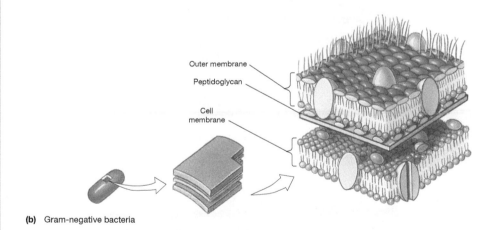

Outer membrane

Peptidoglycan

Cell
membrane

(b) Gram-negative bacteria

FIGURE 17.4 **Cell Walls of Gram-Negative and Gram-Positive Bacteria**

Gram Stain Procedure

1. Collect a clean microscope slide, an inoculation loop, a slide holder, and a broth (liquid) culture of bacteria.

(continued)

2. Transfer two loopfuls of the broth culture onto the surface of the microscope slide. Use the loop to spread the culture into a thin film, and allow the liquid to air dry.

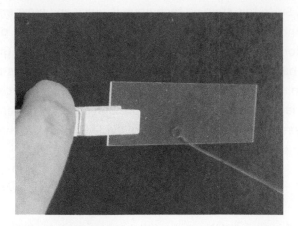

3. Secure one end of your microscope slide into a slide holder.

4. Quickly pass your slide through the flame on a Bunsen burner. *Repeat this step two additional times.* The heat will fix bacteria onto the glass microscope slide; bacteria are also killed by heat during this process.

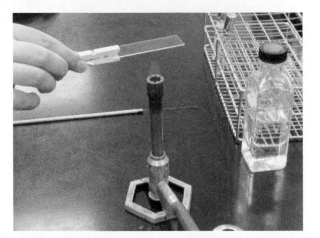

5. Set your slide on top of a staining rack and flood its surface with crystal violet stain. Let the stain sit on the microscope slide for one minute.

6. Gently rinse the excess stain off of your slide with water.

7. Flood the slide with Gram's iodine and allow it to set for one minute.

8. Pour the Gram's iodine off of your microscope slide.

9. While holding your slide at a 45° angle, apply a continuous stream of 95% ethanol for 5 to 7 seconds. **Immediately following this step, rinse the ethanol off with water to stop the decolorization process**.

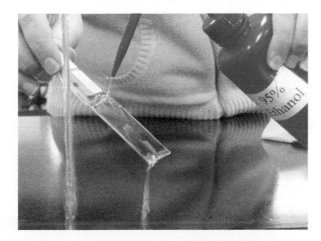

(continued)

10. Flood the slide with Safranin stain and allow it to set for one minute.

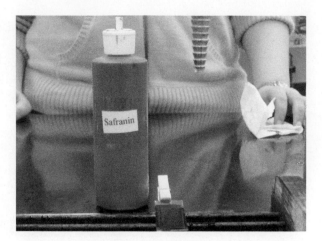

11. Gently rinse the slide with water and blot dry.

12. Observe your microscope slide under the 100x (oil immersion) objective. Refer to *A Closer Look at Oil Immersion* as needed, while performing microscopy.

Review Questions

Are the bacteria on your slide Gram-positive, Gram-negative, or a mixture of the two?

_____

_____

How would you describe the shape(s) and arrangement(s) of bacteria on your microscope slide?

_____

_____

FOOD MICROBIOLOGY

As with most things in life, there are positive and negative sides to food microbiology. Food poisoning definitely falls into the negative category, as most of us know from personal experience! Usually, the symptoms of food poisoning—vomiting, diarrhea, fever, and/or abdominal cramping—are relatively mild and resolve within a day. However, severe cases of food poisoning can lead to extreme dehydration, organ failure, paralysis, and in some cases even death. Pathogenic strains of *Escherichia coli*, *Salmonella typhimurium*, *Shigella sonnei*, and *Staphylococcus aureus* are common culprits in food poisoning cases.

On a positive note, many foods get their signature flavors and textures from microbial activity. A cheeseburger just wouldn't be a cheeseburger without the bun, a slice of cheese, and a few pickle slices. Bacterial species such as *Lactobacillus lactus* give cheese its unique flavor and curd-like texture; this is accomplished as lactose (milk sugar) gets converted into lactic acid. Bacterial species such as *Lactobacillus brevis* give pickles and sauerkraut their tangy flavors by converting sugars into lactic acid. Bread dough rises due to the addition of yeast species such as *Saccharomyces cerevisiae*. As yeast metabolize sugars in the dough, carbon dioxide and ethanol are produced. Although the ethanol quickly evaporates away, bubbles of carbon dioxide get trapped in the dough and make it rise.

STEP BY STEP 17.5 Observing Bacteria in Yogurt

Visualizing THE LAB

Procedure

1. Using a toothpick, place a small drop of plain yogurt on a microscope slide. Spread the yogurt into a thin film with the toothpick.

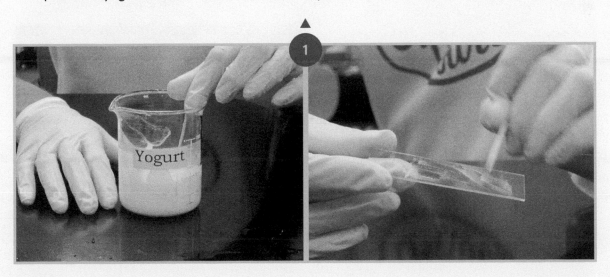

2. Secure one end of your microscope slide in a slide holder.

3. Quickly pass your slide through the flame on a Bunsen burner. *Repeat this step two additional times.* The heat will fix bacteria onto the glass microscope slide; bacteria are also killed by heat during this process.

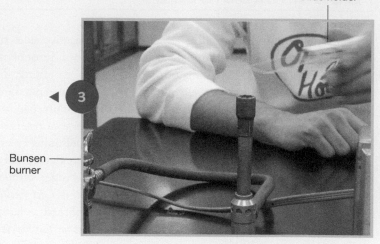

Slide holder

3

Bunsen burner

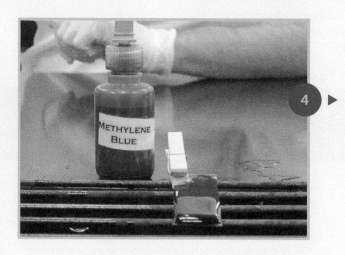

4

4. Set your slide on a staining rack and flood its surface with methylene blue stain.

5. Let the stain sit on the microscope slide for one minute.

6. Gently rinse the slide off with water and blot dry.

7. Observe your microscope slide under the 100x (oil immersion) objective. Refer to *A Closer Look at Oil Immersion*, as needed, while performing microscopy.

6

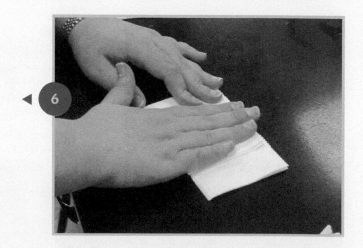

Draw a picture of the bacteria you observed in the yogurt culture. How would you classify the shape(s) and arrangement(s) of these bacteria?

_____

_____

Yogurt gets its tangy flavor when lactose (milk sugar) gets converted into lactic acid. Does lactic acid fermentation ever occur in the human body? If so, where and why does it occur?

_____

_____

EPIDEMIOLOGY

Epidemiologists are medical scientists who study the origin, transmission, distribution, and prevention of infectious diseases. For example, epidemiologists at the Centers for Disease Control and Prevention (CDC, www.cdc.gov) track the spread of infectious disease and simultaneously implement strategies to reduce disease transmission. Initially, epidemiologists collect raw (unprocessed) data on individuals who have contracted a particular disease. This data helps them identify the **index case**, or the first documented case, of an infectious disease within the population. By performing statistical analysis on this data, epidemiologists can also determine 1) the **prevalence** of the disease, which is the percentage of individuals who are infected throughout the population, 2) the groups who are at high risk of contracting and/or dying from this disease, and 3) trends regarding the geographical spread of this disease.

1. Choose a numbered Petri dish with a piece of
    candy inside. (One piece of candy has been
    "infected" with a safe fluorescent powder. The
    other pieces of candy have been coated in
    powdered sugar.) Pick up the candy and smear it
    around your right hand.

2. Drop the candy back into the Petri dish. Do not touch anything with your right
    hand until instructed to do so!

3. Line up at the front of the classroom in numerical order. Student 1 (located
    at one end of the line) firmly shakes hands with student 2. Student 2 then
    shakes hands firmly with student 3, and so on, until all students have partici-
    pated in two handshakes.

Note: To complete the handshaking cycle, the last student in line should shake
hands with student 1.

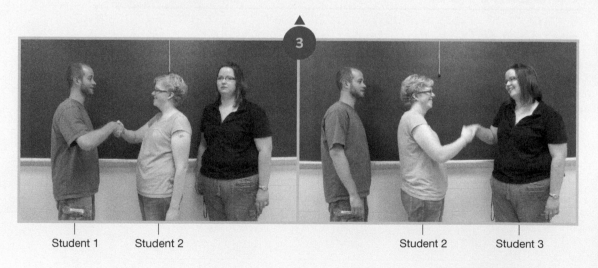

Student 1   Student 2   Student 2   Student 3

4. Turn off the lights in the laboratory.

5. Using a handheld ultraviolet light, the lab instructor will check your right hand for any signs of fluorescence. Fluorescence indicates a positive result for "infection."

◄ 5

6. Wash your hands thoroughly with soap and water.

7. Following hand washing, the lab instructor will reexamine your hands for any signs of fluorescence.

Review Questions for Step by Step 17.6

Determine the index case in this simulated epidemic. How do you know this student is the index case?

_____

_____

Using the equation below, determine the prevalence of this "disease" in the population:

$$\left(\frac{\text{Number of Students "Infected"}}{\text{Total Number of Students}}\right) \times 100 = \text{Prevalence Rate (\%)}$$

Prevalence Rate _____ %

Was any fluorescent powder detected on your hands after hand washing? What does this tell you about your hand washing technique?

_____

_____

Based on the class results, is hand washing an effective method for controlling diseases with direct contact transmission? Which spots on the hands are most frequently missed during hand washing?

_____

_____

ANTIBIOTICS AND ANTIBIOTIC RESISTANCE

When antibiotics are administered to treat bacterial infections, they must selectively kill bacteria without harming the patient's own cells. To do this, antibiotics exploit the structural differences between bacterial and human cells, such as the presence or absence of a cell wall. For example, penicillin targets the cell walls of Gram-positive bacteria. Since its discovery in 1928, penicillin has been used to treat Gram-positive bacterial infections, such as those caused by pathogenic *Staphlylococcus* and *Streptococcus* species.

Antibiotic resistance is currently a hot topic in medicine, research, and even the mainstream media. In order to discuss antibiotic resistance, we must define the terms *susceptibility* and *resistance*. If a particular bacterial strain is **susceptible** to an antibiotic, this means it is vulnerable to the activity of the antibiotic. When a strain of *Staphylococcus aureus* is susceptible to penicillin, for instance, then penicillin should cure infections caused by this pathogen. When a bacterial strain is **resistant** to an antibiotic, this means the bacteria can survive in its presence. Some strains of *Staphylococcus aureus* have become resistant to penicillin; as a result, penicillin is no longer a viable treatment option for all *Staphylococcus aureus* infections.

Genes for antibiotic resistance are often passed from one bacterium to another on small, circular DNA molecules called **plasmids**. Such genes code for proteins that allow the bacteria to withstand—or avoid—the effects of a particular antibiotic. For instance, an antibiotic resistance gene may code for an enzyme that degrades the antibiotic. Figure 17.5 illustrates plasmid structure and several proteins that can impart antibiotic resistance.

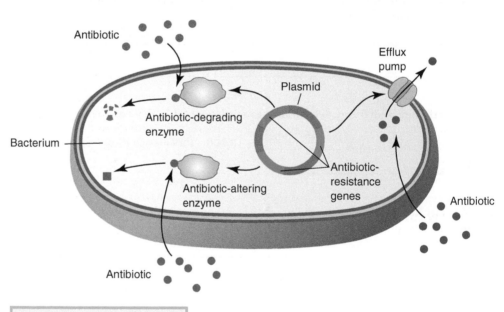

FIGURE 17.5 **Plasmid DNA and Antibiotic Resistance**

The Kirby-Bauer test is used around the world to determine whether bacterial strains are susceptible or resistant to different antibiotics. In the medical field, for instance, this test is used to determine the best course of treatment for severe bacterial infections. During the Kirby-Bauer test, a pure bacterial culture is grown in the presence of antibiotic discs. Following an incubation period, the growth around each antibiotic disc (or lack thereof) is analyzed. As shown in Figure 17.6, a **zone of inhibition** is simply a clearing around the antibiotic disc. The diameter of this clearing—the zone of inhibition—is compared to a standardized chart, and the results indicate whether the bacterial species is susceptible or resistant to the antibiotic.

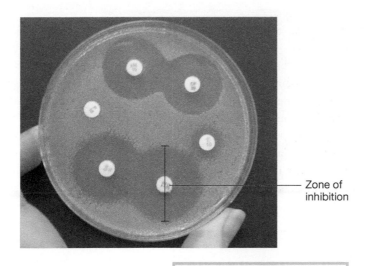

Zone of inhibition

FIGURE 17.6 **Results of the Kirby-Bauer Test**

The terms *susceptible* and *resistant* have already been defined, but there are two additional terms used to categorize Kirby-Bauer results: *moderately susceptible* and *intermediate*. If a bacterial species is **moderately susceptible** to an antibiotic, then this organism should still be responsive to the effects of the antibiotic. When a bacterial species is deemed **intermediate** to an antibiotic, though, then this antibiotic is certainly not ideal for treating infections caused by this organism. Antibiotic resistance can gradually build up over generations; therefore, classifications such as *moderately susceptible* and *intermediate* may indicate a progression toward antibiotic resistance.

1. Collect two demonstration plates for the Kirby-Bauer test.

2. Measure the zone of inhibition that surrounds each antibiotic disc. Record this information in Table 17.1.

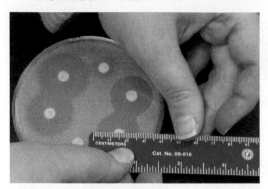

TABLE 17.1   Results of Kirby-Bauer Test

Antibiotic Tested	Zone of Inhibition (Diameter in mm)	Results per Kirby-Bauer Interpretation Chart	Is this antibiotic an effective treatment option for infections caused by this organism?
Bacterial Species 1:			
Bacterial Species 2:			

3. Compare each zone of inhibition to Table 17.2, which contains an interpretation chart for the Kirby-Bauer test. Record the results (Susceptible = S, Moderately Susceptible = MS, Intermediate = I, Resistant = R) in Table 17.1.

4. Determine which antiobiotics (if any) are effective treatment options against the tested organisms. Record the results in Table 17.1.

TABLE 17.2   Interpretation of Kirby-Bauer Results*

Antibiotic	Susceptible	Moderately Susceptible	Intermediate	Resistant
Ampicillin (AM) *Enterobacteriacae* *Staphylococcus* *Enterococci* *Other organisms*	≥ 14 ≥ 29 ≥ 30	≥ 17 22–29	12–13	≤ 11 ≤ 28 ≤ 16 ≤ 21
Augmentin (AMC) *Staphylococcus* *Other organisms*	≥ 20 ≥ 18		14–17	≤ 19 ≤ 13
Cephalothin (CF)	≥18		15–17	≤14
Erythromycin (E)	≥18		14–17	≤13
Gentamycin (GM)	≥15		13–14	≤12
Kanamycin (K)	≥18		14–17	≤13
Naficillin (NF)	≥ 13		11–12	≤ 10
Oxacillin (OX) *Staphylococcus*	≥ 13		11–12	≤10
Penicillin G (P) *Staphylococcus* *Enterococci* *Other organisms*	≥ 19 ≥ 28	≥ 15 20–27		≤ 28 ≤ 14 ≤ 19
Streptomycin (S)	≥ 15		12–14	≤ 11
Tetracycline (TE)	≥ 19		15–18	≤ 14
Trimethoprim (TMP)	≥ 16	11–15		≤ 10
Trimethoprim-sulfamethoxazole (SXT)	≥ 16	11–15		≤ 10
Tobramycin (T)	≥ 15	13–14		≤ 12
Vancomycin (VA)	≥ 12	10–11		≤ 9

*Measurements Given in Millimeters (mm)

REVIEW QUESTIONS FOR LAB 17

1. Are all bacteria harmful to humans? Explain your answer.

_____

_____

2. Label the following bacterial species based on their Gram reactions, cell shapes, and cell arrangements.

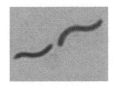   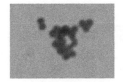

Gram Reaction   _____   _____   _____

Cell Shape and
Arrangement   _____   _____   _____

3. Name three ways that bacteria and humans differ from one another.

_____

_____

4. Name three similarities that are shared by all living organisms, including humans and bacteria.

_____

_____

5. Label an animal cell and a bacterial cell on the image below. How can you tell them apart?

_____

_____

6. _____ study the origin, transmission, distribution, and prevention of infectious diseases.
   a. Oncologists
   b. Podiatrists
   c. Endocrinologists
   d. Epidemiologists

7. Imagine that a small town named Hartford has a population of 500 people. Currently, 100 of the residents are infected with the H1N1 virus. What is the prevalence rate of H1N1 infections in this population?_____

8. Name two aseptic techniques that you used during this lab period.

_____

_____

9. The _____ is the first documented case of an infectious disease within a population.

10. What is a pathogen?

_____

_____

LAB 18:
Evolution

By the End of This Lab, You Should Be Able to Answer the Following Questions:

- How did the structure and function of the human skeleton evolve over time?

- Does biological evolution occur at the individual level or the population level?

- Which structures in human embryos are considered vestigial?

- What lines of scientific evidence are used to create evolutionary (phylogenetic) trees?

- Which bones are homologous in all vertebrates, in spite of their functional differences?

- How do bacterial populations evolve when antibiotics enter the environment?

- What factors drive evolution and speciation?

INTRODUCTION

In 1984, Richard Leakey and his team were hunting for fossils in northern Kenya. **Paleoanthropologists** like Richard Leakey study fossils to learn more about the history of our own species—*Homo sapiens*—as well as the history of our ancient ancestors. During an excavation, it is rare to find enough bones from one **hominid** (a human or human ancestor) to create a partial skeleton for analysis. For this and many other reasons, Leakey's discovery in 1984 amazed and astonished people all over the globe. While digging near Lake Turkana, Leakey's team uncovered a fossilized skeleton (Figure 18.1) from a young hominid boy. Turkana Boy, as he is commonly referred to, lived in Kenya approximately 1.6 million years ago. In addition to the fact that Turkana Boy's skeleton was nearly complete, bone and teeth analysis revealed that he was 7–13 years old at time of death. The fossils of children are particularly useful to paleoanthropologists, since their growth rates can be compared to those of modern-day humans. Although Turkana Boy belonged to a species that has gone extinct (*Homo erectus*), his species is closely related to our own species: *Homo sapiens*. Figure 18.2 delves into the evolutionary relationship between *Homo sapiens*, *Homo erectus*, and other species in the genus *Homo*.

Turkana Boy's Fossilized
Skeleton
Species: *Homo erectus*

Expeditions still continue in the basin of Lake
Turkana. Since 1984, more than 200 hominid
specimens have been discovered in this region.
Turkana Basin Institute: http://turkanabasin.org
Koobi Fora Research Project://www.kfrp.com

FIGURE 18.1 Turkana Boy's
Fossilized Skeleton

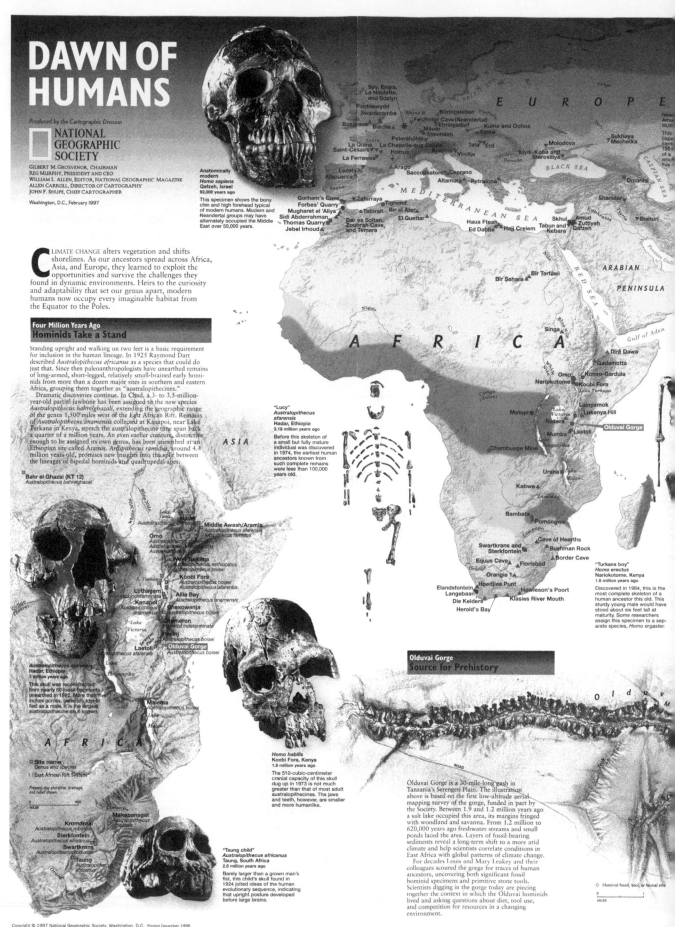

Supplement to NATIONAL GEOGRAPHIC, February 1997

S I B E R I A

SEA OF OKHOTSK

SEA OF JAPAN

Xujiayao
Zhoukoudian

Dingcun
Changwu · Dali · Yiyuan
Lantian Yunxi Nanzhao
Jianshi Yunxian
Longgupo Chaoxian
Tongzi Changyang
Yuanmou Maba

YELLOW SEA
EAST CHINA SEA

HIMALAYA
Brahmaputra
Ganges
▲ Narmada

BAY OF BENGAL

SOUTH CHINA SEA

Plate Carrée Projection
SCALE 1:35,660,000
0   800
MILES

Homo erectus
Zhoukoudian, China
420,000 years ago

Thick browridges and skull walls in this male *erectus* are features shared with African representatives of the species. Fossils and artifacts of "Peking man" have been collected from this cave site since the 1920s.

EUROPE
ASIA
AFRICA
PACIFIC OCEAN
EQUATOR
INDIAN OCEAN
Weber Basin
Java Trench
North Australian Basin
AUSTRALIA

Present-day shoreline
Shoreline 65,000 years ago

☐ Sea ice
☐ Glacier
☐ Tundra
☐ Forest
☐ Grassland, savanna, and open woodland
☐ Loess semidesert
☐ Steppe and desert

0   2,000
MILES

Source: George J. Kukla, Lamont-Doherty Earth Observatory, Columbia University

Sambungmachan
Sangiran · Perning
Trinil
Ngandong

Homo erectus
Trinil, Java, Indonesia
700,000 years ago

Debate over whether it came from a "man-like ape" or an "ape-like man" followed the discovery of this skullcap in 1891. Called Java man, it is the first *erectus* specimen ever collected.

AUSTRALIA

Darling

65,000 Years Ago
Ice Age Routes to New Homes

The map above shows glaciers, vegetation, and shore-lines about 65,000 years ago. Glaciers cover nearly 17 million square miles of earth's surface, and sea levels are more than 400 feet lower than today. Land bridges connect previously separated areas, enabling humans to move into new territories.

Even at this glacial maximum, however, Australia and New Guinea are isolated by the waters of the Java Trench and the North Australian and Weber Basins. The earliest firm evidence of human settlement in Australia dates from about 60,000 years ago, suggesting that some form of boatbuilding had developed by that time,

Design and Olduvai Gorge art: Robert E. Pratt; Research: Jennifer J. Iscol, Kerry Jo Kreiton; Edit: Jonathan E. Kaut; Production: Dianne C. Davis, Neal J. Edwards, Mich A. Laws, Ken Marlow, Stephen P. Wells. Text: M. Lynne Warren, author; Barbara W. McConnell, researcher

Consultants: Leslie Aiello, University College London; Robert J. Blumenschine, Rutgers University; Peter B. deMenocal, Lamont-Doherty Earth Observatory, Columbia University; James Ebert, Ebert & Associates, Inc.; Peter Kershaw, Monash University, Victoria, Australia; Alan Mann, University of Pennsylvania; David Pilbeam, Harvard University; Richard Potts, Smithsonian Institution; Charles Schweger, University of Alberta, Canada; Carl Vondra, Iowa State University; Alan Walker, Pennsylvania State University; Bernard Wood, Liverpool University

Fossil and East Africa art by Rob Wood
Digital relief for main map provided by Jym Terhorst, MountainTop Computing

Today
Prospecting for Hominids

Fossil hominid specimens come from sediments between 10,000 and about five million years old. As the map at right indicates, dense plant growth blocks access to these deposits over much of earth's surface. Finds are most likely in arid, sparsely vegetated regions where erosion has carved through layers of earth. But even in promising areas significant discoveries are exceptional events. Only a tiny fraction of all organisms leave fossil traces anywhere. These constraints suggest that our curiosity about our past will probably always outstrip our ability to collect its remains.

ATLANTIC OCEAN
EUROPE
ASIA
PACIFIC OCEAN
AFRICA
INDIAN OCEAN
EQUATOR
AUSTRALIA

☐ Glacier
☐ Tundra
☐ Mountain vegetation
☐ Coniferous forest
☐ Temperate forest
☐ Tropical deciduous forest and savanna
☐ Rain forest
☐ Steppe and desert
☐ Rocks - 10,000 to five million years old (excludes caves)

0   1,000
MILES

FIGURE 18.2   **Human Evolution**

During this exercise, you will study one aspect of hominid evolution: the evolution of the human skeleton. Refer to Lab 11 (The Skeletal System), as needed, to review the names of the major bones that comprise the human skeleton. Since ancient hominid skeletons are quite rare, a photograph of each skeleton has been provided for you. Compare the structure of your own skeleton (*Homo sapiens sapiens*) to fossilized skeletons from the following hominids: Lucy (*Australopithecus afarensis*), Turkana Boy (*Homo erectus*), and a Neanderthal (*Homo neanderthalensis*).

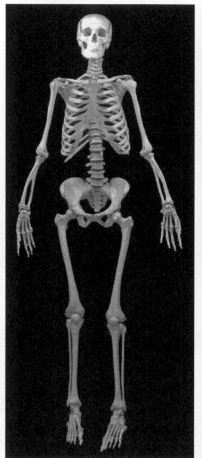
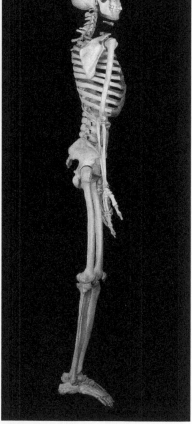
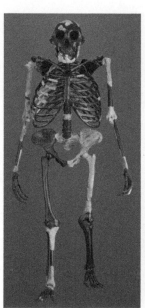

Species: *Australopithecus afarensis* (Lucy)
Fossils Age: ≈3.2 million years old (mya)
Average Female Height: 41" (104 cm) tall
Average Female Weight: 64 lbs (29 kg)
Discovered in 1974 by Donald Johanson and
Maurice Taieb in Hadar, Ethiopia
Photos Courtesy of the Smithsonian Institute

Species: *Homo sapiens sapiens* (female skeleton)
Average Height of an Adult Female: 63.8" (162 cm) tall*
Average Weight of an Adult Female: 164.7 lbs (74.7 kg)*

---

* Source: Anthropometric Reference Data for Children and Adults: United States, 2003–2006. National Center for Health Statistics.

What similarities/differences exist between Lucy's skull and a modern-day human's skull?

Similarities: _____

Differences: _____

What similarities/differences exist between Lucy's ribcage and a modern-day human's ribcage?

Similarities: _____

Differences: _____

What similarities/differences exist between Lucy's pelvis and a modern-day human's pelvis?

Similarities: _____

Differences: _____

What similarities/differences exist between Lucy's limbs and a modern-day human's limbs?

Similarities: _____

Differences: _____

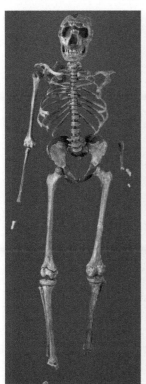
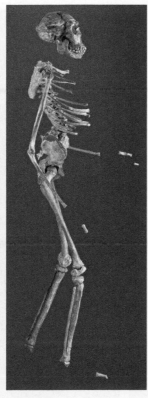
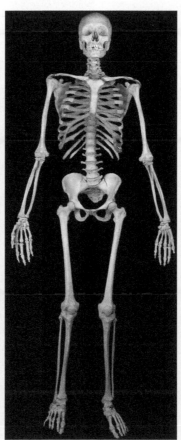

Species: *Homo erectus* (Turkana Boy)
Fossils Age: ≈1.6 million years old (mya)
Height Range for Species: 57" (145 cm) – 73" (185 cm) tall
Weight Range for Species: 88 lbs (40 kg) – 150 lbs (68 kg)
Discovered in 1984 by Kamoya Kimeu and Richard
Leakey in Nariokotome, West Turkana, Kenya
Photos Courtesy of the Smithsonian Institute

Species: *Homo sapiens sapiens* (male skeleton)
Average Height of an Adult Male: 69.4"
(176 cm) tall*
Average Weight of an Adult Male: 194.7 lbs (88.3
kg)*

----

* Source: Anthropometric Reference Data for Children and Adults: United States, 2003–2006. National Center for Health Statistics.

What similarities/differences exist between Turkana Boy's skull and a modern-day human's skull?

Similarities: _____

Differences: _____

What similarities/differences exist between Turkana Boy's ribcage and a modern-day human's ribcage?

Similarities: _____

Differences: _____

What similarities/differences exist between Turkana Boy's pelvis and a modern-day human's pelvis?

Similarities: _____

Differences: _____

What similarities/differences exist between Turkana Boy's limbs and a modern-day human's limbs?

Similarities: _____

Differences: _____

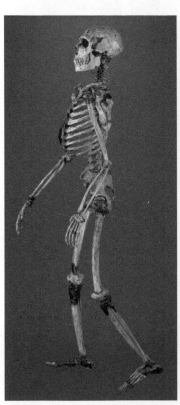
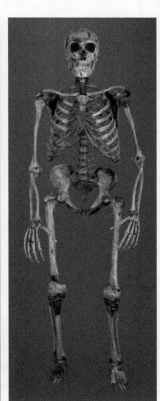
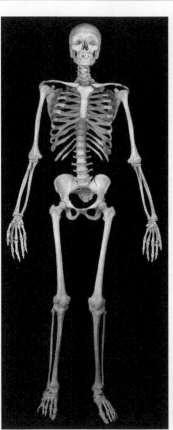
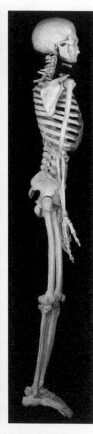

Species: *Homo neanderthalensis*
Composite skeleton reconstruction based on
La Ferrassie 1 and Kebara 2
Fossil Ages: between 50,000 and 70,000 years old
Average Male Height: 65" (164 cm) tall
Average Male Weight: 143 lbs (65 kg)

Species: *Homo sapiens sapiens* (male skeleton)
Average Height of an Adult Male: 69.4" (176 cm) tall*
Average Weight of an Adult Male: 194.7 lbs (88.3 kg)*

---

* Source: Anthropometric Reference Data for Children and Adults: United States, 2003–2006. National Center for Health Statistics.

What similarities/differences exist between a Neanderthal skull and a modern-day human's skull?

Similarities: _____

Differences: _____

What similarities/differences exist between a Neanderthal ribcage and a modern-day human's ribcage?

Similarities: _____

Differences: _____

What similarities/differences exist between a Neanderthal pelvis and a modern-day human's pelvis?

Similarities: _____

Differences: _____

What similarities/differences exist between Neanderthal limbs and a modern-day human's limbs?

Similarities: _____

Differences: _____

EVOLUTION BY MEANS OF NATURAL SELECTION

Why is it that some species (such as *Homo erectus*) go extinct, while other species (such as *Homo sapiens*) come into existence? Before we can address this question, we must define *evolution* in the biological sense. In everyday life, we commonly use the word *evolve* to describe a person's changing attitude, appearance, or philosophy. **Biological evolution**, on the other hand, refers to changes that develop in a *population* as it progresses from one generation to the next. (The term *population* refers to all species members that are close enough geographically to reproduce with one another.) Although the evolution of certain traits can be physically assessed, such as beak length or fur color, these traits must be *genetically* determined to pass from one generation to another. At the genetic level— which is the one that dictates heritable traits—mutations and sexual reproduction increase a population's diversity.

So what exactly drives the evolution of a bacterial, avian, or human population? More than 150 years ago, Charles Darwin proposed one novel answer to this question in the book *On the Origin of Species*: **natural selection**. When resources become scarce in the environment, individuals must compete with one another to secure resources and survive. Although some variations are neutral, meaning that they have no impact on survival, subtle variations in other traits may prove to be advantageous—or detrimental—in the quest for survival:

- Ability (or inability) to obtain food, shelter, and water

- Ability (or inability) to withstand climate changes

- Ability (or inability) to evade predators

- Ability (or inability) to find a mate and reproduce successfully

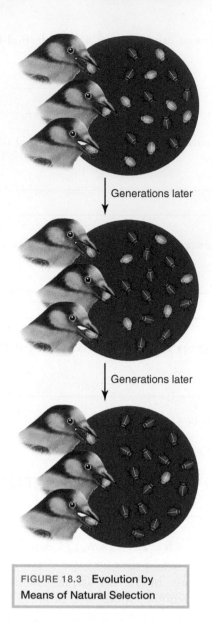

Generations later

Generations later

FIGURE 18.3 **Evolution by Means of Natural Selection**

Even slight variations in height, speed, color, sensory perception, or beak length may have a tremendous impact on survival in nature; an example of this phenomenon is shown in Figure 18.3. Those individuals who are best adapted to the environment will survive, reproduce, and pass their DNA on to future generations. As a result, genetic traits that promote survival inevitably grow more and more prevalent in the population as a whole. Keep in mind, though, that natural selection is not steering populations toward perfection. Ultimately, the environment determines whether a trait is neutral, detrimental, or advantageous at a particular point in time. And since the environment *itself* is constantly changing, populations must also continue evolving in order to adapt and survive.

During this exercise, you will first identify physical variations that exist among members of the same population. How—and why—do you think these variations may affect survival? To create a logical hypothesis, be sure to evaluate the environment that is inhabited by each species.

Deer Mice (*Peromyscus maniculatus*)*

What physical variations exist between these deer mice?

_____

_____

Formulate a hypothesis to describe how these variations may affect survival. Explain your rationale.

_____

_____

How could this hypothesis be tested?

_____

_____

Can you think of an environmental change that would alter your prediction?

_____

_____

*Linnen, C. R., E. P. Kingsely, J. D. Jensen and H. E. Hoekstra 2009. On the origin and spread of an adaptive allele in deer mice. *Science* 325 (5944): 1095–1098.

Giraffes (*Giraffa camelopardalis*)

What physical variations exist between these giraffes?

_____

_____

Formulate a hypothesis to describe how these variations may affect survival. Explain your rationale.

_____

_____

How could this hypothesis be tested?

_____

_____

Can you think of an environmental change that would alter your prediction?

_____

_____

Florida Cottonmouths (*Agkistrodon piscivorus conanti*)

What physical variations exist between these snakes?

_____

_____

Formulate a hypothesis to describe how these variations may affect survival. Explain your rationale.

_____

_____

How could this hypothesis be tested?

_____

_____

Can you think of an environmental change that would alter your prediction?

_____

_____

Male House Finches (*Carpodacus mexicanus*)

What physical variations exist between these male finches?

_____

_____

Formulate a hypothesis to describe how these variations may affect survival and mating. Explain your rationale.

_____

_____

How could this hypothesis be tested?

_____

_____

Can you think of an environmental change that would alter your prediction?

_____

_____

Although it takes many years for humans to develop and reach sexual maturity, the same is not true for viruses and unicellular organisms. Bacteria, for example, can move from one generation to the next in a matter of hours or, in some cases, even minutes. As a result, the evolution of a bacterial population can be examined in a matter of days. During this exercise, you will monitor the evolution of an *Escherichia coli* (*E. coli*) population in response to one environmental change: the addition of an antibiotic. Before you begin this exercise, create a hypothesis to test regarding the effects of antibiotics on bacterial populations.

Hypothesis:

_____

_____

Procedure

1. Collect the following supplies from the materials bench:

   - One Mueller-Hinton (M-H) agar plate with *E. coli* colonies
   - One plate of M-H Streptomycin agar
   - One plate of M-H Tetracycline agar
   - One plate of M-H Penicillin agar
   - One applicator for replica plating

2. Label the *bottom* of each agar plate with your name, the date, and the type of agar it contains.

lid

bottom
of Petri
Plate

3. Draw an arrow pointing toward the edge of each Petri plate; this arrow should be drawn on the bottom portion of the plate.

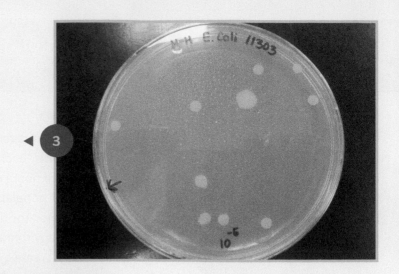

4. Place four strips of double-stick tape along the bottom of the applicator.

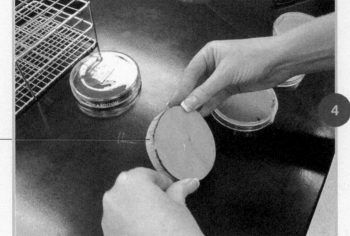

Applicator for replica plating

Note: Refer to Lab17 (Microorganisms and the Human Body) to review aseptic techniques.

5. *Aseptically* attach one sterile velvet circle to the bottom of the applicator. This is best achieved by placing the applicator directly on the velvet and pressing down; the tape will hold the velvet in place.

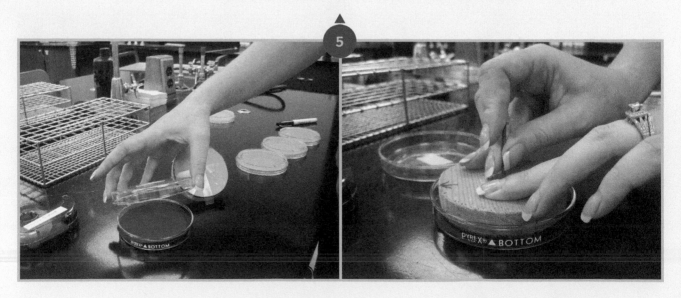

6. Line up the arrow on the applicator with the arrow on the M-H *E. coli* plate.

7. Press the applicator down directly on top of the bacteria. The velvet will pick up cells from each colony on the plate.

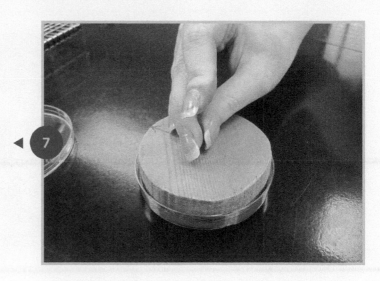

8. Line up the arrow on the applicator with the arrow on the M-H Streptomycin plate. Press the applicator down onto the media.

9. Insert a pair of forceps between the velvet and the applicator to loosen the velvet. Use the forceps to transfer the velvet into a biohazard bag.

10. Repeat steps 4–9 with the M-H Tetracycline plate and the M-H Penicillin plate.

11. Incubate the Petri plates at 37°C for 24–48 hours. Following incubation, these plates may be stored at 4°C until your next lab period.

Next Lab Period

1. Compare the density of growth on each antibiotic plate with the original M-H *E. coli* plate.

2. Using the arrows as a guide, identify *E. coli* colonies from the original plate that successfully produced growth an each antibiotic plate.

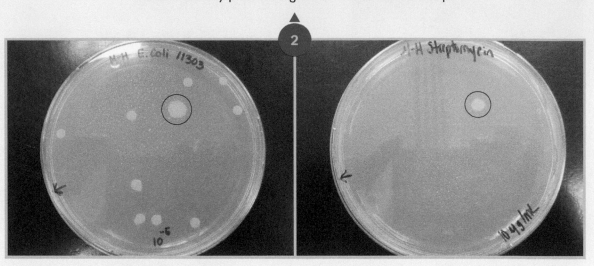

Review Questions for Step by Step 18.3

Was your initial hypothesis supported or rejected by the results of this experiment? If necessary, revise your initial hypothesis in the space provided below.

_____

_____

How did the bacterial population evolve when streptomycin, penicillin, and tetracycline were added to the environment?

_____

_____

In regard to patient care, what measures are taken by medical professionals to minimize the development of antibiotic resistance? What measures can patients take to minimize this risk?

_____

_____

What are potential sources of error in this experiment? What improvements (if any) could be made to this experiment?

_____

_____

THE EVOLUTIONARY TREE OF LIFE

We know that populations evolve over time, but how does evolution lead to the development of new species? How exactly are species distinguished from each other, anyway? By definition, a **species** is a group of organisms that can successfully reproduce with one another. Sometimes, a geographical barrier—such as a river, an ocean, or a mountain range—winds up isolating members of the same species from one another. As a result, two different populations are formed, since these groups can no longer breed with each other. As time passes, each population continues to evolve and adapt to its own unique environment. These differences may accumulate to the point where breeding is impossible, even if the physical barrier is removed. The result of this phenomenon is **speciation**, since the original species diverged into two separate branches on the ever-growing tree of life. Darwin himself pondered finch speciation for years after his voyage to the Galápagos Islands. Although finch species thrived on every island, their beaks had adapted to the unique food sources available in their habitats. Figure 18.4 illustrates the divergence of finch species throughout the Galápagos Islands.

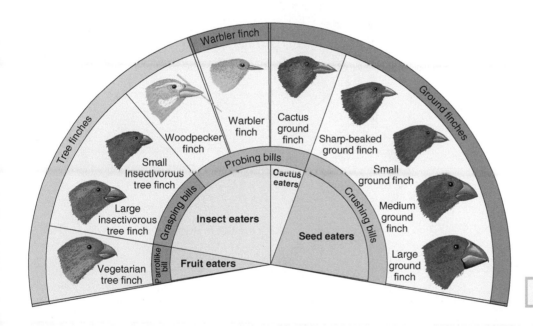

FIGURE 18.4 **Darwin's Finches**

As new species develop over time, other species inevitably go extinct. Darwin spent years pondering how one finch species—the common ancestor—could diverge into every finch species currently living on the Galápagos Islands. By studying the branches—new finch species—that arose from the same common ancestor, Darwin stumbled across another logical question to ponder. What happens if you trace these branches *backwards* for thousands, millions, or even billions of years? Do all living organisms on earth share a common ancestor? Since Darwin's time, scientists have continued refining the evolutionary tree (a.k.a. the **phylogenetic tree**) that unites all forms of life on earth. Scientists use phylogenetic trees to classify organisms, study ancestral lineages, and test new hypotheses related to the theory of evolution. Many scientific disciplines work together to create and refine phylogenetic trees, such as the one displayed in Figure 18.5.

The Timetree of Life

This year marks the 200th anniversary of the birth of Charles Darwin, the author of the most influential book in the history of science: *On the Origin of Species by Means of Natural Selection* (1859). Darwin's work, which provided a mechanism for evolution, transcended science and had great impact on society. A key concept for Darwin was the evolutionary tree, which he first sketched in a notebook in 1837, at age 28. Later, Darwin added the dimension of time and produced the first timetree—an evolutionary tree scaled to time—as the only figure in his book. Darwin referred to the "great tree of life... With its everbranching and beautiful ramifications," and he predicted in a letter to his friend Thomas Huxley, "The time will come I believe, though I shall not live to see it, when we shall have fairly true genealogical trees of each great kingdom of nature."

The study of the morphology of living and extinct species has since helped to build those evolutionary trees. However, it wasn't until recent decades—with the advent of technology for sequencing genes and genomes and methods for analyzing those data—that Darwin's vision of a great tree of life began to emerge in a comprehensive way. Molecules now provide information on both dimensions (branching order and times of divergence) for thousands of species and higher taxa. The timetree of all species is far from complete, but the general patterns largely have been discovered and Darwin's vision is becoming reality. The large circular timetree of 1,610 families shown here is from *The Timetree of Life* (S. Blair Hedges & Sudhir Kumar, editors, Oxford University Press, 2009). It summarizes the current knowledge down to the taxonomic level of family.

The Timetree of Life (2009), edited by S. Blair Hedges and Sudhir Kumar with Foreword by James D. Watson

AUTHORS: A. Louise Allcock, Cajsa Lisa Anderson, Robert J. Asher, John C. Avise, Nadia A. Ayoub, Allan J. Baker, F. Keith Barker, Fabia U. Battistuzzi, Michael J. Benton, Matthew A. Bertone, Debashish Bhattacharya, Jaime E. Blair, Mark Blaxter, Franky Bossuyt, Endri G. Brady, Brigitte Bremer, Christopher A. Brochu, Joseph W. Brown, David C. Cannatella, Mark W. Chase, Arnaud Couloux, Joel Cracraft, Keith A. Crandall, Bryan N. Danforth, Frédéric Delsuc, Rui Diogo, Philip C. J. Donoghue, Christophe J. Douady, Emmanuel J. P. Douzery, Scott V. Edwards, Eduardo Eizirik, Brian D. Ferrell, Félix Forest, John Gatesy, David J. Gower, Felix M. Gradstein, Jeremiah D. Hackett, Cheryl Y. Hayashi, Shuoqing He, S. Blair Hedges, Matthew H. Heinicke, Kassian Ho-Nygrén, Kristin W. Jolu, Jens E. Irving, Rodney L. Honeycutt, Peter Houde, Thomas Jansen, Jungwook Kim, Garey W. Krajewski, Sudhir Kumar, Shigehiro Kuraku, Shiguero Kuratani, Leah Larkin, Annie J. Lindgren, Anne Ludwig, Ole Madsen, Suzanne Magallon, Conrad A. Matthews, Duane O. McKenna, Linda H. Medlin Robert W. Meredith, Kathleen J. Miglia, David P. Mindell, Mosaki Miya, William S. Moore, William J. Murphy, Gavin J. P. Naylor, Thomas J. Near, Angela E. Newton, James G. Oigg, Rinus G. Ote, Zuojiang Peng, Sérgio L. Pereira, Marcos Pérez-Losada, Davide Pisani, Megan L. Porter, Céline Poux, Kathleen M. Pryer, Jean-Claude Ragle, Susanne Renner, Kim Roelants, Alex D. Rogers, Lukas Rüber, Oliver A. Ryder, Jennifer M. Sander, Eric Schuettpelz, H. Bradley Shaffer, A. Jonathan Shaw, Andrew M. Shedlock, Andrew B. Smith, Mark S. Springer, Michael E. Seeper, Jan M. Strugnell, Emma C. Teeling, Michelle O. Trautwein, Marcel van Tuinen, Nicolas Vidal, David R. Vieites, David D. Voha, Marnahea H. Weis, Brian M. Wiegmann, Niklas Wikström, Mark Wilkinson, Shaun L. elinerton, Hwan Su Yoon, Nathan M. Young, Peng Zhang

Oxford University Press: www.oup.com

© 2009 by S. Blair Hedges and Sudhir Kumar
Graphics assistance by Madelyn Owens, Wayne Parkhurst, and Michael Suleski

the TIMETREE of LIFE
by S. BLAIR HEDGES and SUDHIR KUMAR

TIMETREE   www.timetree.org

http://www.timetree.org

FIGURE 18.5   **The Tree of Life**

- Discovering transitional fossils

Tiktaalik: Transitional Fossil from a Limbed Fish

Anchiornis Huxleyi: Fossil from a Small Feathered Dinosaur

- Studying the distribution of fossils based on time and geography
- Discovering that the fossils in **strata** (layers of sedimentary rock) display a complexity gradient that progressed over billions of years

Chart 1

Eonothem / Eon	Erathem / Era	System / Period	Series / Epoch	Geo-chronologic Age (Ma)
Phanerozoic	Cenozoic	Quaternary	Holocene	0.0118
			Pleistocene	2.588
		Neogene	Pliocene	5.332
			Miocene	23.03
		Paleogene	Oligocene	33.9
			Eocene	55.8
			Paleocene	65.5
	Mesozoic	Cretaceous	Upper	99.6
			Lower	145.5
		Jurassic	Upper	161.2
			Middle	175.6
			Lower	199.6
		Triassic	Upper	228.0
			Middle	245.0
			Lower	251.0

Chart 2

Eonothem / Eon	Erathem / Era	System / Period	Series / Epoch	Geo-chronologic Age (Ma)
Phanerozoic	Paleozoic	Permian	Lopingian	251.0 – 260.4
			Guadalupian	270.6
			Cisuralian	299.0
		Carboniferous	Pennsylvanian Upper	306.5
			Pennsylvanian Middle	311.7
			Pennsylvanian Lower	318.1
			Mississippian Upper	326.4
			Mississippian Middle	345.3
			Mississippian Lower	359.2
		Devonian	Upper	385.3
			Middle	397.5
			Lower	416.0
		Silurian	Pridoli	418.7
			Ludlow	422.9
			Wenlock	428.2
			Llandovery	443.7
		Ordovician	Upper	460.9
			Middle	471.8
			Lower	488.3
		Cambrian	Furongian	~501.0
			Series 3	~510.0
			Series 2	~521.0
			Terreneuvian	542.0

Chart 3

Eonothem / Eon	Erathem / Era	System / Period	Geo-chronologic Age (Ma)
Precambrian	Proterozoic	Neo-proterozoic	Ediacaran — 542
			Cryogenian — ~630
			Tonian — 850 / 1000
		Meso-proterozoic	Stenian — 1200
			Ectasian — 1400
			Calymmian — 1600
		Paleo-proterozoic	Statherian — 1800
			Orosirian — 2050
			Rhyacian — 2300
			Siderian — 2500
	Archean	Neo-archean	2800
		Meso-archean	3200
		Paleo-archean	3600
		Eoarchean	Lower limit is not defined

KEY CONTRIBUTIONS FROM MOLECULAR BIOLOGY

- Discovering that ATP is the universal energy currency for cellular activities

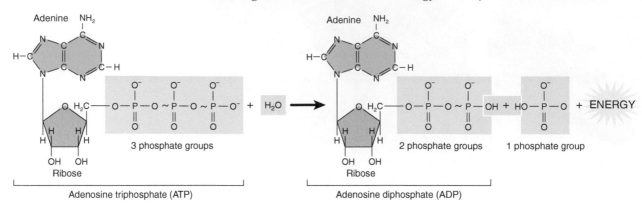

- Discovering that the same four letters (DNA bases) belong to every organism's DNA alphabet

- Discovering that the same 20 letters (amino acids) belong to every organism's protein alphabet

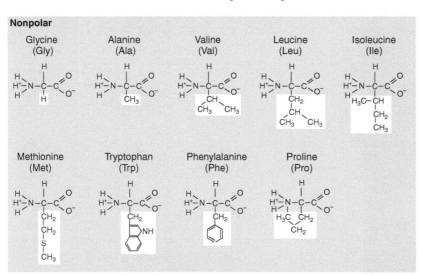

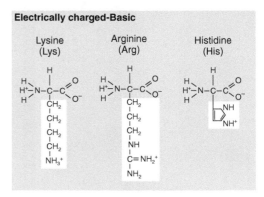

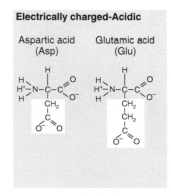

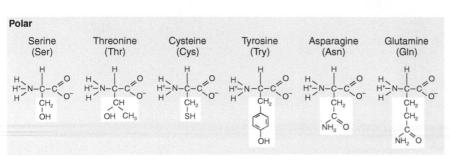

- Refining evolutionary lineages by examining DNA sequences and protein sequences from different organisms

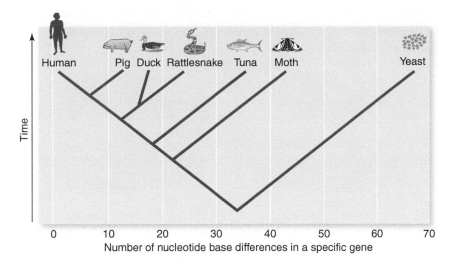

KEY CONTRIBUTIONS FROM CELLULAR BIOLOGY

- Discovering that cells are the basic building blocks, or structural units, of all living organisms
- Identifying similarities and differences in cells from different organisms

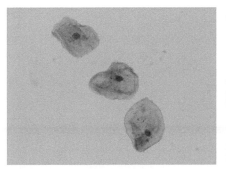

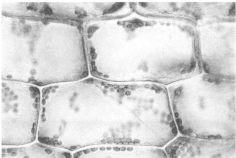

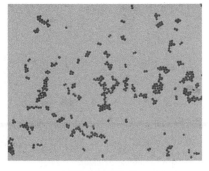

Cheek Cells   Plant Cells   Bacterial Cells

KEY CONTRIBUTIONS FROM GEOLOGY

- Determining how climate changes, weather changes, continental drift, and natural disasters (earthquakes, floods, glaciations, volcanic eruptions, and so on) influenced evolutionary history

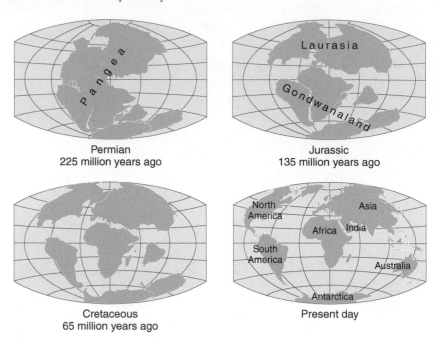

Permian
225 million years ago

Jurassic
135 million years ago

Cretaceous
65 million years ago

Present day

- Dating fossils using a combination of radiometric/radioactive techniques (example: uranium-lead radiometric dating) and stratigraphic techniques (example: cross-dating fossils in strata)

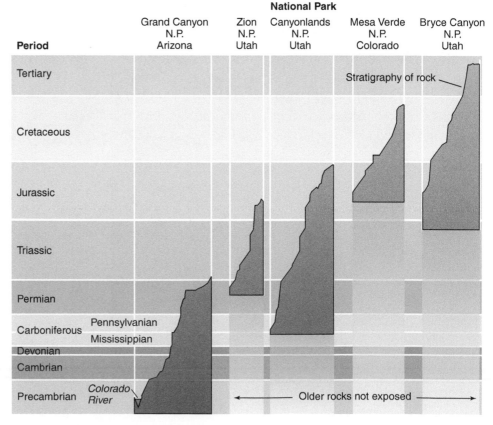

KEY CONTRIBUTIONS FROM ANATOMY

- Identifying **homologous structures**: similar structures that were inherited from a common ancestor

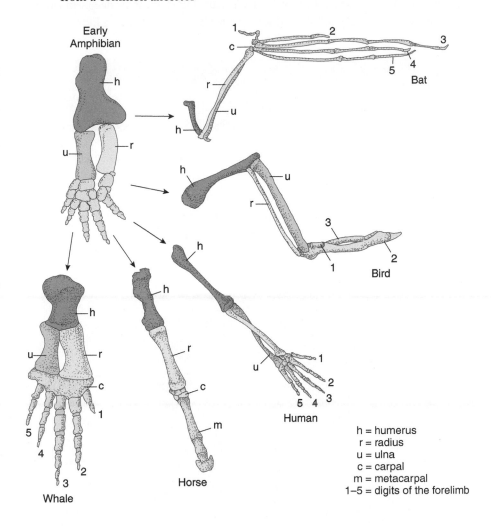

h = humerus
r = radius
u = ulna
c = carpal
m = metacarpal
1–5 = digits of the forelimb

- Identifying **vestigial structures**: structures that were functional in an ancestor, although they no longer function or have a diminished function in the organism of interest

Cloacal spurs (hindlimb vestiges) and pelvic vestiges in pythons and boas

Post-anal tail

Male Anaconda (*Eunectes murinus*)

Human Embryo, Day 30

If you go back far enough in time on a phylogenetic tree, it becomes evident that humans shared ancestors with chimpanzees, gorillas, and orangutans. Go back in time millions of years further, and suddenly humans and chimpanzees share ancestors with mice, worms, and jellyfish. Billions of years ago, humans even shared ancestors with yeast, bacteria, and algae! Evolution goes well beyond a solid, scientific theory; on a personal level, it can be deeply moving to learn how so many differences—and similarities—gradually developed among life on earth.

EXERCISE 18.4   Comparative Embryology

During this exercise, you will examine microscope slides of vertebrate embryos at various stages of development. Depending on the nature of your course, you may be examining frog embryos, chicken embryos, zebra fish embryos, and/or embryos from other model organisms. Compare each microscope slide to the images provided below, and look for the presence of homologous structures and/or vestigial structures.

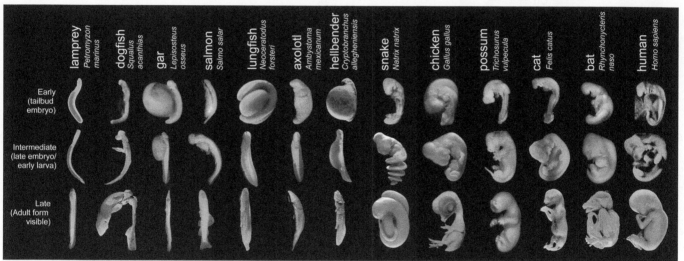

Courtesy of Michael K. Richardson.

Review Questions for Exercise 18.4

What vestigial structures (if any) are present on frog embryos? Chicken embryos? Zebra fish embryos? Human embryos?

_____

_____

What homologies are present among all vertebrates during embryonic development?

_____

_____

What differences arise among vertebrates during embryonic development?

_____

_____

1. Label a vestigial structure on the human embryo that is shown below.

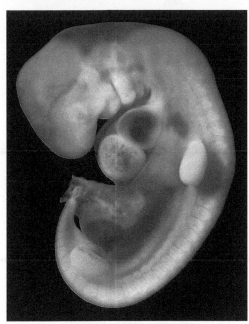

Human Embryo, Day 30

2. Does biological evolution occur at the individual level or the population level? Explain your answer.

_____

_____

_____

3. The forelimb and hindlimb bones of vertebrates are _____ since they descended (with modification) from a common ancestor.

4. Explain how bacterial populations evolve when antibiotics enter the environment.

_____

_____

_____

5. A simplified example of natural selection is shown below. How would you describe this process?

_____

_____

_____

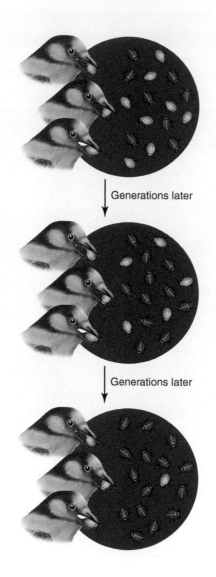

Generations later

Generations later

6. Name three distinct lines of evidence for biological evolution.

_____

_____

_____

7. A _____ is a group of organisms that can successfully reproduce with one another.

8. Based on their environmental conditions, which peppered moth (A or B) appears to have a better chance of survival? Explain.

Example 1: _____

_____

Example 2: _____

_____

Example 1

Example 2

9. Explain how one species can diverge into two separate species over time.

_____

_____

_____

10. Briefly explain how the structure of the human skeleton evolved over time.

_____

_____

_____

LINE ART AND TEXT CREDITS

LAB 1

Figure 1.1: from *Visualizing Human Biology 3rd edition*, by Kathleen Ireland, Reprinted courtesy of John Wiley & Sons, Inc., © 2011.

LAB 3

Figure 3.1: from *Visualizing Human Biology 3rd edition*, by Kathleen Ireland, Reprinted courtesy of John Wiley & Sons, Inc., © 2011. **Figure 3.2:** from *Visualizing Human Biology 3rd edition*, by Kathleen Ireland, Reprinted courtesy of John Wiley & Sons, Inc., © 2011. **Figure 3.3:** from *Principles of Human Anatomy* 11th Edition by Gerald Tortora and Mark T. Nielsen, Reprinted courtesy of John Wiley & Sons, Inc., © 2009. **Figure 3.4** from *Principles of Human Anatomy* 11th Edition by Gerald Tortora and Mark T. Nielsen , Reprinted courtesy of John Wiley & Sons, Inc., © 2009. **Figure 3.5:** from *Principles of Human Anatomy* 11th Edition by Gerald Tortora and Mark T. Nielsen, Reprinted courtesy of John Wiley & Sons, Inc., © 2009. **Figure 3.6:** from *Principles of Human Anatomy* 11th Edition by Gerald Tortora and Mark T. Nielsen, Reprinted courtesy of John Wiley & Sons, Inc., © 2009. **Figure 3.7:** from *Visualizing Human Biology 3rd edition*, by Kathleen Ireland, Reprinted courtesy of John Wiley & Sons, Inc., © 2011. **Figure 3.8:** from *Visualizing Human Biology 3rd edition*, by Kathleen Ireland, Reprinted courtesy of John Wiley & Sons, Inc., © 2011. **Figure 3.9:** from *Visualizing Human Biology 3rd edition*, by Kathleen Ireland, Reprinted courtesy of John Wiley & Sons, Inc., © 2011. **Figure 3.13:** from *Principles of Human Anatomy* 11th Edition by Gerald Tortora and Mark T. Nielsen, Reprinted courtesy of John Wiley & Sons, Inc., © 2009. **Figure 3.14:** from *Principles of Human Anatomy* 11th Edition by Gerald Tortora and Mark T. Nielsen, Reprinted courtesy of John Wiley & Sons, Inc., © 2009. **Figure 3.15:** from *Principles of Human Anatomy* 11th Edition by Gerald Tortora and Mark T. Nielsen, Reprinted courtesy of John Wiley & Sons, Inc., © 2009. **Figure 3.16:** from *Principles of Human Anatomy* 11th Edition by Gerald Tortora and Mark T. Nielsen, Reprinted courtesy of John Wiley & Sons, Inc., © 2009. **Figure 3.17:** from *Visualizing Human Biology 3rd edition*, by Kathleen Ireland, Reprinted courtesy of John Wiley & Sons, Inc., © 2011. **Figure 3.18:** from *Principles of Human Anatomy* 11th Edition by Gerald Tortora and Mark T. Nielsen, Reprinted courtesy of John Wiley & Sons, Inc., © 2009. **Figure 3.19:** from *Principles of Human Anatomy* 11th Edition by Gerald Tortora and Mark T. Nielsen, Reprinted courtesy of John Wiley & Sons, Inc., © 2009. **Figure 3.20:** from *Visualizing Human Biology 3rd edition*, by Kathleen Ireland, Reprinted courtesy of John Wiley & Sons, Inc., © 2011. **Page 3-17:** from *Visualizing Human Biology 3rd edition*, by Kathleen Ireland, Reprinted courtesy of John Wiley & Sons, Inc., © 2011. **Page 3-18:** from *Visualizing Human Biology 3rd edition*, by Kathleen Ireland, Reprinted courtesy of John Wiley & Sons, Inc., © 2011.

LAB 4

Table 4.1: from *Visualizing Human Biology 3rd edition*, by Kathleen Ireland, Reprinted courtesy of John Wiley & Sons, Inc., © 2011.

LAB 5

Figure 5.1: from *Visualizing Human Biology 3rd edition*, by Kathleen Ireland, Reprinted courtesy of John Wiley & Sons, Inc., © 2011. **Figure 5.3:** from *Visualizing Human Biology 3rd edition*, by Kathleen Ireland, Reprinted courtesy of John Wiley & Sons, Inc., © 2011. **Figure 5.4:** from *Visualizing Human Biology 3rd edition*, by Kathleen Ireland, Reprinted courtesy of John Wiley & Sons, Inc., © 2011. **Page 5-21:** from *Visualizing Human Biology 3rd edition*, by Kathleen Ireland, Reprinted courtesy of John Wiley & Sons, Inc., © 2011.

LAB 6

Figure 6.1: from *Visualizing Human Biology 3rd edition*, by Kathleen Ireland, Reprinted courtesy of John Wiley & Sons, Inc., © 2011. **Figure 6.6:** from *Visualizing Human Biology 3rd edition*, by Kathleen Ireland, Reprinted courtesy of John Wiley & Sons, Inc., © 2011. **Figure 6.8:** from *Visualizing Human Biology 3rd edition*, by Kathleen Ireland, Reprinted courtesy of John Wiley & Sons, Inc., © 2011. **Figure 6.11:** from *Principles of Human Anatomy* 11th Edition by Gerald Tortora and Mark T. Nielsen, Reprinted courtesy of John Wiley & Sons, Inc., © 2009. **Page 6-18:** from *Principles of Human Anatomy* 11th Edition by Gerald Tortora and Mark T. Nielsen, Reprinted courtesy of John Wiley & Sons, Inc., © 2009. **Page 6-20:** from *Visualizing Human Biology 3rd edition*, by Kathleen Ireland, Reprinted courtesy of John Wiley & Sons, Inc., © 2011. **Figure 6.12:** from *Visualizing Human Biology 3rd edition*, by Kathleen Ireland, Reprinted courtesy of John Wiley & Sons, Inc., © 2011. **Page 6-31:** from *Visualizing Human Biology 3rd edition*, by Kathleen Ireland, Reprinted courtesy of John Wiley & Sons, Inc., © 2011.

LAB 7

Figure 7.11: from *Visualizing Human Biology 3rd edition*, by Kathleen Ireland, Reprinted courtesy of John Wiley & Sons, Inc., © 2011. **Page 7-23:** from *Visualizing Human Biology 3rd edition*, by Kathleen Ireland, Reprinted courtesy of John Wiley & Sons, Inc., © 2011.

LAB 8

Figure 8.1: from *Visualizing Human Biology 3rd edition*, by Kathleen Ireland, Reprinted courtesy of John Wiley & Sons, Inc., © 2011. **Figure 8.5:** from *Visualizing Human Biology 3rd edition*, by Kathleen Ireland, Reprinted courtesy of John Wiley & Sons,

Inc., © 2011. **Figure 8.6:** from *Visualizing Human Biology 3rd edition*, by Kathleen Ireland, Reprinted courtesy of John Wiley & Sons, Inc., © 2011. **Table 8.2:** from *Visualizing Human Biology 3rd edition*, by Kathleen Ireland, Reprinted courtesy of John Wiley & Sons, Inc., © 2011. **Page 8-25:** from *Visualizing Human Biology 3rd edition*, by Kathleen Ireland, Reprinted courtesy of John Wiley & Sons, Inc., © 2011. **Page 8-26:** from *Visualizing Human Biology 3rd edition*, by Kathleen Ireland, Reprinted courtesy of John Wiley & Sons, Inc., © 2011.

LAB 9

Figure 9.1: from *Visualizing Human Biology 3rd edition*, by Kathleen Ireland, Reprinted courtesy of John Wiley & Sons, Inc., © 2011. **Figure 9.2:** from *Visualizing Human Biology 3rd edition*, by Kathleen Ireland, Reprinted courtesy of John Wiley & Sons, Inc., © 2011. **Figure 9.5:** from *Visualizing Human Biology 3rd edition*, by Kathleen Ireland, Reprinted courtesy of John Wiley & Sons, Inc., © 2011. **Figure 9.6:** from *Principles of Human Anatomy* 11th Edition by Gerald Tortora and Mark T. Nielsen, Reprinted courtesy of John Wiley & Sons, Inc., © 2009. **Figure 9.7:** from *Principles of Human Anatomy* 11th Edition by Gerald Tortora and Mark T. Nielsen, Reprinted courtesy of John Wiley & Sons, Inc., © 2009. **Figure 9.8:** from *Principles of Human Anatomy* 11th Edition by Gerald Tortora and Mark T. Nielsen, Reprinted courtesy of John Wiley & Sons, Inc., © 2009. **Figure 9.9:** from *Principles of Human Anatomy* 11th Edition by Gerald Tortora and Mark T. Nielsen, Reprinted courtesy of John Wiley & Sons, Inc., © 2009. **Figure 9.10:** from *Visualizing Human Biology 3rd edition*, by Kathleen Ireland, Reprinted courtesy of John Wiley & Sons, Inc., © 2011. **Table 9.5:** from *Visualizing Human Biology 3rd edition*, by Kathleen Ireland, Reprinted courtesy of John Wiley & Sons, Inc., © 2011. **Figure 9.12:** "Sexually Transmitted Diseases in America: How Many Cases and at What Cost?", The Henry J. Kaiser Family Foundation, December 1998. **Page 9-30:** from *Visualizing Human Biology 3rd edition*, by Kathleen Ireland, Reprinted courtesy of John Wiley & Sons, Inc., © 2011. **Page 9-31:** from *Visualizing Human Biology 2nd Edition* by Kathleen Ireland, Reprinted courtesy of John Wiley & Sons, Inc., © 2010.

LAB 10

Figure 10.2: from *Psychology 5th Edition*, by Robin Kowalski and Drew Westen, Reprinted courtesy of John Wiley & Sons, Inc., © 2009. **Figure 10.5:** from *Visualizing Human Biology 3rd edition*, by Kathleen Ireland, Reprinted courtesy of John Wiley & Sons, Inc., © 2011. **Figure 10.7:** from *Visualizing Human Biology 3rd edition*, by Kathleen Ireland, Reprinted courtesy of John Wiley & Sons, Inc., © 2011. **Figure 10.8:** from *Principles of Human Anatomy* 11th Edition by Gerald Tortora and Mark T. Nielsen, Reprinted courtesy of John Wiley & Sons, Inc., © 2009. **Figure 10.12:** from *Visualizing Human Biology 3rd edition*, by Kathleen

Ireland, Reprinted courtesy of John Wiley & Sons, Inc., © 2011. **Figure 10.13:** from *Principles of Human Anatomy* 11th Edition by Gerald Tortora and Mark T. Nielsen, Reprinted courtesy of John Wiley & Sons, Inc., © 2009. **Page 10-33:** from *Visualizing Human Biology 3rd edition*, by Kathleen Ireland, Reprinted courtesy of John Wiley & Sons, Inc., © 2011.

LAB 11

Figure 11.1: from *Visualizing Human Biology 3rd edition*, by Kathleen Ireland, Reprinted courtesy of John Wiley & Sons, Inc., © 2011. **Figure 11.2:** from *Visualizing Human Biology 3rd edition*, by Kathleen Ireland, Reprinted courtesy of John Wiley & Sons, Inc., © 2011. **Figure 11.3:** from *Principles of Human Anatomy* 11th Edition by Gerald Tortora and Mark T. Nielsen, Reprinted courtesy of John Wiley & Sons, Inc., © 2009. **Figure 11.4:** from *Visualizing Human Biology 3rd edition*, by Kathleen Ireland, Reprinted courtesy of John Wiley & Sons, Inc., © 2011. **Figure 11.9:** from *Visualizing Human Biology 3rd edition*, by Kathleen Ireland, Reprinted courtesy of John Wiley & Sons, Inc., © 2011. **Table 11.5:** from *Visualizing Human Biology 3rd edition*, by Kathleen Ireland, Reprinted courtesy of John Wiley & Sons, Inc., © 2011. **Page 11-31:** from *Visualizing Human Biology 3rd edition*, by Kathleen Ireland, Reprinted courtesy of John Wiley & Sons, Inc., © 2011. **Page 11-32:** from *Visualizing Human Biology 3rd edition*, by Kathleen Ireland, Reprinted courtesy of John Wiley & Sons, Inc., © 2011.

LAB 12

Figure 12.2: from *Visualizing Human Biology 3rd edition*, by Kathleen Ireland, Reprinted courtesy of John Wiley & Sons, Inc., © 2011. **Figure 12.3:** from *Visualizing Human Biology 3rd edition*, by Kathleen Ireland, Reprinted courtesy of John Wiley & Sons, Inc., © 2011. **Page 12-5:** from *Visualizing Human Biology 3rd edition*, by Kathleen Ireland, Reprinted courtesy of John Wiley & Sons, Inc., © 2011. **Figure 12.4:** from *Visualizing Human Biology 3rd edition*, by Kathleen Ireland, Reprinted courtesy of John Wiley & Sons, Inc., © 2011. **Figure 12.5:** from *Principles of Human Anatomy* 11th Edition by Gerald Tortora and Mark T. Nielsen, Reprinted courtesy of John Wiley & Sons, Inc., © 2009. **Figure 12.6:** from *Visualizing Human Biology 3rd edition*, by Kathleen Ireland, Reprinted courtesy of John Wiley & Sons, Inc., © 2011. **Figure 12.8:** from *Visualizing Human Biology 3rd edition*, by Kathleen Ireland, Reprinted courtesy of John Wiley & Sons, Inc., © 2011. **Figure 12.10:** from *Visualizing Human Biology 3rd edition*, by Kathleen Ireland, Reprinted courtesy of John Wiley & Sons, Inc., © 2011. **Figure 12.12:** from *Visualizing Human Biology 3rd edition*, by Kathleen Ireland, Reprinted courtesy of John Wiley & Sons, Inc., © 2011. **Figure 12.14:** from *Visualizing Human Biology 3rd edition*, by Kathleen Ireland, Reprinted courtesy of John Wiley & Sons, Inc., © 2011. **Table 12.2:** from *Principles of Human Anatomy 12th Edition* by Gerald Tortora and Mark T. Nielsen, Reprinted courtesy of John Wiley & Sons,

Inc., © 2012. **Page 12-40:** from *Visualizing Human Biology 3rd edition*, by Kathleen Ireland, Reprinted courtesy of John Wiley & Sons, Inc., © 2011

LAB 13

Figure 13.1: from Principles of Human Anatomy 11th Edition by Gerald Tortora and Mark T. Nielsen, Reprinted courtesy of John Wiley & Sons, Inc., © 2009. **Figure 13.2:** from *Principles of Human Anatomy 12th Edition* by Gerald Tortora and Mark T. Nielsen, Reprinted courtesy of John Wiley & Sons, Inc., © 2012. **Figure 13.3:** from *Principles of Human Anatomy 12th Edition* by Gerald Tortora and Mark T. Nielsen, Reprinted courtesy of John Wiley & Sons, Inc., © 2012. **Figure 13.4:** from *Visualizing Human Biology 3rd edition*, by Kathleen Ireland, Reprinted courtesy of John Wiley & Sons, Inc., © 2011. **Page 13-4:** from *Principles of Human Anatomy* 12th edition by Gerald Tortora and Mark Nielsen, Reprinted courtesy of John Wiley & Sons, Inc., (c) 2012. **Figure 13.5:** from Biology: Understanding Life 1st edition, by Sandra Alters and Brian Alters, Reprinted courtesy of John Wiley & Sons, Inc., © 2006. **Figure 13.6:** from Principles of Human Anatomy 12th Edition by Gerald Tortora and Mark T. Nielsen, Reprinted courtesy of John Wiley & Sons, Inc., © 2012. **Figure 13.7:** from Principles of Human Anatomy 11th Edition by Gerald Tortora and Mark T. Nielsen, Reprinted courtesy of John Wiley & Sons, Inc., © 2009. **Figure 13.8:** from Principles of Human Anatomy 12th Edition by Gerald Tortora and Mark T. Nielsen, Reprinted courtesy of John Wiley & Sons, Inc., © 2012. **Figure 13.9:** from Principles of Human Anatomy 11th Edition by Gerald Tortora and Mark T. Nielsen, Reprinted courtesy of John Wiley & Sons, Inc., © 2009. **Figure 13.10:** from Principles of Human Anatomy 12th Edition by Gerald Tortora and Mark T. Nielsen, Reprinted courtesy of John Wiley & Sons, Inc., © 2012. **Table 13.2:** from Principles of Human Anatomy 11th Edition by Gerald Tortora and Mark T. Nielsen, Reprinted courtesy of John Wiley & Sons, Inc., © 2009. **Page 13-31:** from Principles of Human Anatomy 12th Edition by Gerald Tortora and Mark T. Nielsen, Reprinted courtesy of John Wiley & Sons, Inc., © 2012. **Page 13-32:** from *Visualizing Human Biology 3rd edition*, by Kathleen Ireland, Reprinted courtesy of John Wiley & Sons, Inc., © 2011 **Page 13-33:** from Principles of Human Anatomy 11th Edition by Gerald Tortora and Mark T. Nielsen, Reprinted courtesy of John Wiley & Sons, Inc., © 2009.

LAB 14

Figure 14.1: from *Visualizing Human Biology 3rd edition*, by Kathleen Ireland, Reprinted courtesy of John Wiley & Sons, Inc., © 2011 **Figure 14.5:** from *Visualizing Human Biology 3rd edition*, by Kathleen Ireland, Reprinted courtesy of John Wiley & Sons, Inc., © 2011. **Figure 14.6:** modified from *Visualizing Human Biology 3rd edition*, by Kathleen Ireland, Reprinted courtesy of John Wiley & Sons, Inc., © 2011. **Page 14-20:** from *Medical Genetics at a Glance, 2nd Edition* by Dorian Pritchard

and Bruce R. Korf, Reprinted courtesy of Wiley-Blackwell, © 2007. **Page 14-21:** from *Medical Genetics at a Glance, 2nd Edition* by Dorian Pritchard and Bruce R. Korf, Reprinted courtesy of Wiley-Blackwell, © 2007. **Page 14-22:** from *Medical Genetics at a Glance, 2nd Edition* by Dorian Pritchard and Bruce R. Korf, Reprinted courtesy of Wiley-Blackwell, © 2007. **Page 14-23:** from *Medical Genetics at a Glance, 2nd Edition* by Dorian Pritchard and Bruce R. Korf, Reprinted courtesy of Wiley-Blackwell, © 2007.

LAB 15

Figure 15.1: from *Cell and Molecular Biology 6th edition* by Gerald Karp, Reprinted courtesy of John Wiley & Sons, Inc., © 2009. **Figure 15.2:** from *Principles of Genetics 5th edition* by D. Peter Snustad and Michael J. Simmons, Reprinted courtesy of John Wiley & Sons, Inc., © 2009. **Table 15.1:** Data from Roberts RJ. Restriction and modification enzymes and their recognition sequences. Nucleic Acids Res. 8 (1): r63–r80. **Figure 15.3:** from Visualizing Human Biology 3rd edition, by Kathleen Ireland, Reprinted courtesy of John Wiley & Sons, Inc., © 2011. **Page 15-14:** from *Cell and Molecular Biology 6th edition* by Gerald Karp, Reprinted courtesy of John Wiley & Sons, Inc., © 2009.

LAB 16

Figure 16.1: Modified from artwork originally created for the National Cancer Institute. Reprinted with permission of the artist, Jeanne Kelly. 2004. **Figure 16.2:** Reprinted by permission from Macmillan Publishers Ltd: Nature Reviews Cancer June 2003 Vol 3 No 6, *"The Pathogenesis of Cancer Metastasis: the 'seed and soil' hypothesis"* revisited by Isiah J. Fidler, 2003. **Figure 16.5:** Modified from artwork originally created for the National Cancer Institute. Reprinted with permission of the artist, Jeanne Kelly. 2004. **Figure 16.6:** from Visualizing Human Biology 3rd edition, by Kathleen Ireland, Reprinted courtesy of John Wiley & Sons, Inc., © 2011. **Table 16.1:** Reprinted courtesy of National Cancer Institute. *"Cancer Staging"* http://www.cancer.gov/cancertopics/factsheet/Detection/staging **Figure 16.7:** from *Medical Genetics at a Glance, 2nd Edition* by Dorian Pritchard and Bruce R. Korf, Reprinted courtesy of Wiley-Blackwell, © 2007. **Figure 16.8:** from *Cell and Molecular Biology 6th edition* by Gerald Karp, Reprinted courtesy of John Wiley & Sons, Inc., © 2009. **Page 16-19:** American Cancer Society. *Cancer Facts and Figures 2010.* Atlanta: American Cancer Society, Inc. **Figure 16.10:** from *Principles of Genetics 5th edition* by D. Peter Snustad and Michael J. Simmons, Reprinted courtesy of John Wiley & Sons, Inc., © 2009.

LAB 17

Figure 17.1: from *Microbiology: Principles and Explorations 7th edition* by Jacquelyn Black, Reprinted courtesy of John Wiley & Sons, Inc., © 2009. **Page 17-7:** modified from *Principles of*

Genetics *5th edition* by D. Peter Snustad and Michael J. Simmons, Reprinted courtesy of John Wiley & Sons, Inc., © 2009. **Page 17-7:** from Visualizing Human Biology 3rd edition, by Kathleen Ireland, Reprinted courtesy of John Wiley & Sons, Inc., © 2011. **Figure 17.2:** from *Microbiology: Principles and Explorations 7th edition* by Jacquelyn Black, Reprinted courtesy of John Wiley & Sons, Inc., © 2009. **Page 17-17:** from *Microbiology: Principles and Explorations 7th edition* by Jacquelyn Black, Reprinted courtesy of John Wiley & Sons, Inc., © 2009. **Figure 17-3:** from *Principles of Human Anatomy* 12th edition by Gerald Tortora and Mark Nielsen, Reprinted courtesy of John Wiley & Sons, Inc., (c) 2012. **Figure 17-4:** from *Microbiology: Principles and Explorations 7th edition* by Jacquelyn Black, Reprinted courtesy of John Wiley & Sons, Inc., © 2009.

LAB 18

Figure 18.4: from Biology: Understanding Life 1st edition, by Sandra Alters and Brian Alters, Reprinted courtesy of John Wiley & Sons, Inc., © 2006. **Page 18-18:** from *Visualizing Earth History 1st edition* by Loren Babcock, Reprinted courtesy of John Wiley & Sons, Inc., © 2009. **Page 18-19:** modified from *Principles of Genetics 5th edition* by D. Peter Snustad and Michael J. Simmons, Reprinted courtesy of John Wiley & Sons, Inc., © 2009. **Page 18-19:** from Visualizing Human Biology 3rd edition, by Kathleen Ireland, Reprinted courtesy of John Wiley & Sons, Inc., © 2011. **Page 18-20:** from Biology: Understanding Life 1st edition, by Sandra Alters and Brian Alters, Reprinted courtesy of John Wiley & Sons, Inc., © 2006. **Page 18-21:** from Principles of Human Evolution, 2nd Edition by Robert Andrew Foley and Roger Lewin, Reprinted courtesy of John Wiley & Sons, Inc., © 2004. **Page 18-22:** Newman, W.L. (1997). Geologic Time. *General Interest Publications of the U.S. Geological Survey* (http://pubs.usgs.gov/gip, 1997). Available online at: http://etc.usf.edu/lit2go/contents/4100/4120/strat.gif. **Page 18-23:** from *Visualizing Earth History 1st edition* by Loren Babcock, Reprinted courtesy of John Wiley & Sons, Inc., © 2009.

PHOTO CREDITS

Visualizing the Lab eye photo: Vincent O'Bryne/Alamy

LAB 1

Page 4 (left): Fawn V. Beckman; page 4 (right): Fawn V. Beckman; page 7: Fawn V. Beckman; page 8: sodapix/Photolibrary; page 10 (top): Fawn V. Beckman; page 10 (center left): Fawn V. Beckman; page 10 (center right): Fawn V. Beckman; page 11 (top left): Fawn V. Beckman; page 11 (top right): Fawn V. Beckman; page 11 (center): Fawn V. Beckman; page 11 (bottom): Fawn V. Beckman.

LAB 2

Page 1 (left): E. Walker/Photo Researchers, Inc; page 1 (right): National Cancer Institute/Photo Researchers, Inc; page 2: Fawn V. Beckman; page 3: Fawn V. Beckman; page 4 Fawn V. Beckman; page 6 (left): Fawn V. Beckman; page 6 (right): Fawn V. Beckman; page 7 (top left): Fawn V. Beckman; page 7 (top right): Fawn V. Beckman; page 7 (bottom left): Fawn V. Beckman; page 7 (bottom right): Fawn V. Beckman; page 8 Fawn V. Beckman; page 9 (top): Fawn V. Beckman; page 9 (bottom left): Fawn V. Beckman; page 9 (bottom right): Fawn V. Beckman; page 10 (top left): Fawn V. Beckman; page 10 (top right): Fawn V. Beckman; page 10 (bottom): Fawn V. Beckman; page 11: Fawn V. Beckman; page 12 (left): Fawn V. Beckman; page 12 (right): Fawn V. Beckman; page 13 (top left): Fawn V. Beckman; page 13 (top right): Fawn V. Beckman; page 13 (bottom): Fawn V. Beckman; page 14 (left): Fawn V. Beckman; page 14 (right): Fawn V. Beckman; page 15 (top): Fawn V. Beckman; page 15 (bottom left): Fawn V. Beckman; page 15 (bottom right): Fawn V. Beckman; page 16 (top): Fawn V. Beckman; page 16 (bottom): Fawn V. Beckman; page 17 (left): Fawn V. Beckman; page 17 (right): Fawn V. Beckman; page 19: Fawn V. Beckman; page 20 (top): Fawn V. Beckman; page 20 (center): Fawn V. Beckman.

LAB 3

Page 2: Rubberball Productions/Getty Images; page 4: Mark Nielsen; page 5 (top): Mark Nielsen; page 5 (bottom): Carolina Biological Supply Co/Visuals Unlimited; page 6: Mark Nielsen; page 8 (top): Mark Nielsen; page 9: Courtesy Michael Ross, University of Florida; page 10: Mark Nielsen; page 11: Mark Nielsen; page 12: Mark Nielsen; page 13: Courtesy Michael Ross, University of Florida; page 14 (top): Dr. Alvin Telser/Visuals Unlimited; page 14 (bottom): Carolina Biological Supply Co/Visuals Unlimited; page 15: Mark Nielsen; page 16: Courtesy Michael Ross, University of Florida; page 17: Rubberball Productions/Getty Images; page 18: Mark Nielsen; page 19 (top): Mark Nielsen; page 19 (center): Mark Nielsen; page 19 (bottom): Mark Nielsen.

LAB 4

Page 5: Fawn V. Beckman; page 6 (left): Fawn V. Beckman; page 6 (right): Fawn V. Beckman; page 7: Fawn V. Beckman; page 8: Fawn V. Beckman; page 10: Fawn V. Beckman; page 11: Fawn V. Beckman; page 15 Fawn V. Beckman; page 16 (top): Fawn V. Beckman; page 16 (bottom): Fawn V. Beckman.

LAB 5

Page 3: 3B Scientific©; page 4 (left): Dissection Shawn Miller; Photograph Mark Nielsen; page 4 (center): Dissection Shawn Miller; Photograph Mark Nielsen; page 4 (right): Dissection Shawn Miller; Photograph Mark Nielsen; page 8 (top left): Fawn V. Beckman; page 8 (top right): Fawn V. Beckman; page 8 (bottom): Fawn V. Beckman; page 10: Fawn V. Beckman; page 11: Fawn V. Beckman; page 13: Fawn V. Beckman; page 16: Fawn V. Beckman; page 19: David M. Martin, M.D./Photo Researchers, Inc.; page 20: ISM/Phototake; page 22: Model, 3B Scientific©. Photo by Fawn V. Beckman.

LAB 6

Page 3 (right): Mark Nielsen; page 4: Martin Rotker/Phototake; page 6: ISM/Phototake; page 7 (top): Dr. Gladden Willis/Visuals Unlimited; page 7 (bottom): Anthony Martinet/Look at Sciences/SPL/Photo Researchers, Inc; page 9: BSIP/Photo Researchers, Inc.; page 10 (left): Dissection Shawn Miller; Photograph Mark Nielsen; page 10 (right): Dissection Shawn Miller; Photograph Mark Nielsen; page 11: Dissection Shawn Miller; Photograph Mark Nielsen; page 13 (top left): Fawn V. Beckman; page 13 (top right): Fawn V. Beckman; page 13 (bottom): Fawn V. Beckman; page 14 (top left): Fawn V. Beckman; page 14 (top right): Fawn V. Beckman; page 19: 3B Scientific©; page 22: CNRI/Photo Researchers, Inc.; page 23 (left): ©Vu/Cabisco/Visuals Unlimited; page 23 (right): W. Ober/Visuals Unlimited; page 24: Fawn V. Beckman; page 25 (top): Fawn V. Beckman; page 25 (bottom left): Fawn V. Beckman; page 25 (bottom right): Fawn V. Beckman; page 26 (top left): Fawn V. Beckman; page 26 (top right): Fawn V. Beckman; page 26 (center): Fawn V. Beckman; page 28: James Cavallini/Photo Researchers, Inc; page 29: James Cavallini/Photo Researchers, Inc; page 30 (top): ISM/Phototake; page 30 (bottom): Max Delson Martins Santos/iStockphoto; page 32 (top): Mark Nielsen; page 32 (bottom): CNRI/Photo Researchers, Inc.

LAB 7

Page 3 (center right): Model, 3B Scientific©. Photo by Fawn V. Beckman; page 3 (center left): Model, 3B Scientific©. Photo by Fawn V. Beckman; page 3 (center right): Model, 3B Scientific©. Photo by Fawn V. Beckman; page 3 (bottom): 3B Scientific©; page 4 (top): Dissection Shawn Miller; Photograph Mark Nielsen; page 4 (center): Dissection Shawn Miller; Photograph Mark Nielsen; page 5: Dissection Shawn Miller; Photograph Mark Nielsen; page 7: Dr. Robert Calentine/Visuals Unlimited; page 8: Carolina Biological Supply Co/Visuals Unlimited; page 9 (center left): Fawn V. Beckman; page 9 (center right): Fawn V. Beckman; page 9 (bottom): Fawn V. Beckman; page 10 (top right): Fawn V. Beckman; page 10 (center): Fawn V. Beckman; page 10 (bottom left): Image courtesy of Nasco; page 10 (bottom right): MedicalRF.com/Getty Images, Inc.; page 11: Pulse Picture Library/CMP Images/Phototake; page 12 (top): Fawn V. Beckman; page 12 (bottom): Fawn V. Beckman; page 13 (top): Fawn V. Beckman; page 13 (center): Fawn V. Beckman; page 13 (bottom): Fawn V. Beckman; page 16 (left): Fawn V. Beckman; page 16 (right): Fawn V. Beckman; page 18 (bottom left): Medical Body Scans/Photo Researchers, Inc; page 18 (bottom right): Dr. John D. Cunningham/Visuals Unlimited; page 20 (top): Zephyr/Photo Researchers, Inc.

LAB 8

Page 3: Fawn V. Beckman; page 4 (top): Photo by Fawn V. Beckman. Model courtesy of Denoyer-Geppert Science Company; copyright 1985; page 4 (bottom left): Fawn V. Beckman; page 4 (bottom right): Model, 3B Scientific©. Photo by Fawn V. Beckman; page 5: Dissection Shawn Miller; Photograph Mark Nielsen; page 6 (top left): Dissection Shawn Miller; Photograph Mark Nielsen; page 6 (top right): Dissection Shawn Miller; Photograph Mark Nielsen; page 6 (bottom): Dissection Shawn Miller; Photograph Mark Nielsen; page 9: Fawn V. Beckman; page 10 (top): Fawn V. Beckman; page 10 (center): Fawn V. Beckman; page 10 (bottom): Fawn V. Beckman; page 11 (top left): Fawn V. Beckman; page 11 (top right): Fawn V. Beckman; page 11 (center left): Fawn V. Beckman; page 11 (bottom right): Fawn V. Beckman; page 12 (top right): Fawn V. Beckman; page 12 (center left): Fawn V. Beckman; page 12 (bottom left): Fawn V. Beckman; page 12 (bottom right): Fawn V. Beckman; page 14: Dr. Dennis Strete; page 15 (left): Mark Nielsen; page 15 (right): Mark Nielsen; page 18 (top right): Fawn V. Beckman; page 18 (bottom left): Fawn V. Beckman; page 19 (center left): Fawn V. Beckman; page 19 (center right): Fawn V. Beckman; page 19 (bottom): Fawn V. Beckman; page 20 (left): Fawn V. Beckman; page 20 (right): Fawn V. Beckman; page 21: Chris Bjornberg/Photo Researchers, Inc; page 22: Dr. Frederick Skvara/Visuals Unlimited; page 24: Medicimage/Phototake; page 26 (top): Dr. Dennis Strete; page 26 (bottom): Fawn V. Beckman.

LAB 9

Page 5 (left): Fawn V. Beckman; page 5 (right): Fawn V. Beckman; page 6 (left): Model, 3B Scientific©. Photo by Fawn V. Beckman; page 6 (right): Model, 3B Scientific©. Photo by Fawn V. Beckman; page 7 (left): Dissection Shawn Miller; Photograph Mark Nielsen; page 7 (right): Dissection Shawn Miller; Photograph Mark Nielsen; page 8 (left): Dissection Shawn Miller; Photograph Mark Nielsen; page 8 (right): Dissection Shawn Miller; Photograph Mark Nielsen;

page 9: Dissection Shawn Miller; Photograph Mark Nielsen; page 12 (top right): Courtesy Michael Ross, University of Florida; page 13: Steve Gschmeissner/Photo Researchers, Inc; page 14: Dr. Keith Wheeler/SPL/Photo Researchers, Inc; page 15 (left): Mark Nielsen; page 15 (right): Mark Nielsen; page 18: Fawn V. Beckman; page 19 (top left): Fawn V. Beckman; page 19 (top right): Fawn V. Beckman; page 19 (center): Fawn V. Beckman; page 19 (bottom): Fawn V. Beckman; page 20 (top): Fawn V. Beckman; page 20 (bottom): Fawn V. Beckman; page 22: David M. Phillips/Photo Researchers, Inc; page 25 (top): Dr P. Marazzi/Photo Researchers, Inc; page 25 (bottom): Dr. Cecil H. Fox/Photo Researchers, Inc; page 26 (top): Garry Watson/Photo Researchers, Inc; page 26 (bottom): Fawn V. Beckman; page 27: Science Photo Library/Photo Researchers, Inc; page 28 (top): Science Source/Photo Researchers, Inc; page 28 (bottom): With kind permission from SOA-AIDS Amsterdam; page 29: Xinhua/Tumpa Mondal/NewsCom; page 31: Dr. Keith Wheeler/SPL/Photo Researchers, Inc; page 32: doc-stock GmbH/Phototake.

LAB 10

Page 4: Mark Nielsen; page 5 (top): Fawn V. Beckman; page 5 (bottom): Fawn V. Beckman; page 6 (center left): Fawn V. Beckman; page 6 (bottom right): Fawn V. Beckman; page 7 (top): Fawn V. Beckman; page 7 (center): Fawn V. Beckman; page 7 (bottom): Fawn V. Beckman; page 13 (top left): Fawn V. Beckman; page 13 (top right): Model, 3B Scientific©. Photo by Fawn V. Beckman; page 13 (bottom left): Fawn V. Beckman; page 13 (bottom right): Fawn V. Beckman; page 14 (top): ©SOMSO Modelle Coburg/Germany, www.somso.com; page 14 (bottom left): ©SOMSO Modelle Coburg/Germany, www.somso.com; page 14 (bottom right): ©SOMSO Modelle Coburg/Germany, www.somso.com; page 15 (top): Dissection Shawn Miller; Photograph Mark Nielsen; page 15 (center): Dissection Shawn Miller; Photograph Mark Nielsen; page 15 (bottom): Dissection Shawn Miller; Photograph Mark Nielsen; page 16: Dissection Shawn Miller; Photograph Mark Nielsen; page 19: Fawn V. Beckman; page 20 (top): Fawn V. Beckman; page 20 (center left): Fawn V. Beckman; page 20 (center right): Fawn V. Beckman; page 20 (bottom): Fawn V. Beckman; page 21: Fawn V. Beckman; page 25 (top): Fawn V. Beckman; page 25 (bottom): Fawn V. Beckman; page 27: Dr. P. Marazzi/Photo Researchers, Inc; page 28 (top): Living Art Enterprises, LLC/Photo Researchers, Inc; page 28 (bottom): Medicimage/Phototake; page 29 (left): ISM/Phototake; page 29 (right): Du Cane Medical Imaging Ltd/Photo Researchers, Inc; page 30 (top right): ISM/Phototake; page 30 (top left): Medical Body Scans/Photo Researchers, Inc; page 30 (center left): Carolina Biological Supply Co/Visuals Unlimited; page 30 (center right): Dr. M. Goedert/Photo Researchers, Inc.; page 31 (top): Brian Bell/Photo Researchers,

Inc; page 31 (bottom): Mark Nielsen; page 32: Fawn V. Beckman; page 33: Dissection Shawn Miller; Photograph Mark Nielsen.

LAB 11

Page 4: Fawn V. Beckman; page 7: Fawn V. Beckman; page 9: Pulse Picture Library/CMP Images/Phototake; page 10: Anatomical Travelogue/Photo Researchers, Inc.; page 11 (left): Fawn V. Beckman; page 11 (right): Fawn V. Beckman; page 12 Dissection Shawn Miller; Photograph Mark Nielsen; page 13 (far left): ISM/Phototake; page 13 (left): ISM/Phototake; page 13 (center): Sourthern Illinois University/Photo Researchers, Inc; page 13 (right): Southern Illinois University/Photo Researchers, Inc; page 14 (top): Fawn V. Beckman; page 14 (bottom): Fawn V. Beckman; page 15 (top left): Fawn V. Beckman; page 15 (top right): Fawn V. Beckman; page 15 (center): Fawn V. Beckman; page 15 (bottom): Fawn V. Beckman; page 16 (top left): Fawn V. Beckman; page 16 (center): ISM/Phototake; page 16 (bottom): Fawn V. Beckman; page 17: ISM/Phototake; page 19: Fawn V. Beckman; page 21: Spencer Grant/Photo Researchers, Inc; page 23 (center left): Fawn V. Beckman; page 23 (center right): Fawn V. Beckman; page 23 (bottom): Dissection Shawn Miller; Photograph Mark Nielsen; page 25 (top): Mark Nielsen; page 25 (bottom left): Barbara Galati/Phototake; page 25 (bottom center): ISM/Phototake; page 25 (bottom right): Scott Camazine/Phototake; page 26 (top left): Fawn V. Beckman; page 26 (top center): Fawn V. Beckman; page 26 (top right): Fawn V. Beckman; page 26 (bottom): Fawn V. Beckman; page 27 (top): Fawn V. Beckman; page 27 (center left): Fawn V. Beckman; page 27 (center right): Barbara Galati/Phototake; page 28: Fawn V. Beckman; page 29: Fawn V. Beckman; page 30 (top): Medicimage/Phototake; page 30 (bottom): ISM/Phototake; page 31 (top): Fawn V. Beckman; page 32: Barbara Galati/Phototake.

LAB 12

Page 2: Dissection Shawn Miller; Photograph Mark Nielsen; page 3: Dissection Shawn Miller; Photograph Mark Nielsen; page 5: Mark Nielsen; page 6: Fawn V. Beckman; page 7 (top left): Fawn V. Beckman; page 7 (top right): Dr. Fred Skvara/Visuals Unlimited; page 7 (bottom): Mark Nielsen; page 8 (top): Dissection Shawn Miller; Photograph Mark Nielsen; page 8 (center): Courtesy James L. Cook, Comparative Orthopaedic Laboratory, University of Missouri; page 9: Courtesy Michael Ross, University of Florida; page 10: Dissection Shawn Miller; Photograph Mark Nielsen; page 12 (top): Fawn V. Beckman; page 12 (center): Fawn V. Beckman; page 12 (bottom): Fawn V. Beckman; page 13 (top): Fawn V. Beckman; page 13 (center): Fawn V. Beckman; page 13 (bottom): Fawn V. Beckman; page 14 (top): Fawn V. Beckman; page 14 (bottom): Fawn V. Beckman; page 20 (center left): Mark Nielsen;

page 20 (center): Mark Nielsen; page 20 (center right): Mark Nielsen; page 20 (bottom left): Mark Nielsen; page 20 (bottom right): Mark Nielsen; page 22: Mark Nielsen; page 23: Mark Nielsen; page 24 (top): Fawn V. Beckman; page 24 (center): Fawn V. Beckman; page 24 (bottom right): Mark Nielsen; page 24 (bottom center): Mark Nielsen; page 24 (bottom left): Mark Nielsen; page 25 (top): VideoSurgery/Photo Researchers, Inc; page 25 (bottom): Fawn V. Beckman; page 26 (top): Fawn V. Beckman; page 26 (bottom): Fawn V. Beckman; page 28 (left): Mark Nielsen; page 28 (right): Mark Nielsen; page 30 (left): Mark Nielsen; page 30 (right): Mark Nielsen; page 31 (top): Mark Nielsen; page 31 (bottom): Fawn V. Beckman; page 32 (top): Mark Nielsen; page 32 (bottom): Fawn V. Beckman; page 33 (top): Mark Nielsen; page 33 (bottom left): Fawn V. Beckman; page 33 (bottom right): Fawn V. Beckman; page 34 (top and bottom): Mark Nielsen; page 35 (just below top): Courtesy Department of Medical Illustration, University of Wisconsin Medical School; page 35 (center): Courtesy Brent Layton; page 35 (top and below center and bottom): Courtesy Department of Medical Illustration, University of Wisconsin Medical School; page 36: ISM/Phototake; page 37 (top): Courtesy Brent Layton; page 37 (bottom): Living Art Enterprises, LLC/Photo Researchers, Inc; page 38: PDSN/Phototake; page 39 (top left): Mark Nielsen; page 39 (top center): Mark Nielsen; page 39 (top right): Fawn V. Beckman; page 40 (top): Courtesy Brent Layton; page 40 (bottom): Mark Nielsen; page 41 (top): ISM/Phototake; page 41 (center): Mark Nielsen; page 42 (top): Fawn V. Beckman; page 42 (center): Fawn V. Beckman; page 42 (bottom): Mark Nielsen; page 43: Mark Nielsen.

LAB 13

Page 5 (top): Courtesy Michael Ross, University of Florida; page 5 (center): Image reprinted with permission from eMedicine.com, 2010. Available at: http://emedicine.medscape.com/article/1189246-overview.; page 6 (top): From Pakurar, AS and Bigbee, JW, Digital Histology: An Interactive CD Atlas and Review Text, 2nd. ed., John Wiley & Sons, Inc., Hoboken; page 6 (bottom): CDC/Dr. Edwin P. Ewing, Jr.; page 7: Courtesy Hiroyouki Sasaki, Yale E. Goldman and Clara Franzini-Armstrong; page 10 (top left): Fawn V. Beckman; page 10 (top center): Fawn V. Beckman; page 10 (top right): Fawn V. Beckman; page 10 (bottom): Fawn V. Beckman; page 11 (top): Fawn V. Beckman; page 11 (center): Fawn V. Beckman; page 11 (bottom): Fawn V. Beckman; page 12: Fawn V. Beckman; page 17 (bottom): John Wilson White; page 18 (top): John Wilson White; page 20: Ferenc Szelepcsenyi/Shutterstock; page 21 (left): 3B Scientific©; page 21 (right): 3B Scientific©; page 22 (left): Fawn V. Beckman; page 22 (right): Fawn V. Beckman; page 23 (left): Fawn V. Beckman; page 23 (right): Fawn V. Beckman; page 24 (top): Dissection Shawn Miller; Photograph Mark Nielsen; page 24 (bottom left): Dissection Shawn Miller; Photograph Mark Nielsen; page 24 (bottom right): Dissection

Shawn Miller; Photograph Mark Nielsen; page 25 (far left): Dissection Shawn Miller; Photograph Mark Nielsen; page 25 (left): Dissection Shawn Miller; Photograph Mark Nielsen; page 25 (center): Dissection Shawn Miller; Photograph Mark Nielsen; page 25 (right): Dissection Shawn Miller; Photograph Mark Nielsen; page 26 (far left): Dissection Shawn Miller; Photograph Mark Nielsen; page 26 (left): Dissection Shawn Miller; Photograph Mark Nielsen; page 26 (center): Dissection Shawn Miller; Photograph Mark Nielsen; page 26 (right): Dissection Shawn Miller; Photograph Mark Nielsen; page 27: Biophoto Associates/Photo Researchers, Inc; page 28: CDC/Dr. Edwin P. Ewing, Jr.; page 29 (top): Courtesy Michael Stadnick, M.D., Radsource; page 29 (bottom): Chris Barry/Phototake; page 30: Deep Light Productions/Photo Researchers, Inc; page 31 (top): Courtesy Hiroyouki Sasaki, Yale E. Goldman and Clara Franzini-Armstrong; page 31 (bottom): Ferenc Szelepcsenyi/Shutterstock; page 32 (top left): Courtesy Michael Ross, University of Florida; page 32 (top right): From Pakurar, AS and Bigbee, JW, Digital Histology: An Interactive CD Atlas and Review Text, 2nd. ed., John Wiley & Sons, Inc., Hoboken; page 33 (left): Dissection Shawn Miller; Photograph Mark Nielsen; page 33 (right): Dissection Shawn Miller; Photograph Mark Nielsen.

LAB 14

Page 3 (top and center): L. Willatt/Photo Researchers, Inc.; page 3 (bottom left): Fawn V. Beckman; page 3 (bottom right): Fawn V. Beckman; page 4 (top left): Fawn V. Beckman; page 4 (top right): Fawn V. Beckman; page 4 (center): Fawn V. Beckman; page 4 (just below center): Fawn V. Beckman; page 4 (bottom): Fawn V. Beckman; page 5: Carlolina Biological Supply Co./Visuals Unlimited; page 6: Liza McCorkle/iStockphoto; page 8: Fawn V. Beckman; page 10 (top): Reprinted from Trends in Genetics, vol. 20:8, Richard A. Sturm, Tony N. Frudakis Eye colour: portals into pigmentation genes and ancestry. Copyright 2004, with permission from Elsevier.; page 10 (center left): Fawn V. Beckman; page 10 (center right): Fawn V. Beckman; page 10 (bottom): Ted Kinsman/Photo Researchers, Inc; page 11 (top): Christopher Futcher/iStockphoto; page 11 (center): Inga Ivanova/iStockphoto; page 13 (top): Fawn V. Beckman; page 13 (center): Fawn V. Beckman; page 13 (bottom): Fawn V. Beckman; page 14 (top): Fawn V. Beckman; page 14 (center): Fawn V. Beckman; page 14 (bottom): Fawn V. Beckman; page 18: L. Willatt/Photo Researchers, Inc.; page 24: ISM/Phototake; page 25: Sovereign, ISM/Phototake; page 26: SPL/Photo Researchers, Inc; page 27: ISM/Phototake; page 28: James King-Holmes/Imperial Cancer Fund/Photo Researchers, Inc.; page 29: ISM/Phototake; page 30: Fawn V. Beckman.

LAB 15

Page 2: Philippe Plailly/Science Photo/Photo Researchers, Inc.; page 5 (top): Fawn V. Beckman; page 5 (center): Fawn V. Beckman; page 5 (bottom): Fawn V. Beckman; page 6 (top left): Fawn V. Beckman; page 6 (top right): Fawn V. Beckman; page 6 (bottom): Fawn V. Beckman; page 7 (top): Fawn V. Beckman; page 7 (center left): Fawn V. Beckman; page 7 (center right): Fawn V. Beckman; page 7 (bottom right): Fawn V. Beckman; page 8 (top left): Fawn V. Beckman; page 8 (top right): Fawn V. Beckman; page 8 (center): Fawn V. Beckman; page 8 (bottom): Fawn V. Beckman; page 9 (top): Fawn V. Beckman; page 9 (center): Fawn V. Beckman; page 9 (bottom): Fawn V. Beckman; page 10 (top left): Copyright ©2010 Edvotek. All rights reserved. www.edvotek.com; page 10 (top right): LookatSciences/Phototake; page 10 (bottom): Fawn V. Beckman; page 11 (top): Fawn V. Beckman; page 11 (center): Fawn V. Beckman; page 11 (bottom): Fawn V. Beckman; page 12: Images courtesy Bio-Rad Laboratories.; (left): image taken with the White Digital Bioimaging System from Vernier Software & Technology, www.vernier.com; (right): image taken with the Blue Digital Bioimaging System from Vernier Software & Technology, www.vernier.com.; page 13: Image taken with the White Digital Bioimaging System from Vernier Software & Technology, www.vernier.com. Image provided courtesy of Bio-Rad Laboratories; page 15: Image taken with the White Digital Bioimaging System from Vernier Software & Technology, www.vernier.com. Image provided courtesy of Bio-Rad Laboratories.

LAB 16

Page 5 (bottom): Fawn V. Beckman; page 13 (top): E. Walker/Photo Researchers, Inc; page 13 (bottom): National Cancer Institute/Photo Researchers, Inc; page 14 (thyroid cancer): Pulse Picture Library/CMP Images/Phototake; page 14 (breast cancer): Scott Camazine/Photo Researchers, Inc; page 14 (cervical cancer): SPL/Photo Researchers, Inc; page 14 (leukemia): Southern Illinois University/Photo Researchers, Inc; page 14 (skin cancer): Biophoto Associates/Photo Researchers, Inc; page 14 (Hodgkins): ISM/Phototake; page 14 (sarcoma): P. Marazzi/Photo Researchers, Inc; page 15 (left): Vu/©Cabisco/Visuals Unlimited; page 15 (right): Vu/© Cabisco/Visuals Unlimited; page 16 (top left): Mark Nielsen; page 16 (top right): ISM/Phototake; page 16 (bottom left): ISM/Phototake; page 16 (bottom right): Dr. Gladden Willis/Visuals Unlimited; page 17 (left): Biophoto Associates/Photo Researchers, Inc.; page 17 (right): CNRI/Photo Researchers, Inc; page 21: From B.N. Ames, J. McCann, and E. Yamasaki, Mutat. Res. 31:347, 1975. Photograph courtesy B.N. Ames.; page 22 (left): Fawn V. Beckman; page 22 (right): Fawn V. Beckman; page 23 (top): Fawn V. Beckman; page 23 (center left): Fawn V. Beckman; page 23 (center right): Fawn V. Beckman; page 24 (top): Fawn V. Beckman; page 24 (center): Fawn V. Beckman; page 24 (bottom): Fawn V. Beckman; page 25 (top): Fawn V. Beckman; page 25 (center): Fawn V. Beckman; page 25 (bottom left): Fawn V. Beckman; page 25 (bottom right):

Fawn V. Beckman; page 26 (top left): Fawn V. Beckman; page 26 (top right): Fawn V. Beckman; page 26 (center left): Fawn V. Beckman; page 26 (center right): Fawn V. Beckman; page 27: UPI Photo/Bill Greenblatt/Landov LLC; page 28 (left): E. Walker/Photo Researchers, Inc; page 28 (right): National Cancer Institute/Photo Researchers, Inc; page 29 (left): Fawn V. Beckman; page 29 (right): Fawn V. Beckman; page 30: ISM/Phototake.

LAB 17

Page 2 (top left): Fawn V. Beckman; page 2 (center left): Fawn V. Beckman; page 2 (center right): Fawn V. Beckman; page 2 (top right): Fawn V. Beckman; page 2 (bottom left): Fawn V. Beckman; page 3 (top left): Fawn V. Beckman; page 3 (top right): Fawn V. Beckman; page 3 (center left): Fawn V. Beckman; page 3 (center right): Fawn V. Beckman; page 3 (bottom): Fawn V. Beckman; page 4: Fawn V. Beckman; page 5 (top): Fawn V. Beckman; page 5 (center left): Fawn V. Beckman; page 5 (center right): Fawn V. Beckman; page 5 (bottom): Fawn V. Beckman; page 6: Fawn V. Beckman; page 9 (center): Fawn V. Beckman; page 9 (bottom): Fawn V. Beckman; page 10 (top): Fawn V. Beckman; page 10 (center): Fawn V. Beckman; page 11: Fawn V. Beckman; page 12 (center left): Fawn V. Beckman; page 12 (center right): Fawn V. Beckman; page 12 (bottom): Fawn V. Beckman; page 13 (top): Fawn V. Beckman; page 13 (center): Fawn V. Beckman; page 13 (bottom): Fawn V. Beckman; page 14 (top): Fawn V. Beckman; page 14 (center): Fawn V. Beckman; page 14 (bottom): Fawn V. Beckman; page 15: Courtesy Robert Pollack; page 17 (far left): Fawn V. Beckman; page 17 (left): Fawn V. Beckman; page 17 (center): Fawn V. Beckman; page 17 (right): Fawn V. Beckman; page 19 (top): Fawn V. Beckman; page 19 (center): Fawn V. Beckman; page 19 (bottom): Fawn V. Beckman; page 20 (top): Fawn V. Beckman; page 20 (center): Fawn V. Beckman; page 20 (bottom): Fawn V. Beckman; page 21 (top): Fawn V. Beckman; page 21 (center): Fawn V. Beckman; page 22 (left): Fawn V. Beckman; page 22 (right): Fawn V. Beckman; page 23 (top): Fawn V. Beckman; page 23 (center): Fawn V. Beckman; page 23 (bottom): Fawn V. Beckman; page 25 (top): Fawn V. Beckman; page 25 (bottom left): Fawn V. Beckman; page 25 (bottom right): Fawn V. Beckman; page 26: Fawn V. Beckman; page 28: Fawn V. Beckman; page 29: Fawn V. Beckman; page 30 (left): Fawn V. Beckman; page 30 (center): Fawn V. Beckman; page 30 (right): Fawn V. Beckman; page 31: Fawn V. Beckman

LAB 18

Page 1 (left): Volker Steger/Photo Researchers, Inc.; page 1 (right): ©Scott Bjelland; pages 2-3: NG Maps; page 4 (far left): Photo by Human Origins Program, Smithsonian Institution; reconstruction by Matt Tocheri and Paul Rhymer; page 4 (left): Chip Clark, Human Origins Program, Smithsonian Institution; page 4 (right): Mark Nielsen; page 4 (far right): Mark Nielsen; page 5 (far left): Chip Clark, Human Origins Program, Smithsonian Institution; page 5 (left): Chip Clark, Human Origins Program, Smithsonian Institution; page 5 (center): Mark Nielsen; page 5 (right): Mark Nielsen; page 6 (far left): Chip Clark, Human Origins Program, Smithsonian Institution; page 6 (left): Chip Clark, Human Origins Program, Smithsonian Institution; page 6 (right): Mark Nielsen; page 6 (far right): Mark Nielsen; page 9: Courtesy Emily Kay, Harvard University; page 10: Stanislav Khrapov/Shutterstock; page 11 (left): Joe McDonald/Visuals Unlimited; page 11 (right): Joe McDonald/Visuals Unlimited; page 12 (left): Robert J. Erwin/Photo Researchers, Inc; page 12 (right): Calvin Larsen/Photo Researchers, Inc; page 13 (left): Fawn V. Beckman; page 13 (right): Fawn V. Beckman; page 14 (top): Fawn V. Beckman; page 14 (center): Fawn V. Beckman; page 14 (bottom left): Fawn V. Beckman; page 14 (bottom right): Fawn V. Beckman; page 15 (top left): Fawn V. Beckman; page 15 (top right): Fawn V. Beckman; page 15 (center): Fawn V. Beckman; page 15 (bottom): Fawn V. Beckman; page 16 (left): Fawn V. Beckman; page 16 (right): Fawn V. Beckman; page 18: Courtesy S. Blair Hedges and Sudhir Kumar; page 19 (top left): Ted Daeschler/Academy of Natural Sciences/VIREO; page 19 (top right): Courtesy Xing Xu, Key Laboratory of Evolutionary Systematics of Vertebrates, Institute of Vertebrate Paleontology & Paleonanthropology, Chinese Academy of Sciences, Beijing; page 21 (left): Fawn V. Beckman; page 21 (center): ISM/Phototake; page 21 (right): Fawn V. Beckman; page 23 (left): Betty & Nathan Cohen/Science Photo Library/Visuals Unlimited; page 23 (right): Omikron/Photo Researchers, Inc; page 24: Image courtesy of Michael K. Richardson; page 25: Omikron/Photo Researchers, Inc; page 27 (bottom left): Perennou Nuridsany/Photo Researchers, Inc; page 27 (bottom right): David Fox/Oxford Scientific/Getty Images, Inc.